CISM International Centre for Mechanical Sciences

Courses and Lectures

Volume 597

Managing Editor

Paolo Serafini, CISM - International Centre for Mechanical Sciences, Udine, Italy

Series Editors

Elisabeth Guazzelli, IUSTI UMR 7343, Aix-Marseille Université, Marseille, France
Franz G. Rammerstorfer, Institut für Leichtbau und Struktur-Biomechanik, TU Wien, Vienna, Wien, Austria
Wolfgang A. Wall, Institute for Computational Mechanics, Technical University Munich, Munich, Bayern, Germany
Bernhard Schrefler, CISM - International Centre for Mechanical Sciences, Udine, Italy

For more than 40 years the book series edited by CISM, “International Centre for Mechanical Sciences: Courses and Lectures”, has presented groundbreaking developments in mechanics and computational engineering methods. It covers such fields as solid and fluid mechanics, mechanics of materials, micro- and nanomechanics, biomechanics, and mechatronics. The papers are written by international authorities in the field. The books are at graduate level but may include some introductory material.

More information about this series at http://www.springer.com/series/76

Jörg Schröder · Paulo de Mattos Pimenta
Editors

Novel Finite Element Technologies for Solids and Structures

Springer

Editors
Jörg Schröder
Faculty of Engineering
University of Duisburg-Essen
Essen, Germany

Paulo de Mattos Pimenta
Polytechnic School
University of São Paulo
São Paulo, Brazil

ISSN 0254-1971 ISSN 2309-3706 (electronic)
CISM International Centre for Mechanical Sciences
ISBN 978-3-030-33522-9 ISBN 978-3-030-33520-5 (eBook)
https://doi.org/10.1007/978-3-030-33520-5

This Springer imprint is published by the registered company Springer Nature Switzerland AG
The registered company address is: Gewerbestrasse 11, 6330 Cham, Switzerland

Preface

Many current engineering applications can be solved by finite element technologies. Nevertheless, for several important problems, the application of standard numerical simulation techniques, as, for example, the Galerkin method, is limited due to drawbacks like numerical stability issues, locking phenomena, and non-smoothness of the solution. In order to improve capabilities and the reliability of numerical simulations, advanced finite element methods are a major part of present research in the field of mechanics and mathematics. Due to the progress in this emerging field, the objective of this course is to present new ideas in the framework of novel finite element discretization schemes. Thereby, the lectures have been focused as well on the mechanical as also on the mathematical background. Here, recent developments in mixed finite element formulations in solid mechanics and on novel techniques for flexible structures at finite deformations have been emphasized. A special focus was aimed at the implementation and automation aspects of these technologies. The presented automation processes pays attention to the application of automatic differentiation technique, combined with the symbolic problem description, automatic code generation, and code optimization. The combination of these approaches leads to highly efficient numerical codes, which are fundamental for reliable simulations of complicated engineering problems. The presented modeling techniques cover a huge range of advanced finite element techniques. The special topics of this course have been: The isogeometric concept applied to solid and shell structures, novel C^1-continuous finite element technologies for Kirchhoff–Love shell models and Bernoulli–Euler beams, robust mixed, and discontinuous Galerkin methods for solids, plates, and shells including strong material anisotropies, the Virtual Finite Element Method, and concepts of robust preconditioning techniques for large-scale problems. Furthermore, the course introduces the theory and application of AceGen: A multi-language and multi-environment tool for highly efficient numerical code generation. These techniques encounter a wide range of applications from elasticity, viscoelasticity, plasticity, and viscoplasticity in classical engineering disciplines, as, for instance, civil and mechanical engineering, as well as in modern branches as biomechanics and multiphysics. The CISM course on "Novel Finite Element

Technologies for Solids and Structures", held in Udine from September 18 to 22, 2017, was addressed to master's students, doctoral students, postdocs, and experienced researchers in engineering, applied mathematics, and material science, which are interested in conducting research on the topics of advanced mixed Galerkin FEM, structural finite element methods, mathematical analysis as well as formulations and applications of these methods to finite strain or coupled problems.

It is our pleasure to thank the lecturers of the CISM course Sven Klinkel (Aachen, Germany), Jože Korelc (Ljubljana, Slovenia), Paulo de Mattos Pimenta (São Paulo, Brasil), Joachim Schöberl (Vienna, Austria), Jörg Schröder (Essen, Germany), Fleurianne Bertrand (Berlin, Germany) as well as the additional contributors to these CISM lecture notes Margarita Chasapi (Aachen, Germany), Bernhard Kober (Essen, Germany), Teja Melink (Ljubljana, Slovenia), Marcel Moldenhauer (Essen, Germany), Gerhard Starke (Essen, Germany), and Nils Viebahn (Essen, Germany). We furthermore thank all the participants who made the course a success. Finally, we extend our thanks to the Rectors, the Board, and the staff of CISM for the excellent support and kind help.

Essen, Germany Jörg Schröder
São Paulo, Brazil Paulo de Mattos Pimenta

Contents

Engineering Notes on Concepts of the Finite Element Method for Elliptic Problems

Jörg Schröder

Abstract In this contribution, we discuss some basic mechanical and mathematical features of the finite element technology for elliptic boundary value problems. Originating from an engineering perspective, we will introduce step by step of some basic mathematical concepts in order to set a basis for a deeper discussion of the rigorous mathematical approaches. In this context, we consider the boundedness of functions, the classification of the smoothness of functions, classical and mixed variational formulations as well as the H^{-1}-FEM in linear elasticity. Another focus is on the analysis of saddle point problems occurring in several mixed finite element formulations, especially on the solvability and stability of the associated discretized versions.

1 Introduction

This chapter deals with some fundamental concepts needed for the understanding of the mathematical background of the finite element method (FEM). Starting from a one-dimensional boundary value problem, we motivate the formulation of an abstract minimization problem in order to generalize the problems occurring in the numerical approximation of elliptic boundary value problems. The presented general explanations originate from an engineering point of view and are consulted of the mathematical framework needed for a deeper understanding. Of course, there are a variety of excellent textbooks dealing with this topic, from the engineering as well as from the mathematical point of view. Textbooks with a more mechanical motivation are (amongst many others) e.g. Hughes (1987), Wriggers (2008), Auricchio et al. (2004), Gockenbach (2006), Berdichevsky (2009), Becker et al. (1981), and Oden and Carey

J. Schröder (✉)
Department of Civil Engineering, Faculty of Engineering, Institute of Mechanics, University of Duisburg-Essen, Essen, Germany
e-mail: j.schroeder@uni-due.de

J. Schröder and P. de Mattos Pimenta (eds.), *Novel Finite Element Technologies for Solids and Structures*, CISM International Centre for Mechanical Sciences 597,
https://doi.org/10.1007/978-3-030-33520-5_1

(1983); representatives with a mathematical background are e.g. Braess (1997), Boffi et al. (2013), Oden and Reddy (1976), Brenner and Scott (2002), and Ern and Guermond (2013).

2 Introductory Example and Propaedeutic Remarks

Let $\mathcal{B} \subset \mathbb{R}^d$ be the body of interest parametrized in $x \in \mathbb{R}^d$ with $d = 1, 2, 3$. The boundary $\partial\mathcal{B}$ of $\mathcal{B}$ is decomposed into $\partial\mathcal{B}_N$ and $\partial\mathcal{B}_D$, where Neumann and Dirichlet boundary conditions are prescribed, respectively. They satisfy

$$\partial\mathcal{B} = \partial\mathcal{B}_N \cup \partial\mathcal{B}_D \quad \text{and} \quad \partial\mathcal{B}_N \cap \partial\mathcal{B}_D = \emptyset\,. \tag{1}$$

The boundary value problem is typically defined by a set of partial differential equations (PDEs) on the open domain $\mathcal{B}$ and boundary conditions.

For simplicity we start with the simple one-dimensional ($d = 1$) boundary value problem

$$-\,(EAu'(x))' + K_s u(x) = f(x) \quad \text{in} \quad x \in \mathcal{B} = (0, l)\,, \tag{2}$$

with Young's modulus $E > 0$, cross section $A > 0$, and continuous elastic support $K_s > 0$, with units $[E] = \text{N/m}^2$, $[A] = \text{m}^2$, $[K_s] = \text{N/m}^2$, $[u] = \text{m}$, $[f] = \text{N/m}$, see Fig. 1. At $x = 0$ a Dirichlet and at $x = l$ a Neumann boundary condition is applied:

$$u(0) = 0 \quad \text{and} \quad EA\,u'(l) = t_l\,, \tag{3}$$

respectively. For the following explanations, we assume that the solution $u(x)$ and the distributed loading $f(x)$ are sufficiently regular.

Analytical solution. The general solution of (2) for constant EA and $f(x) = f_0 + \Delta f\, x/l$ is based on the ansatz

$$u(x) = e^{\lambda x} \quad \rightsquigarrow \quad u'(x) = \lambda e^{\lambda x} \quad \rightsquigarrow \quad u''(x) = \lambda^2 e^{\lambda x}\,. \tag{4}$$

Substituting these expressions into the homogeneous part of (2), denoted by $\tilde{u}(x)$, yields

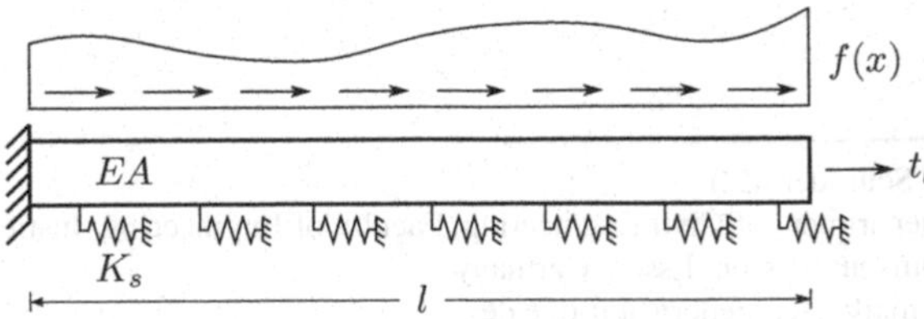

Fig. 1 Bar with continuous elastic support

$$(-EA\,\lambda^2 + K_s)e^{\lambda x} = 0 \quad \rightsquigarrow \quad \lambda_{1,2} = \pm\sqrt{\frac{K_s}{EA}} =: \pm\alpha\,, \tag{5}$$

i.e., the solution is of the form

$$\begin{aligned}\tilde{u}(x) &= \tilde{c}_1\, e^{+\alpha x} + \tilde{c}_2\, e^{-\alpha x} \\ &= \underbrace{(\tilde{c}_1 + \tilde{c}_2)}_{c_1} \underbrace{\frac{e^{+\alpha x} + e^{-\alpha x}}{2}}_{\cosh(\alpha x)} + \underbrace{(\tilde{c}_1 - \tilde{c}_2)}_{c_2} \underbrace{\frac{e^{+\alpha x} - e^{-\alpha x}}{2}}_{\sinh(\alpha x)}\,.\end{aligned} \tag{6}$$

Adding the particular solution $f_0/K_s + (\Delta f \cdot x)/(K_s\, l)$ yields

$$u(x) = c_1 \cosh(\alpha x) + c_2 \sinh(\alpha x) + \frac{f_0}{K_s} + \frac{\Delta f}{K_s}\frac{x}{l}\,. \tag{7}$$

Evaluating the boundary conditions yields the analytical expressions for the constants c_1 and c_2:

$$\begin{aligned} u(0) &= c_1 + \frac{f_0}{K_s} = 0 \\ \rightarrow c_1 &= \frac{-f_0}{K_s}\,, \\ u'(l) &= c_2\,\alpha \cosh(\alpha\, l) + \frac{\Delta f}{K_s}\frac{1}{l} = t_l \\ \rightarrow \quad c_2 &= \left(t_l - \frac{\Delta f}{K_s}\frac{1}{l}\right)\frac{1}{\alpha} e^{-\alpha l}(1 + \tanh(\alpha\, l))\,. \end{aligned} \tag{8}$$

A **weak formulation** of the boundary value problem is obtained by multiplying (2) with a test function δu and partial integration:

$$\int_0^l (EAu'\delta u' + K_s\, u\, \delta u)\mathrm{d}x = \int_0^l f\ \delta u\, \mathrm{d}x + t_l\, \delta u(l) \quad \forall\ \delta u \in V\,, \tag{9}$$

where V is a suitable space of functions, e.g., a Hilbert space. All test functions $\delta u \in V$ have to vanish at the Dirichlet boundary condition $\delta u(0) = 0$. In the variational formulation u is an element of the class of trial functions V_{trial}. The collection of both functions are denoted as admissible functions; in this simple case, V and V_{trial} coincide and have to satisfy

$$V = V_{trial} = \left\{u(x) : \int_0^l (u^2 + (u')^2)\,\mathrm{d}x < \infty, u(0) = 0\right\}. \tag{10}$$

The idea of approximation methods is to compute an approximate solution $u_h \in V_h$ in the finite-dimensional subspace $V_h \subset V$, based on

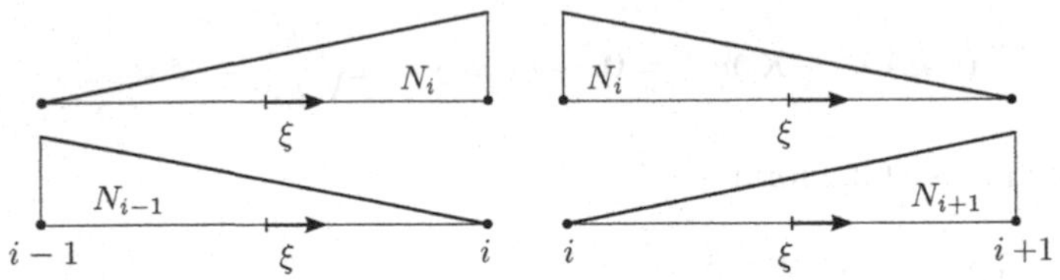

Fig. 2 Linear ansatz functions of neighboring elements, $\xi \in \left[-\frac{l^e}{2}, \frac{l^e}{2}\right]$

$$\int_0^l (EAu_h'\delta u_h' + K_s u_h\, \delta u_h)\mathrm{d}x = \int_0^l f\ \delta u_h\, \mathrm{d}x + t_l\, \delta u_h(l)\quad \forall\, \delta u_h \in V_h\,. \tag{11}$$

In order to do this within the **finite element method**, we have to subdivide the domain in $\mathrm{num_{ele}}$ subsections, here we choose individual finite elements with (for simplicity reasons) unit length $l^e = l/\mathrm{num_{ele}}$. On this individual elements we define a set of ansatz functions, i.e., shape functions $N_i | i = 1, \ldots k$ with local support. We use piecewise polynomial functions which are globally C^0 continuous, as depicted in Fig. 2.

In this case, the continuity can be easily enforced by sharing the degrees of freedom at the interface between two neighboring elements. With this definitions we approximate the individual fields on element level as follows:

$$u_h = \underline{\boldsymbol{N}}^e\, \underline{\boldsymbol{d}}^e, \quad \delta u_h = \underline{\boldsymbol{N}}^e\, \underline{\delta \boldsymbol{d}}^e, \quad u_h' = \underline{\boldsymbol{B}}^e\, \underline{\boldsymbol{d}}^e, \quad \delta u_h' = \underline{\boldsymbol{B}}^e\, \underline{\delta \boldsymbol{d}}^e \tag{12}$$

with the matrix of shape functions $\underline{\boldsymbol{N}}^e$, the matrix containing the derivatives of the shape functions $\underline{\boldsymbol{B}}^e$, and the vectors of nodal (virtual) degrees of freedom $(\underline{\delta \boldsymbol{d}}^e)\, \underline{\boldsymbol{d}}^e$ of the element $e \in \{1, 2, \ldots, \mathrm{num_{ele}}\}$:

$$\underline{\boldsymbol{d}}^e = \begin{bmatrix} d_1^e \\ d_2^e \end{bmatrix}, \quad \underline{\delta \boldsymbol{d}}^e = \begin{bmatrix} \delta d_1^e \\ \delta d_2^e \end{bmatrix}, \quad \underline{\boldsymbol{N}}^e = \begin{bmatrix} N_1 \\ N_2 \end{bmatrix}, \quad \underline{\boldsymbol{B}}^e = \begin{bmatrix} N_{1,x} \\ N_{2,x} \end{bmatrix}. \tag{13}$$

After substituting these approximations equation (11) is reformulated into

$$\begin{aligned} \sum_{e=1}^{\mathrm{num_{ele}}} \underline{\delta \boldsymbol{d}}^{eT} \underbrace{\int_{l^e} (EA\, \underline{\boldsymbol{B}}^{eT} \underline{\boldsymbol{B}}^e + K_s \underline{\boldsymbol{N}}^{eT}\, \underline{\boldsymbol{N}}^e)\mathrm{d}x}_{\underline{\boldsymbol{k}}^e}\, \underline{\boldsymbol{d}}^e &= \\ \sum_{e=1}^{\mathrm{num_{ele}}} \underbrace{\underline{\delta \boldsymbol{d}}^{eT} \int_{l^e} f\ \underline{\boldsymbol{N}}^{eT}\, \mathrm{d}x + t_l\, \underline{\delta \boldsymbol{d}}^e(l)}_{\underline{\delta \boldsymbol{d}}^{eT} \underline{\boldsymbol{r}}^e} \\ \rightarrow \sum_{e=1}^{\mathrm{num_{ele}}} \underline{\delta \boldsymbol{d}}^{eT} \{\underline{\boldsymbol{k}}^e\, \underline{\boldsymbol{d}}^e - \underline{\boldsymbol{r}}^e\} &= 0\,. \end{aligned} \tag{14}$$

Assembling the element matrices,

$$\underline{\boldsymbol{K}} = \mathop{\mathsf{A}}_{e=1}^{\text{num}_{\text{ele}}} \underline{\boldsymbol{k}}^e, \quad \underline{\boldsymbol{R}} = \mathop{\mathsf{A}}_{e=1}^{\text{num}_{\text{ele}}} \underline{\boldsymbol{r}}^e, \tag{15}$$

yields

$$\delta\underline{\boldsymbol{D}}^T\{\underline{\boldsymbol{K}}\,\underline{\boldsymbol{D}} - \underline{\boldsymbol{R}}\} = 0 \;\; \forall\, \delta\underline{\boldsymbol{D}} \quad \rightarrow \quad \underline{\boldsymbol{K}}\,\underline{\boldsymbol{D}} = \underline{\boldsymbol{R}}\,, \tag{16}$$

where $\underline{\boldsymbol{K}}$ is the global element stiffness matrix, $\underline{\boldsymbol{R}}$ the global right-hand side, $\underline{\boldsymbol{D}}$ the global vector of unknowns, and $\delta\underline{\boldsymbol{D}}$ the global vector of virtual node displacements. The FEM solution is depicted for a various number of elements num_{ele} based on a constant K_s in Fig. 3. Figure 4 compares the approximation with $\text{num}_{\text{ele}} = 16$ to the analytical solution considering different values of K_s.

Generalizations: In order to formulate an **abstract minimization problem** we define a quadratic energy functional $J(u)$, e.g., the total potential energy,

$$J(u) := \frac{1}{2}a(u, u) - L(u)\,, \tag{17}$$

with the symmetric form $a(u, v) = a(v, u)$. We assume that $a(u, u)$ is positive definite, i.e.,

$$a(v, v) > 0 \;\; \forall\, v \neq 0\,. \tag{18}$$

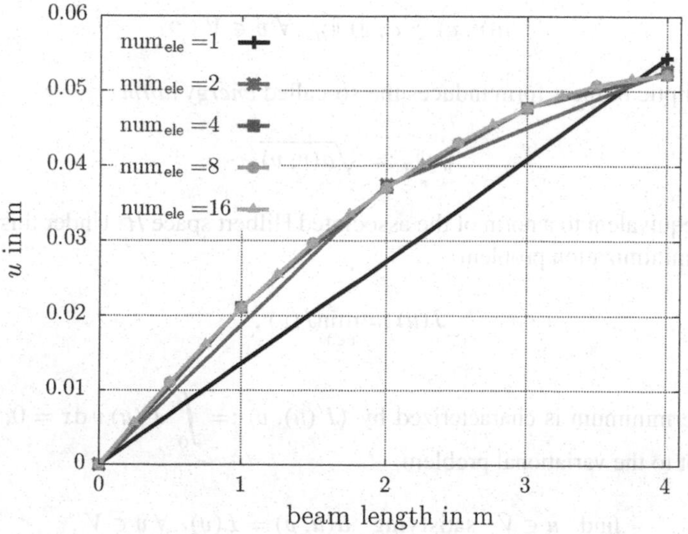

Fig. 3 Approximate FEM solution for $\text{num}_{\text{ele}} = \{1, 2, 4, 8, 16\}$ ($K_s = 10^4$ kN/m^2, $E = 210 \cdot 10^3$ kN/m^2, $l = 4$ m, $t_l = 250$ kN, $f_0 = \Delta f = 10^3$ kN/m)

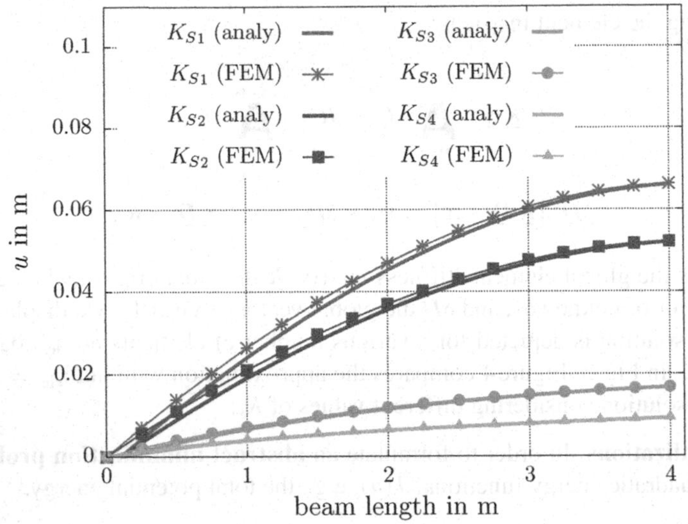

Fig. 4 Comparison of analytical and approximate FEM solution ($\text{num}_{\text{ele}} = 16$) for various values of $K_{Si} = \{10^3, 10^4, 10^5, 210 \cdot 10^3\}$ kN/m^2, $E = 210 \cdot 10^3$ kN/m^2, $t_l = 250$ kN, $f_0 = \Delta f = 10^3$ kN/m, $l = 4$ m

A bilinear form $a : V \times V \to \mathbb{R}$ is called H-elliptic (or simply elliptic) if there exists a constant $c_\alpha > 0$ such that

$$a(v, v) \geq c_\alpha \, \|v\|_H^2 \quad \forall \, v \in V \, . \tag{19}$$

The H-elliptic bilinear form induces the so-called *energy norm*

$$\|v\|_a := \sqrt{a(v, v)} \, , \tag{20}$$

which is equivalent to a norm of the associated Hilbert space H. Under this assumptions the minimization problem

$$J(u) = \min_{v \in V} J(v) \, , \tag{21}$$

where the minimum is characterized by $\langle J'(u), v \rangle := \displaystyle\int_0^l J'(u) \, v \, \mathrm{d}x = 0$, which is equivalent to the variational problem

$$\text{find} \quad u \in V \quad \text{satisfying} \quad a(u, v) = L(v) \quad \forall \, v \in V \, . \tag{22}$$

Identifying our model problem with the abstract formulation yields the quadratic functional

$$a(u, u) = \int_0^l EA\,(u'(x))^2\,\mathrm{d}x + \int_0^l K_s\,(u(x))^2\,\mathrm{d}x \tag{23}$$

and the linear functional

$$L(u) = \int_0^l u(x)\,f(x)\,\mathrm{d}x + t_l\,u(l) \tag{24}$$

with the Dirichlet boundary condition $u(0) = 0$, to be satisfied by the function $u(x)$, and the Neumann boundary condition $EA\,u'(l) = t_l$, appearing as a natural boundary condition in the functional.

Modus operandi. There are several direct methods for the computation of the approximate solution. Beyond this, there are several mathematical frameworks for the qualitative analysis of the existence and uniqueness of solutions. Beside well-known direct methods for the treatment of established models, described by partial differential equations, this topic is rather important for the derivation of new models in continuum thermodynamics. In this contribution, we want to motivate the main ideas of this scientific branch.

A functional is called **bounded from below** on the space V, if there exist a constant $c \in \mathbb{R}$, such that

$$J(u) \geq c \quad \forall u \in V\,. \tag{25}$$

This requirement can be violated if the functional is not bounded by below on V or if it is bounded by below but its minimum is not reached on V, for a physical interpretation see Berdichevsky (2009), Chap. 5.

We assume that in the quadratic functional $a(u, u)$ of our model problem (23) $u(x)$ is a differentiable function and that the integrals exists. We conclude with the meaningful engineering constants

$$EA > 0 \quad \text{and} \quad K_s > 0\,, \tag{26}$$

that $a(u, u)$ is obviously nonnegative and therefore bounded from below. A linear functional $L(u) =< l, u >$ is bounded (from above) if for $C_L > 0$

$$\|L(u)\| = |L(u)| \leq C_L\|u\| \quad \forall u \in V\,. \tag{27}$$

Now we have to answer the question, if the functional (17) is bounded from below or not. Let us consider our model problem. Our functional (17) can now be estimated, compare Braess (1997) Chap. 2.5, by

$$
\begin{aligned}
J(u) &\geq \frac{1}{2} c_\alpha \|u\|^2 - \|l\| \, \|u\| \\
&= \frac{1}{2} c_\alpha \left(\|u\|^2 - \frac{2}{c_\alpha} \|l\| \, \|u\| + \frac{\|l\|^2}{c_\alpha^2} - \frac{\|l\|^2}{c_\alpha^2} \right) \\
&= \frac{1}{2} c_\alpha \left(\|u\| - \frac{\|l\|}{c_\alpha} \right)^2 - \frac{\|l\|^2}{2c_\alpha} \\
&\geq -\frac{\|l\|^2}{2c_\alpha} \, .
\end{aligned}
\tag{28}
$$

Obviously, the functional is bounded from below.

Lax–Milgram Theorem (existence of classical solutions); Let V' be a Hilbert space, $a : V \times V \to \mathbb{R}$ a *continuous* and *H-elliptic* bilinear form defined on V, $L \in V'$ any *continuous* linear functional. Subject to these conditions there exists a *unique* solution

$$u \in V$$

such that

$$a(u, v) = L(v) \quad \forall \, v \in V \, . \qquad \blacksquare$$

Reminder, the properties of the bilinear and linear form are:

- the bilinear form has to be continuous (bounded from above), i.e., there exists a constant $C_a \in \mathbb{R}^+$[1] such that

$$|a(w, v)| \leq C_a \, \|w\|_V \|v\|_V \quad \forall \, w, v \in V \, ,$$

- the bilinear form has to be H-elliptic, i.e., there exists a constant $c_a \in \mathbb{R}^+$ such that

$$a(v, v) \geq c_a \, \|v\|_V^2 \quad \forall \, v \in V \, ,$$

- the linear functional L is continuous, i.e., there exists a constant $C_L \in \mathbb{R}^+$ such that

$$|L(v)| \leq C_L \, \|v\|_V \quad \forall \, v \in V \, .$$

Note: From the continuity of the bilinear form $a(\cdot, \cdot)$, discussed in exercise 1, we obtain $|a(u, u)| \leq C_a \, \|u\|^2$. The continuity of the linear form yields $|L(u)| \leq C_L \, \|u\|$. From the H-ellipticity of the bilinear form $a(\cdot, \cdot)$, see (19), we deduce

$$c_a \, \|u\|^2 \leq a(u, u) \leq C_a \, \|u\|^2 \, . \tag{29}$$

Exploiting the continuity of the linear form allows with $C \in \mathbb{R}^+$ for

$$c_a \, \|u\|^2 \leq a(u, u) \leq C_a \, \|u\|^2 \leq C \langle l, u \rangle \tag{30}$$

[1] nonnegative real values $\mathbb{R}_0^+$, positive real values $\mathbb{R}^+ = \mathbb{R}_0^+ \backslash 0$.

and we deduce from $\langle l, u\rangle \leq \|l\| \, \|u\|$

$$c_a \|u\|^2 \leq C \, \|l\| \, \|u\| \quad \text{and} \quad c_a \, \|u\| \leq C \, \|l\| \tag{31}$$

and

$$\|u\| \leq \frac{C}{c_a} \|l\| \quad \forall \, v \in V \, . \tag{32}$$

In this sense the bounded linear functional is generated by the continuity and H-ellipticity of the bilinear form $a(\cdot, \cdot)$.

Approximate solutions: In general it is cumbersome or even impossible to find exact solutions $u \in V$, therefore we are interested in **approximate solution concepts**. Applying **Ritz method** we seek a solution $u_h \in V_h$ with the discrete subspace $V_h \subset V$, i.e.,

$$J(u_h) = \min_{v_h \in V_h} J(v_h) \, . \tag{33}$$

The Ritz approach, based on our technical assumptions, is equivalent to the **Galerkin method** of the variational counterpart

$$\text{find} \quad u_h \in V_h \quad \text{satisfying} \quad a(u_h, v_h) = l(v_h) \quad \forall \, v_h \in V_h \, , \tag{34}$$

where $a(u_h, v_h)$ is a bilinear functional (linear in both arguments).

Exercise 1 Show that the bilinear form of our model problem in Eq. (9)

$$a(u, v) = \int_0^l EA \, u' v' \, \mathrm{d}x + \int_0^l K_s \, u \, v \, \mathrm{d}x \tag{35}$$

with $EA \in (0, \infty)$ and $K_s \in (0, \infty)$, is continuous!

Remark: Definition of *continuous bilinear forms* $a : U \times V \to \mathbb{R}$ on linear normed spaces U and V: A bilinear form $a(\cdot, \cdot)$ is a continuous bilinear form, if there exists a constant $C_a \in \mathbb{R}^+$ such that

$$|a(u, v)| \leq C_a \|u\| \|v\| \quad \forall \, u \in U, v \in V \, . \tag{36}$$

In anticipation of the following chapters we introduce the norm

$$\|u\|^2_{H^1} = (u, u)_{H^1} = \int_0^l (u^2 + (u')^2) \, \mathrm{d}x = \|u\|^2_{L^2} + \|u'\|^2_{L^2} \, ; \tag{37}$$

obviously we obtain the inequalities

$$\|u\|^2_{L^2} \leq \|u\|^2_{H^1} \quad \text{and} \quad \|u'\|^2_{L^2} \leq \|u\|^2_{H^1} \, . \tag{38}$$

Solution. In order to show that the bilinear form is continuous, consider

$$\begin{aligned}\tilde{a}(u,v) = \frac{1}{EA}a(u,v) &= \int_0^l u'\,v'\,\mathrm{d}x + \int_0^l \frac{K_s}{EA}\,u\,v\,\mathrm{d}x \\ &\leq \left|\int_0^l u'\,v'\,\mathrm{d}x\right| + \int_0^l \frac{K_s}{EA}\,|u\,v|\,\mathrm{d}x \\ &\leq \left|\int_0^l u'\,v'\,\mathrm{d}x\right| + \frac{K_s}{EA}\int_0^l |u|\,|v|\,\mathrm{d}x \\ &= |(u',v')_{L^2}| + \frac{K_s}{EA}\,(|u|,|v|)_{L^2}\,.\end{aligned} \tag{39}$$

Applying the Cauchy–Schwarz inequality yields

$$|(u',v')_{L^2}| + \frac{K_s}{EA}\,(|u|,|v|)_{L^2} \leq \|u'\|_{L^2}\,\|v'\|_{L^2} + \frac{K_s}{EA}\,\|u\|_{L^2}\,\|v\|_{L^2}\,. \tag{40}$$

Using the inequalities (38) yields the final estimation

$$\begin{aligned}\tilde{a}(u,v) &\leq \|u\|_{H^1}\|v\|_{H^1} + \frac{K_s}{EA}\,\|u\|_{H^1}\,\|v\|_{H^1} \\ &= (1+\frac{K_s}{EA})\,\|u\|_{H^1}\,\|v\|_{H^1}\,.\end{aligned} \tag{41}$$

For our bilinear form we write

$$a(u,v) \leq C_a\,\|u\|_{H^1}\|v\|_{H^1} \quad \text{with} \quad C_a = (EA + K_s)\,. \tag{42}$$

Thus, $a(u, v)$ is continuous, or in other words it is bounded by above. ♠

3 Classification of the Smoothness of Functions

In this section we discuss the classification of functions and their derivatives with respect to Hilbert spaces. For this we first set a few notations, a more detailed summary is given in Appendix A.

$$L^2(\mathcal{B}) = \left\{u : \|u\|^2_{L^2(\mathcal{B})} = \int_{\mathcal{B}} |u|^2 \mathrm{d}v \; < \; +\infty\right\} \tag{43}$$

characterizes the space of square integrable functions on $\mathcal{B}$. At this point, it seems to be meaningful to give some remarks concerning the *Riemann integral* and the *Lebesgue integral*. The Riemann integral has some disqualifications if we would like to use it for a satisfactory theory of PDEs.

In order to obtain a satisfactory theory of PDEs, we must—for technical reasons—integrate certain singular functions. If functions are regular enough to integrate them they are called *Lebesgue measurable*. An introduction to this topic is given in Royden (1968). For $m \in \mathbb{N}_0$[2] we define

$$H^m(\mathcal{B}) = \left\{ u : D^{\boldsymbol{\alpha}} u \in L^2(\mathcal{B}) \quad \forall \quad |\boldsymbol{\alpha}| \leq m \right\} , \tag{44}$$

with the multi-index notation for the derivatives of u, with the 3-tuple of nonnegative integers

$$\boldsymbol{\alpha} = (\alpha_1, \alpha_2, \alpha_3) \qquad \text{and} \qquad |\boldsymbol{\alpha}| = (\alpha_1 + \alpha_2 + \alpha_3) . \tag{45}$$

Thus the α-th derivative of u with respect to (x_1, x_2, x_3) is defined by

$$D^{\boldsymbol{\alpha}} u = \frac{\partial^{\alpha_1+\alpha_2+\alpha_3} u}{\partial x_1^{\alpha_1}\, \partial x_2^{\alpha_2} \ldots \partial x_3^{\alpha_n}} = \frac{\partial^{|\boldsymbol{\alpha}|} u}{\partial x_1^{\alpha_1}\, \partial x_2^{\alpha_2} \ldots \partial x_3^{\alpha_n}} \tag{46}$$

Explanatory examples:

$$\begin{aligned} &D^{(0,0,0)} u = u; \qquad D^{(1,0,0)} u = \frac{\partial u}{\partial x_1}; \qquad D^{(0,0,1)} u = \frac{\partial u}{\partial x_3}; \\ &D^{(1,1,0)} u = \frac{\partial^2 u}{\partial x_1\, \partial x_2}; \qquad D^{(3,2,1)} u = \frac{\partial^6 u}{\partial x_1^3\, \partial x_2^2\, \partial x_3} . \end{aligned} \tag{47}$$

The introduction of the Sobolev spaces $H^m(\mathcal{B})$ allows for the quantification of the smoothness (regularity) of functions. Let $C^m(\mathcal{B})$ be the linear space of functions u with continuous derivatives $D^{|\boldsymbol{\alpha}|}u$ of the order $0 \leq |\boldsymbol{\alpha}| \leq m$. The Sobolev spaces $H^m(\mathcal{B})$ are related with the $C^k(\overline{\mathcal{B}})$ spaces by the **Sobolev embedding theorem:** Let $\overline{\mathcal{B}} = \mathcal{B} \cup \partial\mathcal{B}$ be a bounded domain with a Lipschitz boundary. Every function in $H^m(\mathcal{B})$ belongs to $C^k(\overline{\mathcal{B}})$ if

$$m > k + 1 \quad \text{for } \mathcal{B} \subset \mathbb{R}^2, \qquad m > k + 3/2 \quad \text{for } \mathcal{B} \subset \mathbb{R}^3 .$$ ■

It should be noted that the embedding is continuous:

$$H^m(\mathcal{B}) \subseteq C^k(\overline{\mathcal{B}}) .$$

Furthermore we introduce the notation

$$H_0^1(\mathcal{B}) := \{u \in H^1(\mathcal{B}),\, u|_{\partial\mathcal{B}} = 0\},\; H_{0,D}^1(\mathcal{B}) := \{u \in H^1(\mathcal{B}),\, u|_{\partial\mathcal{B}_D} = 0\} . \tag{48}$$

[2] positive integers $\mathbb{N}_+ = \{1, 2, 3, \ldots\}$, nonnegative integers $\mathbb{N}_0 = \{0, 1, 2, 3, \ldots\} = \mathbb{N}_+ \cup \{0\}$.

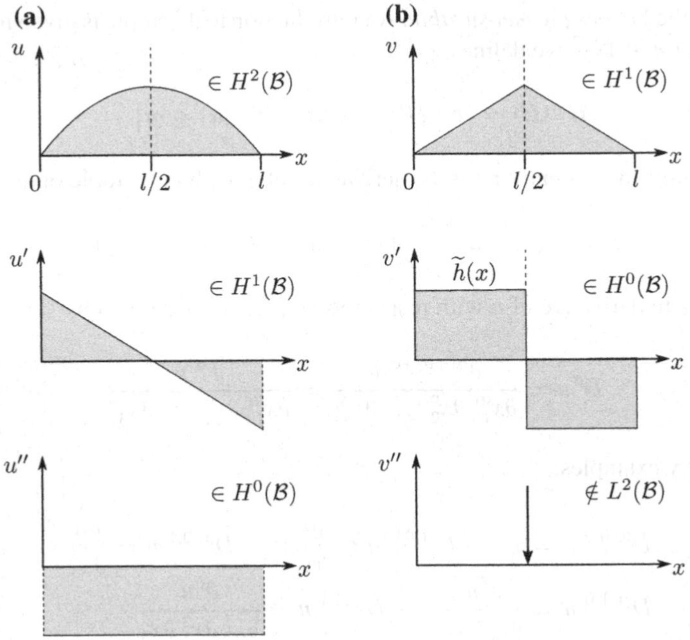

Fig. 5 Regularity of functions u and v and their derivatives on $\mathcal{B} = [0, l]$

3.1 One-Dimensional Example

To discuss the smoothness of functions, we consider the two functions depicted in Fig. 5.
For the function u in Fig. 5a we observe

$$u \in C^2(\mathcal{B}) \,, \tag{49}$$

because u is twice continuously differentiable and $u \in H^2(\mathcal{B})$. In contrast, the function v depicted in Fig. 5b is

$$v \in C^0(\mathcal{B}) \,, \tag{50}$$

because already its first derivative is not continuous. Obviously, $v \in H^1(\mathcal{B})$, due to the fact that its first derivative is square integrable,[3] i.e., $v' \in L^2(\mathcal{B})$. Although the *classical* derivative of $v(x)$ does not exist at $x = l/2$ we can define the weak derivative of v. Consider

[3]The derivatives occurring in $H^m(\mathcal{B})$ have to be interpreted as *weak* or *generalized* derivatives. Classical derivatives are functions defined pointwise on an interval. A weak derivative need only to be locally integrable. If the function is sufficiently smooth, e.g., $v \in C^m(\overline{\mathcal{B}})$, then its weak derivatives $D^\alpha u$ coincide with the classical ones for $|\alpha| \leq m$.

$$\int_0^l v\,\eta'\,\mathrm{d}x = \int_0^{l/2} v\,\eta'\,\mathrm{d}x + \int_{l/2}^l v\,\eta'\,\mathrm{d}x\,, \tag{51}$$

with the infinitely differentiable function η, satisfying $\eta(0) = \eta(l) = 0$. Integration by part, i.e.,

$$\int_a^b v\,\eta'\,\mathrm{d}x = v\,\eta\big|_a^b - \int_a^b v'\,\eta\,\mathrm{d}x\,, \tag{52}$$

yields

$$\begin{aligned}\int_0^l v\,\eta'\,\mathrm{d}x &= v(^l/_2)\,\eta(^l/_2) - \int_0^{l/2} v'\,\eta\,\mathrm{d}x - \eta(^l/_2)\,v(^l/_2) - \int_{l/2}^l v'\,\eta\,\mathrm{d}x \\ &= -\left\{\int_0^{l/2} v'\,\eta\,\mathrm{d}x + \int_{l/2}^l v'\,\eta\,\mathrm{d}x\right\}.\end{aligned} \tag{53}$$

The function $v'(x)$ in (53) is denoted as the weak derivative of $v(x)$.

Let us now consider the function in Fig. 6 which is a delta function representing a force acting at a point. This function (distribution) is not square integrable. Before we are able to quantify its smoothness it has to be integrated. In order to generalize the discussion we introduce the *antiderivative* D^{-1}, by means of

$$D(D^{-1}u) = u \quad \text{with} \quad D := \frac{\mathrm{d}}{\mathrm{d}x}\,. \tag{54}$$

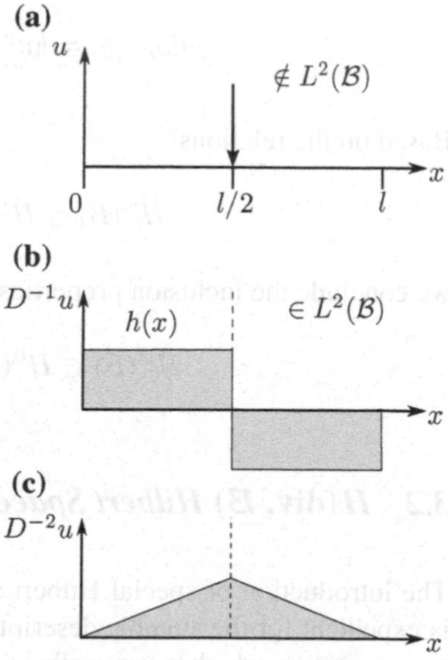

Fig. 6 Antiderivatives of function u

The meaning of this operator becomes clear if we consider again the function $\widetilde{h}(x)$ in Fig. 5b:

$$v' = Dv = \widetilde{h}(x)\,.$$

The calculation of the antiderivative

$$D^{-1}v' = D^{-1}(Dv) = \int \widetilde{h}(x)\,\mathrm{d}x$$

yields the hat function depicted in Fig. 5b up to a constant. Switching back to our function (distribution) shown in Fig. 6a: Evaluating the antiderivative of the delta function $\delta(l/2)$ leads to

$$D^{-1}u = h(x)\,, \tag{55}$$

then we conclude that $h(x)$ is a square integrable function. In other words its first antiderivative, i.e., its "first integral", is in $L^2(\mathcal{B})$. Therefore we define

$$u \in H^{-1}(\mathcal{B})\,. \tag{56}$$

The question is: What are negative Sobolev spaces?
Let m be a positive integer, then the negative Sobolev space $H^{-m}(\mathcal{B})$ is defined as the dual of $H_0^m(\mathcal{B})$, i.e.,

$$H^{-m}(\mathcal{B}) = (H_0^m(\mathcal{B}))'\,. \tag{57}$$

The associated norm, exemplarily for $m = 1$ is

$$\|u\|_{H^{-1}(\mathcal{B})} = \|u\|_{-1,\mathcal{B}} = \min_{v\in H_0^1(\mathcal{B})\backslash 0} \frac{(u,v)_{0,\mathcal{B}}}{\|v\|_{1,\mathcal{B}}}\,. \tag{58}$$

Based on the relations

$$H_0^m(\mathcal{B}) \subset H^m(\mathcal{B}) \subset H^0(\mathcal{B}) = L^2(\mathcal{B}) \tag{59}$$

we conclude the inclusion properties

$$H^m(\mathcal{B}) \subset H^0(\mathcal{B}) = L^2(\mathcal{B}) \subset H^{-m}(\mathcal{B})\,. \tag{60}$$

3.2 $H(\mathrm{div}, \mathcal{B})$ Hilbert Spaces

The introduction of special Hilbert spaces related to vector or tensor valued fields is expedient for the suitable description of many engineering problems. A frequent representative, which is especially of importance in the field of elasticity, heat conduction, flow problems, etc. is the $H(\mathrm{div}, \mathcal{B})$ space which demands a $L^2(\mathcal{B})$-measurable weak divergence. The corresponding space in introduced by

$$H(\mathrm{div}, \mathcal{B}) = \left\{ \boldsymbol{u} \in (L^2(\mathcal{B}))^d \wedge \mathrm{div}\, \boldsymbol{u} \in L^2(\mathcal{B}) \right\} , \tag{61}$$

whereas d denotes the dimension of vector $\boldsymbol{u}$.

4 Variational Formulations of Linear Elasticity

In the following chapters, we concentrate on the formulation of elasticity. Let $\mathcal{B} \subset \mathbb{R}^3$ be the body of interest, parametrized in $\boldsymbol{x} \in \mathbb{R}^3$, $\boldsymbol{\sigma}$ the second-order stress tensor, $\boldsymbol{\varepsilon} = \nabla_s \boldsymbol{u}$ the symmetric second-order strain tensor, $\boldsymbol{u}$ the displacement field, $\boldsymbol{f}$ the given body force per unit volume, $\mathbb{C}$ the fourth-order elasticity tensor and $\bar{\boldsymbol{t}}$ the prescribed Neumann boundary conditions. Then the governing equations in linear elasticity are given by[4]

$$\begin{aligned}
&\text{balance of momentum:} && \mathrm{div}\, \boldsymbol{\sigma} + \boldsymbol{f} = \boldsymbol{0} \\
&\text{constitutive law:} && \boldsymbol{\sigma} = \mathbb{C} : \boldsymbol{\varepsilon} \\
&\text{kinematical condition:} && \boldsymbol{\varepsilon} = \nabla_s \boldsymbol{u} = \tfrac{1}{2}(\mathrm{grad}\, \boldsymbol{u} + \mathrm{grad}^T \boldsymbol{u}) \\
&\text{balance of angular momentum:} && \boldsymbol{\sigma} = \boldsymbol{\sigma}^T \\
&\text{Dirichlet boundary condition:} && \boldsymbol{u} = \boldsymbol{0} \text{ on } \partial\mathcal{B}_D \\
&\text{Neumann boundary condition:} && \boldsymbol{\sigma} \cdot \boldsymbol{n} = \bar{\boldsymbol{t}} \text{ on } \partial\mathcal{B}_N
\end{aligned} \tag{62}$$

4.1 Classical (Bubnov-)Galerkin Formulation

A direct substitution of $(62)_2$–$(62)_4$ into $(62)_1$ leads to a variational formulation where solely the displacements are solved in a weak form. Multiplication with a test function $\delta\boldsymbol{u}$ and integration over the domain leads to the problem of seeking $\boldsymbol{u}$ such that

$$\int_{\mathcal{B}} (\mathrm{div}[\mathbb{C} : \nabla_s \boldsymbol{u}] + \boldsymbol{f}) \cdot \delta\boldsymbol{u}\, \mathrm{d}v = 0 \quad \forall\, \delta\boldsymbol{u}. \tag{63}$$

Integration by parts and the insertion of the important test function property $\delta\boldsymbol{u} = \boldsymbol{0}$ on $\partial\mathcal{B}_D$ leads to the formulation of seeking $\boldsymbol{u} \in H^1_{0,D}(\mathcal{B})$ of

$$\int_{\mathcal{B}} (\nabla_s \delta\boldsymbol{u} : \mathbb{C} : \nabla_s \boldsymbol{u} - \delta\boldsymbol{u} \cdot \boldsymbol{f})\, \mathrm{d}v - \int_{\partial\mathcal{B}_N} \delta\boldsymbol{u} \cdot \bar{\boldsymbol{t}}\, \mathrm{d}a = 0 \quad \forall\, \delta\boldsymbol{u} \in H^1_{0,D}(\mathcal{B}). \tag{64}$$

It can be recognized, that for the latter weak formulation, the corresponding function space of the trial function $\boldsymbol{u}$ and the test function $\delta\boldsymbol{u}$ coincide, which is the classical

[4] Note that a restriction to homogeneous Dirichlet boundary conditions is only of technical nature and does not constitute a loss of generality, see, e.g., Braess (1997).

characteristic of the (Bubnov-)Galerkin method. The solution of (64) is equivalent to the minimizer $\boldsymbol{u} \in H^1_{0,D}(\mathcal{B})$ of the potential energy

$$\Pi(\boldsymbol{u}) = \int_{\mathcal{B}} \frac{1}{2}\nabla_s \boldsymbol{u} : \mathbb{C} : \nabla_s \boldsymbol{u}\, \mathrm{d}v - \int_{\mathcal{B}} \boldsymbol{u} \cdot \boldsymbol{f}\, \mathrm{d}v - \int_{\mathcal{B}_N} \boldsymbol{u} \cdot \boldsymbol{t}\, \mathrm{d}a \qquad (65)$$

and constitutes the basis of the well-known displacement based FEM for linear elasticity.

4.2 Alternative Methods

In the previously discussed approach, Green's theorem has been applied to shift a derivative from the trial to the test functions. Particularly in the framework of finite elements, this is the prevalent approach. However, various alternative approximation techniques are available. In these formulations, the space of the approximative solution is distinct to the space of test functions. A first example is represented by the variational problem in Eq. (63) which is e.g. the basis of collocation methods. In the corresponding discrete formulations, the approximative solution is sought in a subspace of $H^2(\mathcal{B})$ whereas the admissible test space corresponds to $L^2(\mathcal{B})$.

In contrast to these formulations where broken (discontinuous) test spaces are appropriate, both derivatives may be transferred to the test spaces. This is executed by means of successive application of Green's theorem. The corresponding formulation is established as the so-called H^{-1}-method, proposed by Rachford et al. (1974). In this H^{-1}-Galerkin method different subspaces for the space of trial functions (approximation functions) U_k and the space of test functions (weighting functions) W_h are used, i.e.,

$$U_k \neq W_h. \qquad (66)$$

In Kendall and Wheeler (1976), the authors adopted the procedure to a Crank–Nicolson–H^{-1}-Galerkin procedure and investigates single space variables in parabolic problems. The ansatz was recapitulated in Thomée (2006), Chap. 16. A discussion of negative norm error estimates for semi-discrete Galerkin-type Finite Element formulations for nonheterogenous parabolic equations is given in Thomée (1980). A more recent approach can be found in Goebbels (2015).
The basis of the approach is the theory of distributional differential equations. Let u be a distribution, v a test function and the equation of interest is the second-order ordinary differential equation

$$\mathrm{D}^2 u(x) = f(x) \qquad x \in \mathbb{R} \qquad (67)$$

with appropriate boundary conditions.

Based on this assumptions we can set up a family of distributional differential equations, see Oden and Reddy (1976), page 365 ff:

$$\begin{aligned} -\langle \mathrm{D}^2 u, v\rangle &= \langle f, v\rangle \quad \forall v \in \mathcal{D}(\mathbb{R}) \\ \langle \mathrm{D}u, \mathrm{D}v\rangle &= \langle f, v\rangle \quad \forall v \in \mathcal{D}(\mathbb{R}) \\ -\langle u, \mathrm{D}^2 v\rangle &= \langle f, v\rangle \quad \forall v \in \mathcal{D}(\mathbb{R}) \\ \langle \mathrm{D}^{-1} u, \mathrm{D}^3 v\rangle &= \langle f, v\rangle \quad \forall v \in \mathcal{D}(\mathbb{R}) \end{aligned} \tag{68}$$

with the space of distribution $\mathcal{D}(\mathbb{R})$. From the distributional point of view all equations can be interpreted as equivalent. Equation $(68)_1$ is the basis for collocation methods, $(68)_2$ for the classical Galerkin method and $(68)_3$ for the H^{-1}-method.

A descriptive explanation of the H^{-1}-method on the basis of a one-dimensional boundary value problem

$$-\langle u, \mathrm{D}^2 v\rangle = \langle f, v\rangle \quad \text{on } x \in (0, l) \tag{69}$$

with $u(0) = u(l) = 0$ is discussed in Sect. 5.2.

4.3 *Mixed Variational Frameworks for Linear Elasticity*

Considering again the governing equations in linear elasticity in Eq. (62) it is apparent that the direct substitution of $(62)_2$–$(62)_4$ into $(62)_1$ is not mandatory. Alternatively it is possible to solve another set of equations of (62) in a weak sense. Here, especially two common variants are considered in the following.

Hellinger–Reissner Formulation: The stress–displacement based formulation solves $(62)_1$ and $(62)_2$ in a weak sense, seeking $\boldsymbol{\sigma} \in H(\mathrm{div}, \mathcal{B})$ and $\boldsymbol{u} \in H^1_{0,D}(\mathcal{B})$ such that

$$\begin{aligned} &\int_{\mathcal{B}} (\mathrm{div}\,\boldsymbol{\sigma} + \boldsymbol{f}) \cdot \delta\boldsymbol{u} \,\mathrm{d}v = 0 && \forall\, \delta\boldsymbol{u} \in L^2(\mathcal{B})\,, \\ &\int_{\mathcal{B}} \left(\nabla_s \boldsymbol{u} - \boldsymbol{\sigma} : \mathbb{C}^{-1}\right) : \delta\boldsymbol{\sigma} \,\mathrm{d}v = 0 && \forall\, \delta\boldsymbol{\sigma} \in L^2(\mathcal{B})\,. \end{aligned} \tag{70}$$

On this basis, two additional variational formulations can be achieved which differ in their corresponding solution spaces.

Application of integration by parts in $(70)_1$ leads to the so-called *primal* version of the Hellinger–Reissner formulation. This yields a the saddle point problem seeking for $\boldsymbol{\sigma} \in L^2(\mathcal{B})$ and $\boldsymbol{u} \in H^1_{0,D}(\mathcal{B})$ such that

$$\begin{aligned} &\int_{\mathcal{B}} (\boldsymbol{\sigma} : \nabla_s \delta\boldsymbol{u} - \boldsymbol{f} \cdot \delta\boldsymbol{u}) \,\mathrm{d}v - \int_{\partial\mathcal{B}_N} \delta\boldsymbol{u} \cdot \boldsymbol{t} \,\mathrm{d}a = 0 && \forall\, \delta\boldsymbol{u} \in H^1_{0,D}(\mathcal{B})\,, \\ &\int_{\mathcal{B}} \left(\nabla_s \boldsymbol{u} - \boldsymbol{\sigma} : \mathbb{C}^{-1}\right) : \delta\boldsymbol{\sigma} \,\mathrm{d}v = 0 && \forall\, \delta\boldsymbol{\sigma} \in L^2(\mathcal{B})\,. \end{aligned} \tag{71}$$

Equivalently, this problem can be described by the potential

$$\Pi^{\mathrm{HR}}(\boldsymbol{\sigma}, \boldsymbol{u}) = \int_{\mathcal{B}} \left(-\frac{1}{2}\boldsymbol{\sigma} : \mathbb{C}^{-1} : \boldsymbol{\sigma} + \boldsymbol{\sigma} : \nabla_s \boldsymbol{u} - \boldsymbol{f} \cdot \boldsymbol{u} \right) \mathrm{d}v - \int_{\partial \mathcal{B}_N} \boldsymbol{u} \cdot \bar{\boldsymbol{t}} \, \mathrm{d}A \,. \tag{72}$$

The expression

$$\boldsymbol{\sigma} : \nabla_s \boldsymbol{u} - \frac{1}{2}\boldsymbol{\sigma} : \mathbb{C}^{-1} : \boldsymbol{\sigma} = \boldsymbol{\sigma} : \nabla_s \boldsymbol{u} - \psi^{\star}(\boldsymbol{\sigma}) \tag{73}$$

represents the free energy $\psi(\boldsymbol{\varepsilon})$ in terms of the complementary potential $\psi^{\star}(\boldsymbol{\sigma})$, i.e., $\psi(\boldsymbol{\varepsilon}) = \boldsymbol{\sigma} : \nabla_s \boldsymbol{u} - \psi^{\star}(\boldsymbol{\sigma})$.

In contrast, integration by parts may be applied to $(70)_2$, which yields the so-called dual Hellinger–Reissner formulation, seeking the saddle point $\boldsymbol{\sigma} \in H(\mathrm{div}, \mathcal{B})$ and $\boldsymbol{u} \in L^2(\mathcal{B})$ such that

$$\begin{aligned} &\int_{\mathcal{B}} (\mathrm{div}\, \boldsymbol{\sigma} + \boldsymbol{f}) \cdot \delta \boldsymbol{u} \, \mathrm{d}v = 0 && \forall\, \delta \boldsymbol{u} \in L^2(\mathcal{B}) \,, \\ &\int_{\mathcal{B}} \left(\boldsymbol{\sigma} : \mathbb{C}^{-1} : \delta \boldsymbol{\sigma} + \boldsymbol{u} \cdot \mathrm{div}\, \delta \boldsymbol{\sigma} \right) \mathrm{d}v = 0 && \forall\, \delta \boldsymbol{\sigma} \in H(\mathrm{div}, \mathcal{B}) \,. \end{aligned} \tag{74}$$

It should be remarked, that in this formulation the traction boundary condition $(62)_6$ has to be incorporated into the solution space of the stresses, since they do not appear in the weak form. In addition the stress symmetry condition has to be enforced.

Hu-Washizu Functional – Three Field Formulation: A third option is the independent interpolation of all variables entering the elasticity problem. This formulation solves $(62)_1$–$(62)_3$ in a weak sense. The optimization problem is: seek a saddle point $\boldsymbol{\varepsilon} \in L^2(\mathcal{B})$, $\boldsymbol{\sigma} \in L^2(\mathcal{B})$ and $\boldsymbol{u} \in H^1_{0,D}(\mathcal{B})$ such that

$$\begin{aligned} &\int_{\mathcal{B}} (\mathbb{C} : \boldsymbol{\varepsilon} - \boldsymbol{\sigma}) : \delta \boldsymbol{\varepsilon} \, \mathrm{d}v = 0 && \forall\, \delta \boldsymbol{\varepsilon} \in L^2(\mathcal{B}) \,, \\ &\int_{\mathcal{B}} (\nabla_s \boldsymbol{u} - \boldsymbol{\varepsilon}) : \delta \boldsymbol{\sigma} \, \mathrm{d}v = 0 && \forall\, \delta \boldsymbol{\sigma} \in L^2(\mathcal{B}) \,, \\ &\int_{\mathcal{B}} (\boldsymbol{\sigma} : \nabla_s \delta \boldsymbol{u} - \boldsymbol{f} \cdot \delta \boldsymbol{u}) \, \mathrm{d}v - \int_{\partial \mathcal{B}} \boldsymbol{t} \cdot \delta \boldsymbol{u} \, \mathrm{d}a = 0 && \forall\, \delta \boldsymbol{u} \in H^1_0(\mathcal{B}) \,. \end{aligned} \tag{75}$$

An equivalent potential formulation can be given by

$$\Pi^{\mathrm{HW}}(\boldsymbol{\varepsilon}, \boldsymbol{\sigma}, \boldsymbol{u}) = \int_{\mathcal{B}} (\psi(\boldsymbol{\varepsilon}) - \boldsymbol{\sigma} : (\nabla_s \boldsymbol{u} - \boldsymbol{\varepsilon}) - \boldsymbol{u} \cdot \boldsymbol{f}) \, \mathrm{d}v - \int_{\partial \mathcal{B}_N} \boldsymbol{u} \cdot \bar{\boldsymbol{t}} \, \mathrm{d}a. \tag{76}$$

with $\psi(\boldsymbol{\varepsilon}) = {}^1\!/\!{}_2\, \boldsymbol{\varepsilon} : \mathbb{C} : \boldsymbol{\varepsilon}$.

5 Finite Element Method

The finite element method constitutes the most prevalent discretization technique for the approximation of boundary value problems in the field of computational mechanics. As discussed in the previous chapters, the solution of the variational equations are in their corresponding Sobolev space. For the numerical treatment this solution space is restrained to a finite-dimensional space, in the following called finite element space and is denoted by the subscript h, e.g., V_h.

5.1 Conforming and Non-conforming Finite Elements

In case of a *conforming* discretization, the finite element space is a discrete subspace of the corresponding Sobolev space. Considering the problem of linear elasticity with the displacements as the only unknown, we seek $\boldsymbol{u} \in H^1_{0,D}(\mathcal{B})$

$$\int_{\mathcal{B}} (\nabla_s \delta \boldsymbol{u} : \mathbb{C} : \nabla_s \boldsymbol{u} - \delta \boldsymbol{u} \cdot \boldsymbol{f})\, \mathrm{d}v - \int_{\partial\mathcal{B}} \delta \boldsymbol{u} \cdot \boldsymbol{t}\, \mathrm{d}a = 0 \quad \forall\, \delta \boldsymbol{u} \in H^1_{0,D}(\mathcal{B})\,. \tag{77}$$

A conforming discretization of $\boldsymbol{u}_h \in V_h$ demands in this case

$$V_h \subset H^1_{0,D}(\mathcal{B})\,. \tag{78}$$

It can be shown that $\boldsymbol{u}_h$ of a conforming finite element converges monotonically to $\boldsymbol{u}$ with increasing mesh density, if it is in addition able to represent the rigid body displacements and the constant strain states, see Bathe (1996).

Standard $H^1(\mathcal{B})$ conforming finite elements on triangles $\mathbb{P}_k$ or quadrilaterals $\mathbb{Q}_k$ are assigned with $k+1$ nodes on each edge of the element, see Fig. 7. Continuity of the approximated variable is enforced, when these nodes are shared with the adjacent elements.

$H(\mathrm{div}, \mathcal{B})$ conforming elements can be constructed, for example, with help of the Raviart–Thomas functions. In case of triangles the $\mathbb{RT}_k$ elements have $k+1$ vector-valued sampling points on each edge and in addition $k\,(k+1)$ vector-valued sampling points in the interior of the element, as exemplary depicted in Fig. 8.

In contrast, the finite element space of *non-conforming* elements is not a subspace of the appropriate Sobolev solution space and convergence is not obvious. Due to the non-conforming discretization, an additional error is introduced and has to be controlled. However, this leads to additional flexibility in the design of the finite element. The simplest non-conforming element is the Crouzeix–Raviart element, see Crouzeix and Raviart (1973). Here, we only assign k nodes on each edge and generally do not have continuity across inter-element boundaries; thus, these are non-conforming elements. The Crouzeix–Raviart finite elements are depicted in Fig. 9.

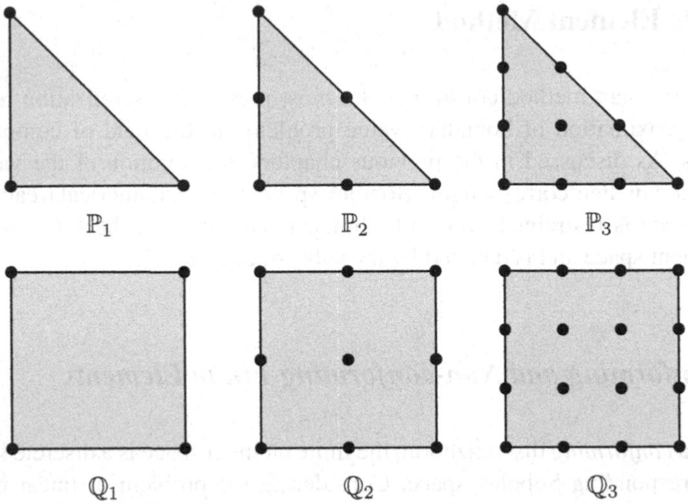

Fig. 7 Examples of $\mathbb{P}_k$ and $\mathbb{Q}_k$ elements

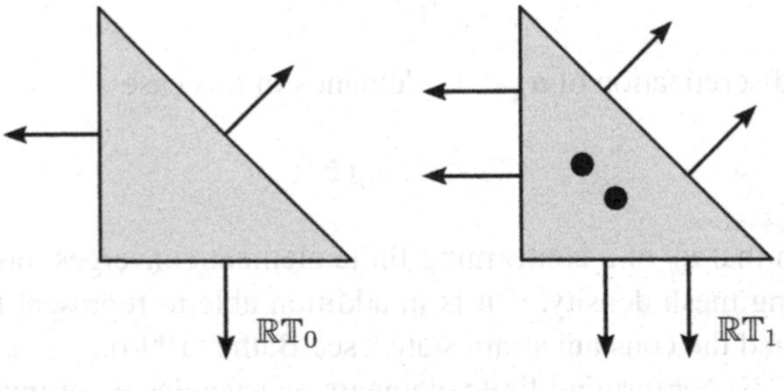

Fig. 8 Examples for $\mathbb{RT}_k$ elements with $k \geq 0$; $\dim \mathbb{RT}_k = (k+1)(k+3)$

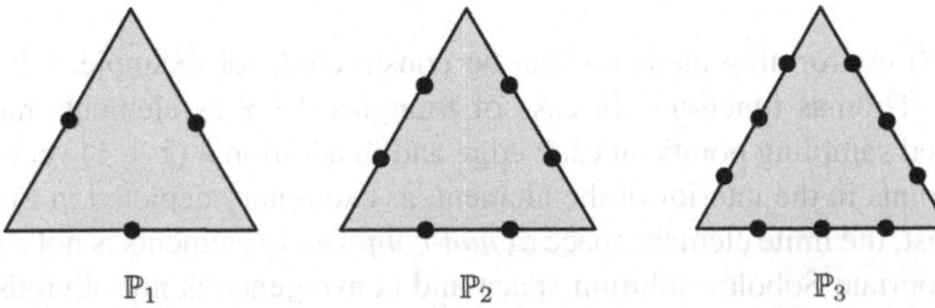

Fig. 9 Non-conforming Crouzeix–Raviart $\mathbb{P}_k$-finite elements

5.2 Example of H^{-1}-FEM for 1D Elliptic Problem

In order to approach the H^{-1}-method we analyze a one-dimensional truss element, in analogy to the one examined in Rachford et al. (1974):

$$L u = \big(\mathrm{EA}(x)\, u(x)'\big)' = -f(x), \qquad x \in \mathcal{B} = (0, 1)\,. \tag{79}$$

where $EA(x)u(x)$ characterize the normal force in the straight bar with the longitudinal stiffness

$$\mathrm{EA}(x) = \alpha^{-1} + \alpha(x - \bar{x})^2 \quad \text{and} \quad \alpha > 0\,, \tag{80}$$

where α and $\bar{x}$ are constant parameters and the right- hand side is given by

$$f(x) = 2\,(\,1 + \alpha(x - \bar{x})(\,\arctan \alpha(x - \bar{x}) + \arctan \alpha\bar{x}\,)\,)\,. \tag{81}$$

The Dirichlet boundary conditions are defined by

$$u(0) = u(1) = 0\,. \tag{82}$$

The closed-form solution of this problem is given by

$$u(x) = (1 - x)(\arctan \alpha(x - \bar{x}) + \arctan \alpha\bar{x}) \tag{83}$$

and is explicitly depicted for two different sets of α and $\bar{x}$ in Fig. 10c and 11c. Considering the plots of the longitudinal stiffness and the applied load in Fig. 10, where the parameter are chosen as $\alpha = 5$ and $\bar{x} = 0.5$. I would like to draw the reader's attention to the low stiffness in the middle of the domain. In the case $\alpha = 1000$ the domain responds to this with a rapid, jump like rising displacement.

Variational approach: The solution in terms of a variational weak form is obtained via

$$\int_{\mathcal{B}} \big(\mathrm{EA}(x)\, u'(x)\big)'\, v \,\mathrm{d}x + \int_{\mathcal{B}} f(x)\, v \,\mathrm{d}x = 0. \tag{84}$$

A reformulation using integration by parts and exploiting $v(0) = v(1) = 0$ yields

$$\int_{\mathcal{B}} \mathrm{EA}(x)\, u'(x)\, v'(x)\,\mathrm{d}x - \int_{\mathcal{B}} f(x)\, v(x)\,\mathrm{d}x = 0. \tag{85}$$

The classical FE discretization with $u_h \in U_h \subset H_0^1$ and $u_h \in \mathbb{P}_1$ or $u_h \in \mathbb{P}_2$, yields an approximation of the displacements as illustrated in Figs. 12 and 13. This standard displacement FEM ansatz even with second-order interpolation is inaccurate in an extreme edge case. It is also worth mentioning that the normal force computed from these element are also very inaccurate in comparison to the analytical solution.

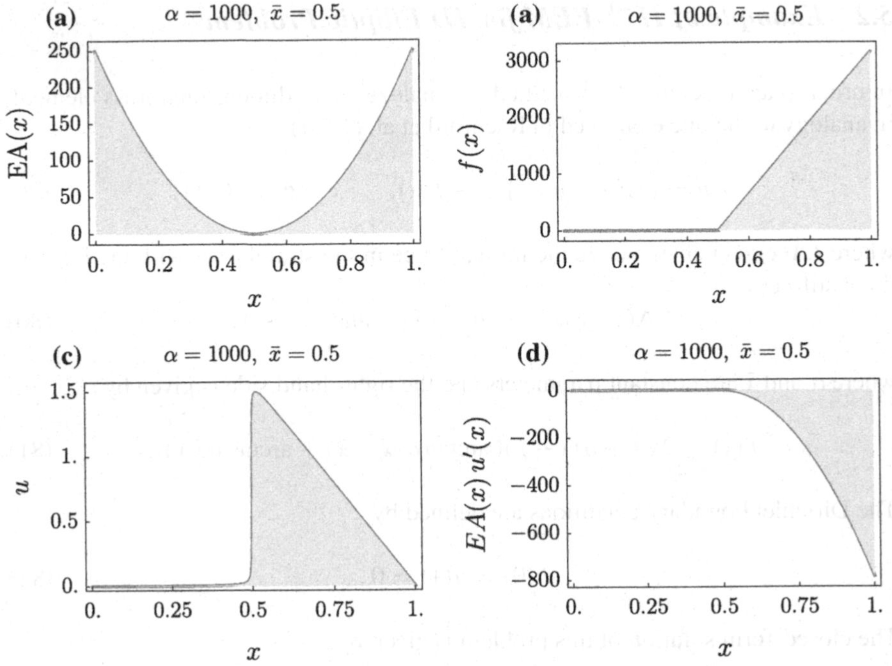

Fig. 10 Distributions of **a** stiffness EA(x), **b** load $f(x)$, **c** analytical solution for the displacements $u(x)$, and **d** longitudinal force distribution EA$(x)u'(x)$ over $\mathcal{B}$ for $\alpha = 1000$ and $\bar{x} = 0.5$

Repeated application of integration by parts leads to another weak form, which constitutes the basis of the H^{-1}-FE approach

$$\int_{\mathcal{B}} \mathrm{EA}(x)\, u(x)\, \eta''(x)\, \mathrm{d}x + \int_{\mathcal{B}} f(x)\, \eta(x)\, \mathrm{d}x = 0. \tag{86}$$

In this case the natural discretization is of the form $u_h \in U_h \subset H^0(\mathcal{B})$, i.e., it is possible to choose discontinuous approximations of u_h, denoted by $u_h \in \mathrm{d}\mathbb{P}$. This reduces the coupling between the elements. The associated subspace consists of all piecewise polynomial functions in C^k. Simultaneously the continuity requirements regarding the test space V_h are increased. Let $\mathcal{B}_h$ denote the discretization of $\mathcal{B}$, with

$$\mathcal{B}_h = \bigcup_e \mathcal{B}^e \quad \text{with} \quad B^j = [x_{j-1}, x_j] \quad \text{and} \quad h_j = x_j - x_{j-1}. \tag{87}$$

By setting $r \geq 1$ and $-1 \leq k \leq r-2$ we define the trial space

$$U_h = \left\{\, u_h \in C^k(\mathcal{B}_h) : u_h|_{\mathcal{B}^e} \in \mathbb{P}^{r-1} \text{ for } e = 1, \ldots, \mathrm{num_{ele}} \,\right\} \tag{88}$$

and the test space

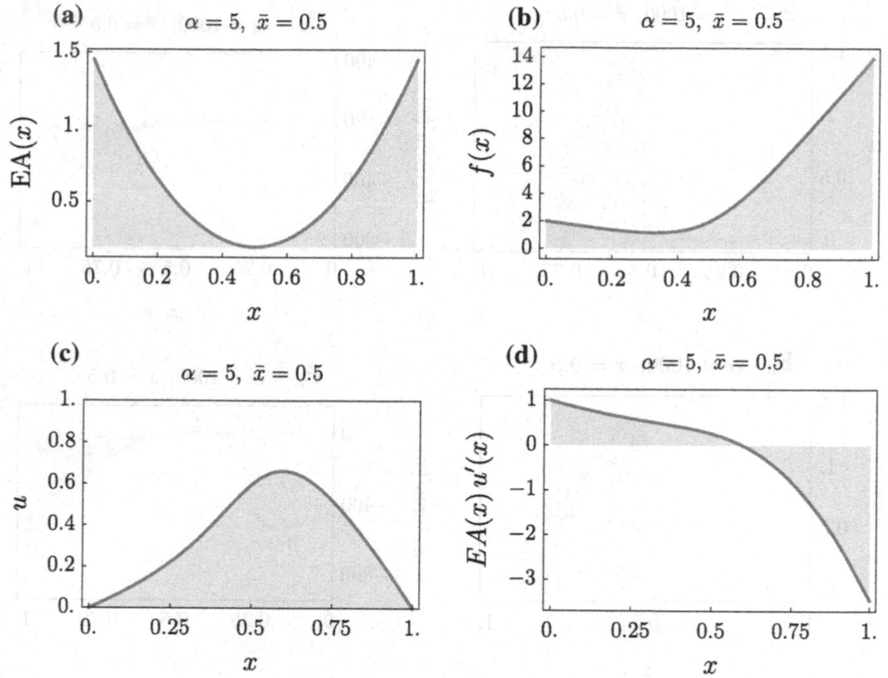

Fig. 11 Distributions of **a** stiffness $\mathrm{EA}(x)$, **b** load $f(x)$, **c** analytical solution for the displacements $u(x)$, and **d** longitudinal force distribution $\mathrm{EA}(x)u'(x)$ over $\mathcal{B}$ for $\alpha = 5$ and $\bar{x} = 0.5$

$$V_h = \left\{ v_h \in C^{k+2}(\mathcal{B}_h) : v_h|_{\mathcal{B}^e} \in \mathbb{P}^{r+1} \text{ for } e = 1, \ldots, \text{num}_{\text{ele}},\ \eta(0) = \eta(1) = 0 \right\}. \tag{89}$$

For $k = -1$ the trial space U_h exhibits discontinuities at the node of the partition, where the functions in the test space V_h are continuously differentiable. Thus we have

$$U_h = \left\{ u_h \in C^{-1}(\mathcal{B}_h) : u_h|_{\mathcal{B}^e} \in \mathbb{P}^{r-1} \text{ for } e = 1, \ldots, \text{num}_{\text{ele}} \right\}, \tag{90}$$

whereas $C^{-1}(\mathcal{B}_h)$ considers all functions whose antiderivative is in $C^0(\mathcal{B}_h)$, which means we do not require continuity at the nodal points. For convenience we define this space by

$$U_h = \left\{ u_h \in \mathrm{d}\mathbb{P}^{r-1} \text{ for } e = 1, \ldots, \text{num}_{\text{ele}} \right\} \tag{91}$$

to enforce that the trial functions are discontinuous at the exterior nodes. This leads to the corresponding space for the test functions as

$$V_h = \left\{ \eta \in C^1(\mathcal{B}_h) : \eta|_{\mathcal{B}^e} \in \mathbb{P}^{r+1} \text{ for } e = 1, \ldots, \text{num}_{\text{ele}} \right\}. \tag{92}$$

The numerical results obtained from this discretization are given in Fig. 14 for the displacements and for the stresses. The discontinuity in the displacements is clearly

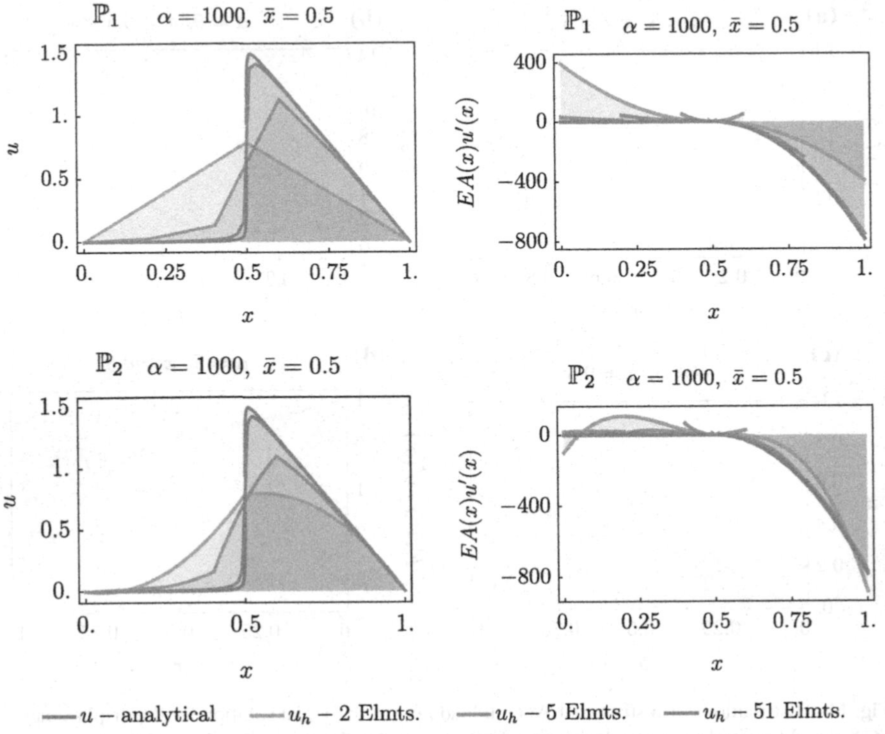

Fig. 12 Illustration of numerical solutions for $u(x)$ and $EA(x)u'(x)$ with $\alpha = 1000$ and $\bar{x} = 0.5$ using classical finite elements with $u \in \mathbb{P}_1$ (top) and $u \in \mathbb{P}_2$ (bottom)

visible, especially for the coarse discretization. The method, however, shows significant advantage for this model problem in comparison to the standard FE method. This is even more significant in terms of the normal force. The interested reader is referred to the error plots of each solution space in Figs. 15, 16, 17 and 18, with respect to both considered loading cases.

6 Analysis of Mixed Finite Elements

For the existence, uniqueness, and approximation of saddle point, problems arise from Lagrangian multipliers see Brezzi (1974). The following explanations are mainly based on the excellent treatises of Auricchio et al. (2004) and Boffi et al. (2013).

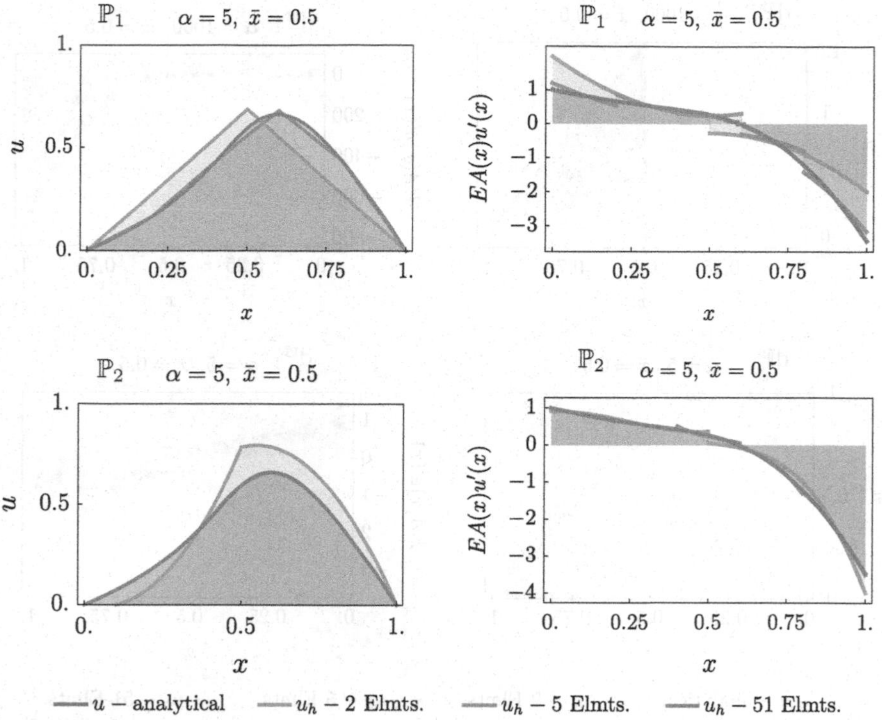

Fig. 13 Illustration of numerical solutions for $u(x)$ and $EA(x)u'(x)$ with $\alpha = 5$ and $\bar{x} = 0.5$ using classical finite elements with $u \in \mathbb{P}_1$ (top) and $u \in \mathbb{P}_2$ (bottom)

6.1 Theoretical Framework

The idea of mixed methods is based on the introduction of Lagrangian multipliers in order to relax several constraints denoted by $constr(v) = 0$, e.g., the incompressibility condition $\operatorname{div} \boldsymbol{u} = 0$. Let's start from the constrained minimization problem

$$\min_{v \in V} \left\{ J(v) \quad \text{subjected to} \quad constr(v) = 0 \right\} . \tag{93}$$

This can be reformulated by means of a Lagrangian functional of the form

$$\begin{aligned} \mathcal{L}(v, q) &= J(v) + b(v, q) \\ &= \tfrac{1}{2} a(v, v) - L(v) + b(v, q) , \end{aligned} \tag{94}$$

with

$$b(v, q) = \int_{\mathcal{B}} q \, constr(v) \, \mathrm{d}v , \tag{95}$$

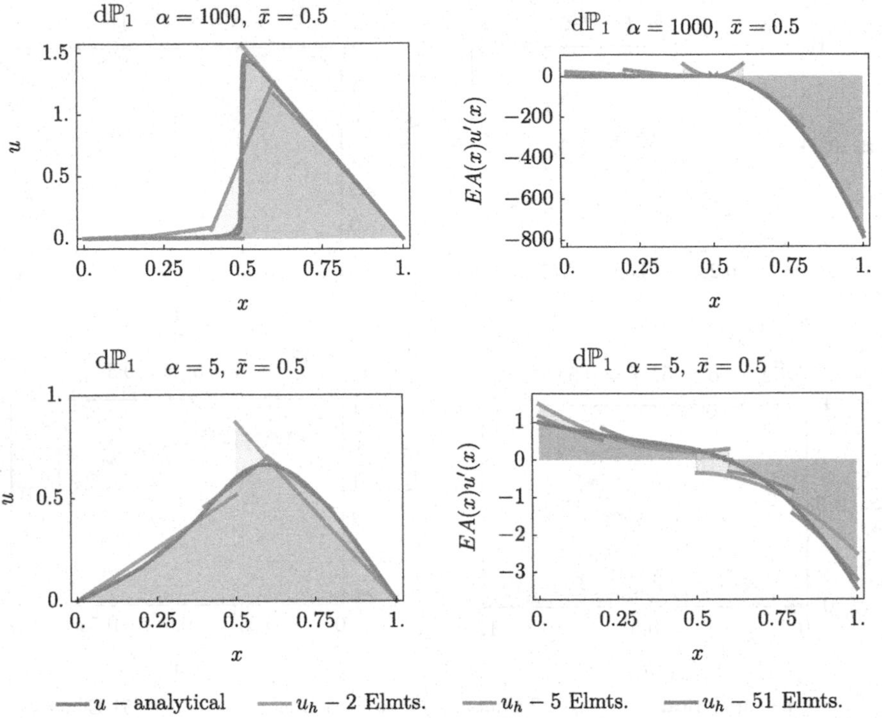

Fig. 14 Numerical solution of $u(x) \in d\mathbb{P}_1$ and of $EA(x)u'(x)$ for $\alpha = 1000$, $\alpha = 5$ and $\bar{x} = 0.5$

where q denotes the Lagrange multiplier. The solution of this abstract optimization problem is (u, p) if the condition

$$\mathcal{L}(u, q) \leq \mathcal{L}(u, p) \leq \mathcal{L}(v, p) \qquad \forall\, v \in V,\ \forall\, q \in \Sigma \tag{96}$$

is fulfilled, V and Σ are suitable Hilbert spaces.

$$a : V \times V \to \mathbb{R} \quad \text{and} \quad b : V \times \Sigma \to \mathbb{R} \tag{97}$$

are continuous bilinear forms, and $L(v) : V \to \mathbb{R}$ is a continuous linear form. It should be noted that the classical Lax–Milgram Lemma cannot be applied. In fact we should apply the so-called Banach–Nečas–Babuška theorem also known as the generalized Lax–Milgram theorem, see, e.g., Ern and Guermond (2013). In summary the variational formulation has a unique solution if

1. The continuous linear form $a(\cdot, \cdot)$ is coercive on

$$K = \{v \in V : b(q, v) = 0 \ \ \forall q \in \Sigma\},$$

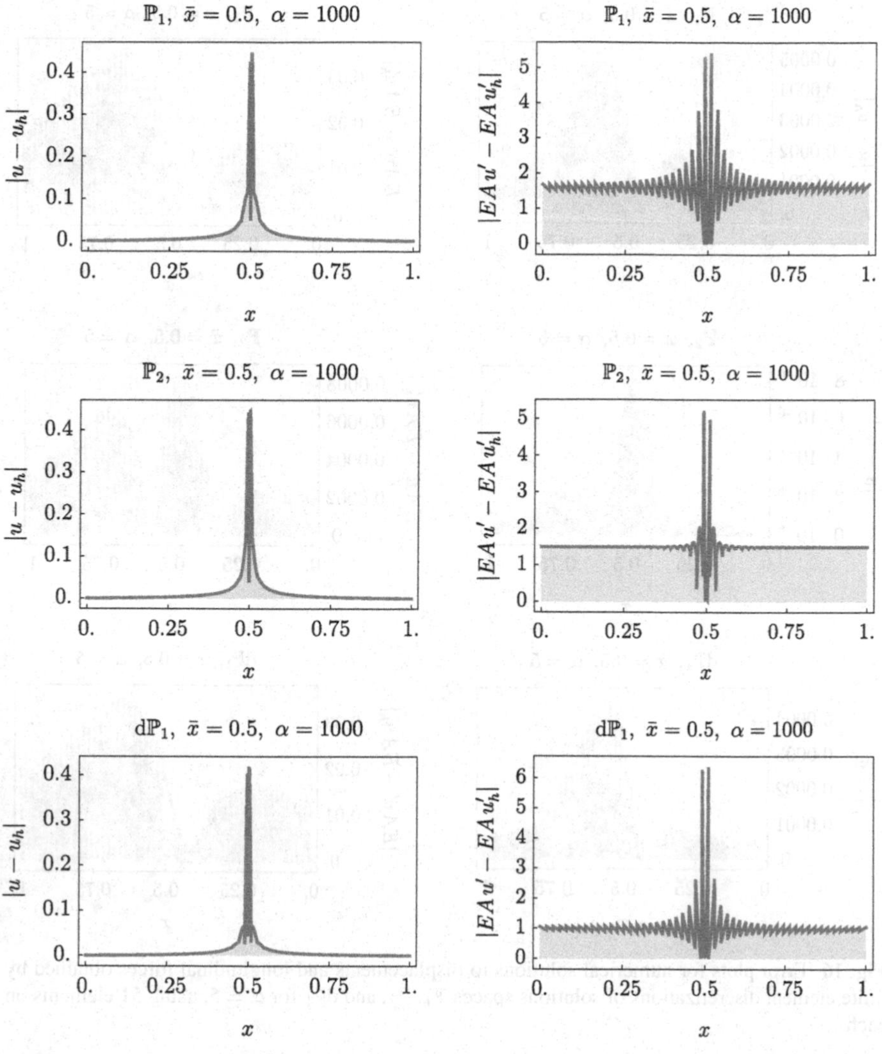

Fig. 15 Error plots for numerical solutions to displacements and longitudinal forces obtained by finite element discretizations of solutions spaces $\mathbb{P}_1$, $\mathbb{P}_2$, and $\mathrm{d}\mathbb{P}_1$ for $\alpha = 1000$, using 51 elements on each

i.e., there exist an $\alpha \in \mathbb{R}_+$, such that

$$a(v, v) \geq \alpha \|v\|_V^2 \quad \forall\, v \in K,$$

and

2. the inf-sup condition, also known as LBB-condition (Ladyzhenskaya–Babuška–Brezzi), is verified, i.e., there exists a $\beta \in \mathbb{R}^+$, such that

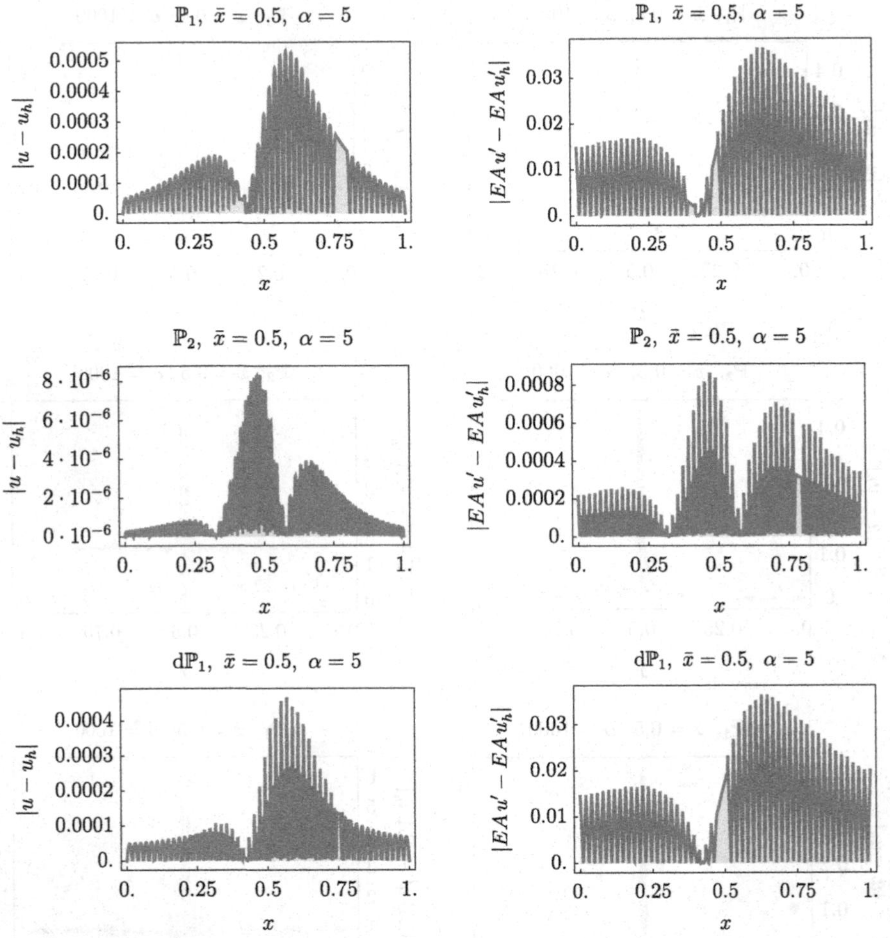

Fig. 16 Error plots for numerical solutions to displacements and longitudinal forces obtained by finite element discretizations of solutions spaces $\mathbb{P}_1$, $\mathbb{P}_2$, and $\mathrm{d}\mathbb{P}_1$ for $\alpha = 5$, using 51 elements on each

$$\inf_{q \in \Sigma \setminus 0} \sup_{v \in V \setminus 0} \frac{b(v,q)}{\|q\|_\Sigma \, \|v\|_V} \geq \beta \,.$$

Furthermore, there exists the a priori estimate for the solution

$$\|u\|_V \leq \frac{1}{\alpha} \|f\|_{V'} + \frac{1}{\beta} \left(1 + \frac{C}{\alpha}\right) \|q\|_{\Sigma'} \tag{98}$$

and

$$\|p\|_\Sigma \leq \frac{1}{\beta} \left(1 + \frac{C}{\alpha}\right) \|f\|_{V'} \,. \tag{99}$$

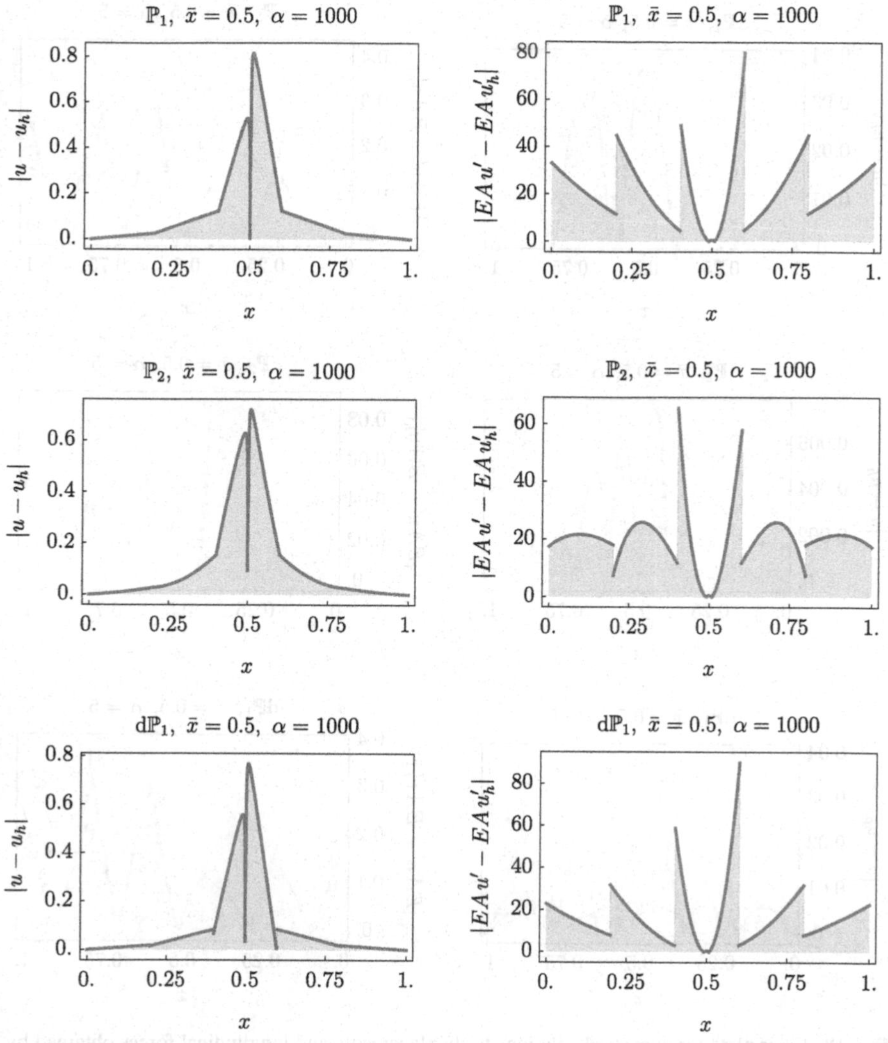

Fig. 17 Error plots for numerical solutions to displacements and longitudinal forces obtained by finite element discretizations of solutions spaces $\mathbb{P}_1$, $\mathbb{P}_2$, and $\mathrm{d}\mathbb{P}_1$ for $\alpha = 1000$, using 5 elements on each

6.2 Treatment of Saddle Point Problems, Sensitization

The discrete mixed problem is given by the matrix representation

$$\underbrace{\begin{pmatrix} \boldsymbol{A} & \boldsymbol{B}^T \\ \boldsymbol{B} & \boldsymbol{0} \end{pmatrix}}_{\widehat{\boldsymbol{K}}} \underbrace{\begin{pmatrix} \boldsymbol{d}_u \\ \boldsymbol{d}_p \end{pmatrix}}_{\widehat{\boldsymbol{d}}} = \underbrace{\begin{pmatrix} \boldsymbol{r}_u \\ \boldsymbol{r}_p \end{pmatrix}}_{\widehat{\boldsymbol{r}}}, \tag{100}$$

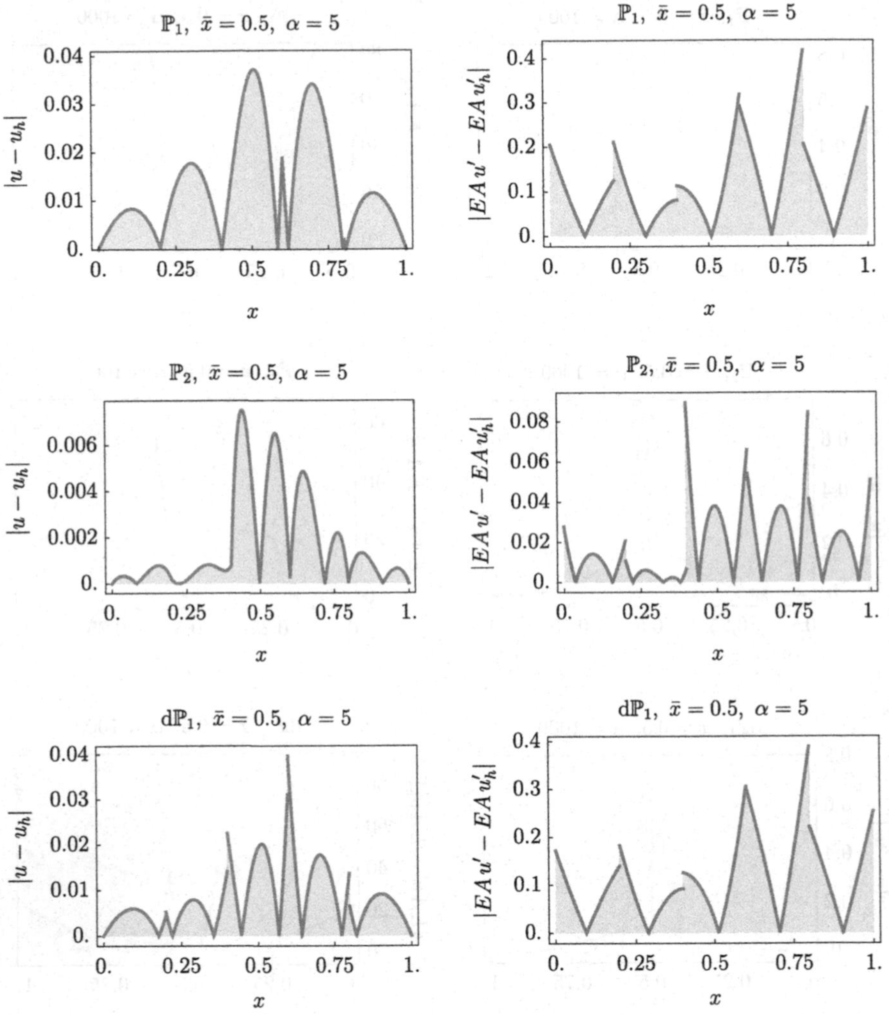

Fig. 18 Error plots for numerical solutions to displacements and longitudinal forces obtained by finite element discretizations of solutions spaces $\mathbb{P}_1$, $\mathbb{P}_2$, and $\mathrm{d}\mathbb{P}_1$ for $\alpha = 5$, using 5 elements on each

with $\boldsymbol{A} \in \mathbb{R}^{n\times n}$, $\boldsymbol{B} \in \mathbb{R}^{m\times n}$, $\boldsymbol{B}^T \in \mathbb{R}^{n\times m}$, $(\boldsymbol{d}_u, \boldsymbol{r}_u) \in \mathbb{R}^n$, $(\boldsymbol{d}_p, \boldsymbol{r}_p) \in \mathbb{R}^m$, $\widehat{\boldsymbol{K}} \in \mathbb{R}^{(n+m)\times(n+m)}$, and $(\widehat{\boldsymbol{d}}, \widehat{\boldsymbol{r}}) \in \mathbb{R}^{n+m}$.

For the **solvability** of (100) we postulate that the system has a unique solution for every right-hand side $\boldsymbol{r}_u$ and $\boldsymbol{r}_p$. Obviously, this condition is fulfilled if $\widehat{\boldsymbol{K}}$ is nonsingular. In other words, we must have a continuous dependency of the solution upon the right-hand side. Therefore, the existence of a constant c, satisfying

$$\|\boldsymbol{d}_u\|_? + \|\boldsymbol{d}_p\|_? \leq c\,(\|\boldsymbol{r}_u\|_? + \|\boldsymbol{r}_p\|_?)\,, \tag{101}$$

is required. However, the existence of c does not depend on the chosen norms, because in finite dimensions all norms are equivalent. Indeed the numerical values will depend on the dimension of the system. As examples consider $\boldsymbol{u} \in \mathbb{R}^n$ with the equivalent norms

$$\|\boldsymbol{u}\|_1 := \sum_i |u_i| \quad \text{and} \quad \|\boldsymbol{u}\|_2 := \sqrt{\sum_i |u_i|^2}\,. \tag{102}$$

For $n < \infty$ there exist the two positive constants c_1 and c_2 satisfying

$$c_1\|\boldsymbol{u}\|_2 \leq \|\boldsymbol{u}\|_1 \leq c_2\|\boldsymbol{u}\|_2 \quad \text{with optimal values} \quad c_1 = 1, c_2 = \sqrt{n}\,. \tag{103}$$

For $n \to \infty$ the latter inequality becomes unbounded from above.

In addition to the solvability condition we are interested in an estimate of the **stability** of (100): In general we consider a sequence of discrete saddle point problems with increasing mesh densities $h \to 0$ and therefore with increasing dimensions. Let $k = 0, 1, 2, 3, ..$ denote a sequence of discretizations with increasing mesh densities, i.e., we consider

$$\begin{pmatrix} \boldsymbol{A}_k & \boldsymbol{B}_k^T \\ \boldsymbol{B}_k & \boldsymbol{0} \end{pmatrix} \begin{pmatrix} \boldsymbol{d}_u^k \\ \boldsymbol{d}_p^k \end{pmatrix} = \begin{pmatrix} \boldsymbol{r}_u^k \\ \boldsymbol{r}_p^k \end{pmatrix} \tag{104}$$

with $\boldsymbol{A}_k \in \mathbb{R}^{n_k \times n_k}$, $\boldsymbol{B}_k \in \mathbb{R}^{m_k \times n_k}$, ..., where the dimensions n_k, m_k increase with the sequence of k. In addition to the solvability condition we are interested in an estimate

$$\|\boldsymbol{d}_u^k\|_? + \|\boldsymbol{d}_p^k\|_? \leq c\,(\|\boldsymbol{r}_u^k\|_? + \|\boldsymbol{r}_p^k\|_?) \tag{105}$$

with a constant c independent on k, i.e., independent of the increasing mesh densities. For a meaningful analysis we have to specify the norms entering (105) carefully. For the stability requirement this choice is rather important, because constants appearing in the relations between equivalent (discrete) norms depend on the dimension of the problem, which goes to infinity with $h \to 0$. Obviously, the stability is a concept which has to be applied to a sequence of discretized boundary value problems.

Incidental remark: In order to get an impression on the well-known dependency of some characteristic values of a discretized system on the mesh density we perform an **eigenvalue analysis** of a one-dimensional elasticity problem, minimizing the discrete energy potential functional on $\mathcal{B} \in [0, 1] = \mathcal{B}_h = \bigcup_e \mathcal{B}^e$

$$\Pi(u_h) = \sum_{\mathcal{B}^e} \int_{\mathcal{B}^e} \left(\frac{1}{2} EA(u_h')^2 - f\,u_h\right) \mathrm{d}x\,. \tag{106}$$

In order to find the minimum we compute $\delta\Pi(u_h, \delta u_h) = 0$, with

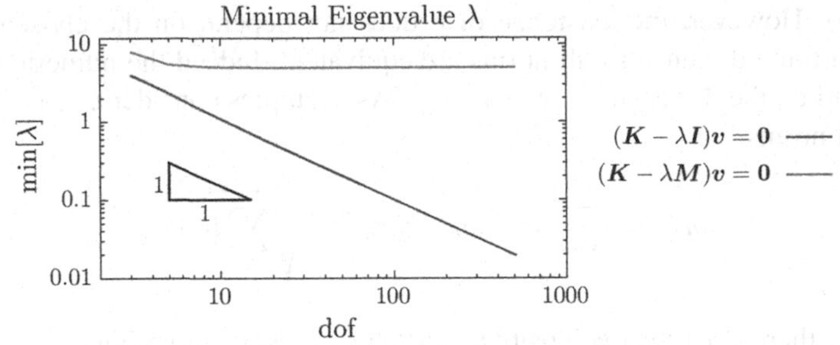

Fig. 19 Minimal eigenvalue depicted over number of unknowns (dof)

$$\delta\Pi(u_h, \delta u_h) = \sum_{\mathcal{B}^e} \int_{\mathcal{B}^e} \left(\delta u_h' \, EA \, u_h' - \delta u_h f\right) \mathrm{d}x \,. \tag{107}$$

We investigate the evolution of the minimal eigenvalue $\lambda_{\min}$ for the eigenvalue problems

$$(\boldsymbol{K}_k - \lambda_k \boldsymbol{I}_k)\boldsymbol{v}^k = \boldsymbol{0} \quad \text{and} \quad (\boldsymbol{K}_k - \lambda_k \boldsymbol{M}_k)\boldsymbol{v}^k = \boldsymbol{0} \,, \tag{108}$$

where $\boldsymbol{K}_k$ is the stiffness matrix, λ_k the associated eigenvalue to the eigenvector $\boldsymbol{v}^k$, $\boldsymbol{I}_k$ the identity and $\boldsymbol{M}_k$ the mass matrix. We consider a truss clamped at the edges, i.e., the Dirichlet boundary conditions $u(0) = u(1) = 0$, a Young's modulus $E = 1$, a cross section $A = 1$. Figure 19 shows the evolution of the minimal eigenvalues with respect to mesh refinement.

Obviously, the eigenvalue problem $(108)_1$ exhibits a decrease of the amplitude of the lowest eigenvalue with increasing mesh density (from $h = 1/2 \rightarrow h = 1/500$) whereas the formulation $(108)_2$ seems to offer a lower bound for $\min(\lambda_k)$. ■

6.3 A Saddle Point Problem-Finite-Dimensional Case

Let us concentrate on the saddle point problem for a linear incompressible material behavior. Starting from the general strong form of elasticity given in (62) and substituting the pressure as an additional unknown field as $p = \lambda \, \mathrm{tr}(\boldsymbol{\varepsilon}(\boldsymbol{u}))$ leads for the incompressible case $\lambda \rightarrow \infty$ to

$$\begin{aligned} &\mathrm{div}(2\,\mu\,\boldsymbol{\varepsilon}(\boldsymbol{u}) + p\boldsymbol{I}) + \boldsymbol{f} = \boldsymbol{0}\,, \\ &\mathrm{tr}(\boldsymbol{\varepsilon}(\boldsymbol{u})) = 0\,, \\ &\boldsymbol{u} = 0 \quad \text{on} \quad \partial\mathcal{B}_D\,, \\ &\boldsymbol{\sigma} \cdot \boldsymbol{n} = \bar{\boldsymbol{t}} \quad \text{on} \quad \partial\mathcal{B}_N\,. \end{aligned} \tag{109}$$

The solution of this problem is similar to the saddle point of the potential

$$\Pi(\boldsymbol{u}, p) = \int_{\mathcal{B}} (\mu \, \nabla_s \boldsymbol{u} : \nabla_s \boldsymbol{u} + p \, \operatorname{div}(\boldsymbol{u}) - \boldsymbol{f} \cdot \boldsymbol{u}) \, \mathrm{d}V - \int_{\partial \mathcal{B}_N} \bar{\boldsymbol{t}} \cdot \boldsymbol{u} \, \mathrm{d}A \tag{110}$$

The variational approach and a finite element discretization leads to the finite-dimensional saddle point problem

$$\begin{aligned}
\sum_{e=1}^{\mathrm{num_{ele}}} \Big\{ \underbrace{\int_{\mathcal{B}^e} 2\mu \, \nabla_s \delta \boldsymbol{u}_h : \nabla_s \boldsymbol{u}_h \, \mathrm{d}v}_{a(\delta \boldsymbol{u}_h, \boldsymbol{u}_h)} + \underbrace{\int_{\mathcal{B}^e} p_h \, \operatorname{div}(\delta \boldsymbol{u}_h) \mathrm{d}v}_{b(\delta \boldsymbol{u}_h, p_h)} \Big\} - \underbrace{\int_{\mathcal{B}^e} \boldsymbol{f}_h \cdot \delta \boldsymbol{u}_h \, \mathrm{d}v}_{f(\boldsymbol{u}_h)} &= 0 \\
\sum_{e=1}^{\mathrm{num_{ele}}} \underbrace{\int_{\partial \mathcal{B}^e} \delta p_h \, \operatorname{div}(\boldsymbol{u}_h) \mathrm{d}v}_{b(\boldsymbol{u}_h, \delta p_h)} &= 0 \, .
\end{aligned} \tag{111}$$

The equivalent problem is the stationarity requirement of the discrete Lagrange functional

$$\mathcal{L}_h(\boldsymbol{d}_u, \boldsymbol{d}_p) = \frac{1}{2} \boldsymbol{d}_u^T \, \boldsymbol{A}_h \, \boldsymbol{d}_u - \boldsymbol{d}_u^T \boldsymbol{f}_h + \boldsymbol{d}_p^T \, \boldsymbol{B}_h \, \boldsymbol{d}_u \, , \tag{112}$$

i.e., $\delta_{d_u} \mathcal{L}_h = 0$ and $\delta_{d_p} \mathcal{L}_h = 0$, which leads to

$$\begin{aligned}
\delta_{d_u} \mathcal{L}_h &= \delta \boldsymbol{d}_u^T \{ \boldsymbol{A}_h \boldsymbol{d}_u - \boldsymbol{f}_h + \boldsymbol{B}_h^T \boldsymbol{d}_p \} \qquad \forall \, \delta \boldsymbol{d}_u \, , \\
\delta_{d_p} \mathcal{L}_h &= \delta \boldsymbol{d}_p^T \{ \boldsymbol{B}_h \, \boldsymbol{d}_u \} \qquad \forall \, \delta \boldsymbol{d}_p \, ,
\end{aligned} \tag{113}$$

where the field quantities have been substituted by the approximations

$$\boldsymbol{u}_h = \mathbb{N}_u \, \boldsymbol{d}_u \, , \quad \delta \boldsymbol{u}_h = \mathbb{N}_u \, \delta \boldsymbol{d}_u \, , \quad p_h = \mathbb{N}_p \, \boldsymbol{d}_p \, , \quad \delta p_h = \mathbb{N}_p \, \delta \boldsymbol{d}_p \, , \tag{114}$$

with $\mathbb{N}$ and $\boldsymbol{d}$ denoting the shape functions and nodal values corresponding to the displacements, pressure and its virtual counterparts. The solution of this set of algebraic equation follows from

$$\begin{bmatrix} \boldsymbol{A}_h & \boldsymbol{B}_h^T \\ \boldsymbol{B}_h & \boldsymbol{0} \end{bmatrix} \begin{bmatrix} \boldsymbol{d}_u \\ \boldsymbol{d}_p \end{bmatrix} = \begin{bmatrix} \boldsymbol{f}_h \\ \boldsymbol{0} \end{bmatrix} , \tag{115}$$

with $\boldsymbol{A}_h \in \mathbb{R}^{n \times n}$, $\boldsymbol{A}_h = \boldsymbol{A}_h^T$ and positive definite, $\boldsymbol{B}_h \in \mathbb{R}^{m \times n}$, $\boldsymbol{f}_n \in \mathbb{R}^n$ and $m < n$. The physical interpretation of $m < n$ is obvious, there must be less constraints than "free" variables. Obviously, we have the "identical" structure as we obtain from equation (111). Now we have to ensure that (115) is solvable for all right-hand sides $\boldsymbol{f}_h$, following the remarks of Devendran et al. (2009). This is of course the fact if the whole matrix is invertible, i.e., nonsingular. Let us consider the congruent transformation, known as Sylvester's law of inertia

$$\begin{bmatrix} \boldsymbol{A}_h & \boldsymbol{B}_h^T \\ \boldsymbol{B}_h & \boldsymbol{0} \end{bmatrix} = \begin{bmatrix} \boldsymbol{I} & \boldsymbol{0} \\ \boldsymbol{B}_h \boldsymbol{A}_h^{-1} & \boldsymbol{I} \end{bmatrix} \begin{bmatrix} \boldsymbol{A}_h & \boldsymbol{0} \\ \boldsymbol{0} & -\boldsymbol{B}_h \boldsymbol{A}_h^{-1} \boldsymbol{B}_h^T \end{bmatrix} \begin{bmatrix} \boldsymbol{I} & \boldsymbol{A}_h^{-1} \boldsymbol{B}_h^T \\ \boldsymbol{0} & \boldsymbol{I} \end{bmatrix} , \tag{116}$$

this transformation preserves the number of positive and negative eigenvalues (but not their numerical values). However, our system has full rank if the Schur complement

$$S_h = -B_h A_h^{-1} B_h^T \quad (= S_h^T) \tag{117}$$

is nonsingular. In this case S_h is invertible and we can solve system (115). The full rank requirement is equivalent to

$$d_p^T B_h A_h^{-1} B_h^T d_p > 0 \quad \forall d_p \in \mathbb{R}^m \backslash 0, \tag{118}$$

i.e., the Schur complement is negative definite. Due to the assumption that A_h and therefore A_h^{-1} is positive definite we argue

$$B_h^T d_p = 0 \quad \text{iff} \quad d_p = \mathbf{0}\,. \tag{119}$$

This means that the kernel of B^T, i.e.,

$$\mathrm{Ker}(B_h^T) := \left\{ d_p \in R^m : B_h^T d_p = \mathbf{0} \right\}, \tag{120}$$

is trivial, i.e., the image of B^T is

$$\mathrm{Im}(B_h^T) = \mathbb{R}^m, \tag{121}$$

in other words $B_h^T \in \mathbb{R}^{n \times m}$, with $m < n$ has full column rank. If this conditions are fulfilled the system (115) is invertible, i.e.,

$$\begin{bmatrix} A_h & B_h^T \\ B_h & \mathbf{0} \end{bmatrix}^{-1} = \begin{bmatrix} A_h^{-1}(I - B_h^T S_h^{-1} B_h A_h^{-1}) & A_h^{-1} B_h^T S_h^{-1} \\ S_h^{-1} B_h A_h^{-1} & S_h^{-1} \end{bmatrix}. \tag{122}$$

Let $\beta^2 > 0$ denote the smallest singular value of B_h. The condition that the smallest eigenvalue β, is greater than zero is directly related to the inf-sup condition of saddle point problems, which states

$$\inf_{d_p \in \mathbb{R}^m \backslash 0} \ \sup_{d_u \in \mathbb{R}^n \backslash 0} \frac{d_p^T B_h^T d_u}{\|d_p\| \, \|d_u\|} \geq \beta^2 > 0 \tag{123}$$

or equivalently

$$\max_{d_u \in \mathbb{R}^m \backslash 0} \frac{d_p^T B_h^T d_u}{\|d_u\|} \geq \beta^2 \, \|d_p\| \quad \forall d_p \in \mathbb{R}^m. \tag{124}$$

The independency of the mesh size, as discussed for Eq. (105), demands here β to be bounded above zero for $h \to 0$.

Furthermore, we obtain the bounds

$$
\begin{aligned}
\|\boldsymbol{d}_u\|_{A_h} &\leq \|\boldsymbol{f}_h\|_{A_h^{-1}} \leq \frac{1}{\alpha}\|\boldsymbol{f}_h\| \\
\|\boldsymbol{d}_p\| &\leq \frac{1}{\beta}\|\boldsymbol{f}_h\|_{A^{-1}} \leq \frac{1}{\alpha\,\beta}\|\boldsymbol{f}\|\,,
\end{aligned}
\tag{125}
$$

with the energy norm $\|\boldsymbol{d}_u\|_A = \sqrt{\boldsymbol{d}_u^T \boldsymbol{A}\,\boldsymbol{d}_u}$. Obviously, if β is small the bound for $\boldsymbol{d}_p$ gets large.

Numerical Inf-Sup Test The numerical inf-sup test was proposed by Chapelle and Bathe (1993). In order to evaluate the inf-sup constant we use the fact that it is equivalent to the square root of the smallest eigenvalue of

$$
\left(\boldsymbol{B}_h \boldsymbol{M}_{u,h}^{-1} \boldsymbol{B}_h^T - \Lambda\,\boldsymbol{M}_{p,h}\right) \boldsymbol{d}_p = \boldsymbol{0}\,. \tag{126}
$$

with the global mass matrices $\boldsymbol{M}_{u,h}$, $\boldsymbol{M}_{p,h}$ as

$$
\boldsymbol{M}_{u,h} = \sum_{e=1}^{\text{num}_{\text{ele}}} \int_{\mathcal{B}^e} \mathbb{B}_u^T \mathbb{B}_u \,\mathrm{d}v \quad \text{and} \quad \boldsymbol{M}_{p,h} = \sum_{e=1}^{\text{num}_{\text{ele}}} \int_{\mathcal{B}^e} \mathbb{N}_p^T \mathbb{N}_p \,\mathrm{d}v\,, \tag{127}
$$

whereas, $\mathbb{B}_u$ contains spatial derivatives of the shape functions such that it holds $\boldsymbol{\varepsilon}_h = \mathbb{B}_u \boldsymbol{d}_u$. For exemplary purposes, the inf-sup stability is investigated by means of an inf-sup test on the example of the well-known Q_1P_0 and T_2P_0, representing elements with a discontinuous pressure approximation, and the T_2P_1, representing an element with a continuous pressure approximation, see Hood and Taylor (1974). The considered boundary value problem is a simple supported rectangle in 2D and brick in 3D, whereas a consecutive number of mesh refinements is considered. The statement on the inf-sup criterium of the considered elements are well known and the formal proofs can be found in the literature, e.g., Boffi et al. (2009). The T_2P_1 element is a famous representative of the Taylor–Hood family, which is well known to be inf-sup stable. In contrast the Q_1P_0 formulation is a text book example for an element which does not satisfy the inf-sup criterium neither in the two-dimensional nor in the three-dimensional case. Interestingly the T_2P_0 formulation fulfills the inf-sup condition in the two-dimensional case but fails in three dimensions. The depicted results in Fig. 20 approve numerically the statements on the inf-sup stability of the elements: In 2D the T_2P_0 and the T_2P_1 elements have an approximately constant $\Lambda > 0$, whereas Λ tends to zero for the Q_1P_0 element. In 3D, only the T_2P_1 seems to have a bounded value for Λ. Of course, a purely numerical check of the LBB condition is not sufficient, but it gives a first impression of the properties of the solution. To be save, a rigorous mathematical proof is needed.

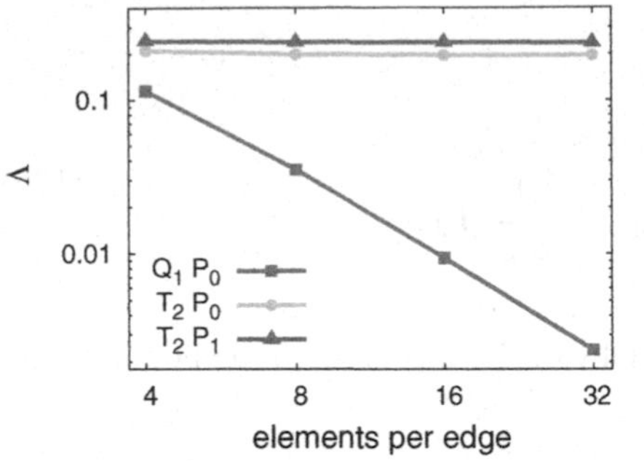

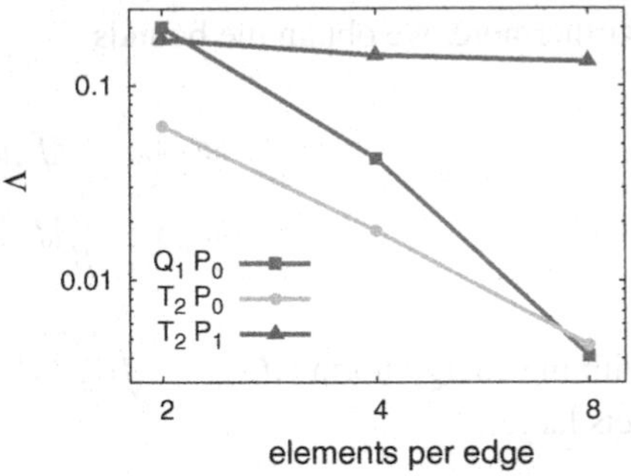

Fig. 20 Inf-Sup test results in 2D (left) and 3D (right). The inf-sup test is satisfied if Λ is bounded above zero, independent of the element size

Acknowledgements We thank the DFG for the financial support within the SPP 1748 *Reliable Simulation Techniques in Solid Mechanics. Development of Non-standard Discretization Methods, Mechanical and Mathematical Analysis*, project *Novel finite elements— Mixed, Hybrid and Virtual Element formulations* (Projectnumber: 255432295) (SCHR 570/23-2). I would also like to thank Nils Viebahn for helpful discussions and his help with the manuscript and Sascha Maassen and Rainer Niekamp for the implementation of the H^{-1}procedure and accompanying discussions.

A Sobolev and Hilbert Spaces

In the following we will use the Sobolev and Hilbert Spaces, they are based on the space of square integrable functions on $\mathcal{B}$:

$$L^2(\mathcal{B}) = \left\{u : \|u\|^2_{L^2(\mathcal{B})} = \int_{\mathcal{B}} |u|^2 \mathrm{d}v < +\infty\right\}. \tag{128}$$

Let $s \geq 0$ be a real number, the standard notation for a Sobolev space is $H^s(\mathcal{B})$ and $H^s(\partial\mathcal{B})$ with the inner products and norm

$$(u, u)_{s,\mathcal{B}}, \quad (u, u)_{s,\partial\mathcal{B}} \quad \text{and} \quad \|u\|_{s,\mathcal{B}}, \quad \|u\|_{s,\partial\mathcal{B}}, \tag{129}$$

respectively. For $s = 0$ the space $H^0(\mathcal{B})$ represents the Hilbert space $L^2(\mathcal{B})$ of all square integrable functions, i.e.,

$$L^2(\mathcal{B}) = H^0(\mathcal{B}) = \{u \in L^2(\mathcal{B})\}. \tag{130}$$

If s is a *positive integer* the spaces $H^s(\mathcal{B})$ consist of all square integrable functions whose derivatives up to the order s are also square integrable, i.e.,

$$H^s(\mathcal{B}) = \{u + \sum_{\alpha=1}^{s} \mathrm{D}^\alpha u \in L^2(\mathcal{B})\}. \tag{131}$$

Here we shall use the semi-norms

$$|u|_{k,\mathcal{B}} := \sqrt{\sum_{\alpha=k} |\mathrm{D}^\alpha u|^2_{L^2(\mathcal{B})}}, \qquad k = 0, 1, \ldots, s, \tag{132}$$

and the norm

$$\|u\|_{s,\mathcal{B}} := \sqrt{\sum_{k \le s} |u|^2_{k,\mathcal{B}}}. \tag{133}$$

Critism: This expression for the norm does not take into account a typical length scale l of the problem, i.e., we are adding, for example, a square integrable function $|u|^2_{L^2(\mathcal{B})}$ and its square integrable derivative $|u'|^2_{L^2(\mathcal{B})}$. Without any physically meaningful parameters these expression is hardly to interpret. This could be avoided by using the expression

$$\|u\|_{s,\mathcal{B}} := \sqrt{\sum_{k \le s} l^{\mathrm{d}k}\ |u|^2_{k,\mathcal{B}}}, \tag{134}$$

where d characterizes the dimension of $\mathcal{B} \subset \mathbb{R}^{\mathrm{d}}$, Boffi et al. (2013).

With D^α as the α-st weak differential operator. Thus the often used spaces $H^1(\mathcal{B})$ and $H^1_0(\mathcal{B})$ are defined by

$$H^1(\mathcal{B}) = \{u + \mathrm{D}^1 u \in L^2(\mathcal{B})\}, \tag{135}$$

and

$$H^1_0(\mathcal{B}) = \{u \in H^1(\mathcal{B})\ :\ u = 0 \text{ on } \partial\mathcal{B}\}. \tag{136}$$

For completeness we introduce the spaces $H^2(\mathcal{B})$ and $H^2_0(\mathcal{B})$ defined by

$$H^2(\mathcal{B}) = \{u + \mathrm{D}^1 u + \mathrm{D}^2 u \in L^2(\mathcal{B})\}, \tag{137}$$

and

$$H^2_0(\mathcal{B}) = \{u \in H^2(\mathcal{B})\ :\ u = 0 \text{ and } \frac{\partial u}{\partial n} = 0 \text{ on } \partial\mathcal{B}_u\}. \tag{138}$$

For negative superscripts, i.e., $H^{-s}(\mathcal{B})$ with $s > 0$, the spaces are identified with the duals of $H^s_0(\mathcal{B})$:

$$H^{-s}(\mathcal{B}) = (H^s_0(\mathcal{B}))'. \tag{139}$$

For example, the norm associated to $H^{-1}(\mathcal{B})$, which is the dual of $H^1_0(\mathcal{B})$, is defined as

$$\|u\|_{-1,\mathcal{B}} = \min_{v \in H_0^1(\mathcal{B})\backslash 0} \frac{(u, v)_{0,\mathcal{B}}}{\|v\|_{1,\mathcal{B}}} . \tag{140}$$

The norm associated to $H^{-1/2}(\partial\mathcal{B})$, the dual of $H_0^{1/2}(\partial\mathcal{B})$, is defined as

$$\|u\|_{-1/2,\partial\mathcal{B},0} = \min_{v \in H^{1/2}(\partial\mathcal{B})\backslash 0} \frac{(u, v)}{\|v\|_{1/2,\partial\mathcal{B}}} . \tag{141}$$

The Hilbert space $H_0^m(\mathcal{B})$ is a closed subspace of $H^m(\mathcal{B})$; furthermore is $H_0^0(\mathcal{B}) = L_2(\mathcal{B})$.

$$\begin{aligned} \ldots H^{-2}(\mathcal{B}) \supseteq H^{-1}(\mathcal{B}) \supseteq L_2(\mathcal{B}) \supseteq H_0^1(\mathcal{B}) \supseteq H_0^2(\mathcal{B}) \ldots \\ \ldots \|u\|_{-2,\mathcal{B}} \leq \|u\|_{-1,\mathcal{B}} \leq \|u\|_{0,\mathcal{B}} \leq \|u\|_{1,\mathcal{B}} \leq \|u\|_{2,\mathcal{B}} \ldots \end{aligned} \tag{142}$$

For tensorial Sobolev spaces, e.g., the three-dimensional tensor product space

$$H^s(\mathcal{B}) \times H^s(\mathcal{B}) \times H^s(\mathcal{B}) \tag{143}$$

we use the abbreviation

$$[H^s(\mathcal{B})]^3 = \prod_{i=1}^{3} H^s(\mathcal{B}) \quad \text{and analogously} \quad [L^2(\mathcal{B})]^3 = \prod_{i=1}^{3} L^2(\mathcal{B}) . \tag{144}$$

Let $\boldsymbol{u} \in \mathbb{R}^3$ and set the Hilbert space

$$H(\text{div}; \mathcal{B}) = \left\{ \boldsymbol{u} \in [L^2(\mathcal{B})]^3 \ : \ \text{div}\, \boldsymbol{v} \in L^2(\mathcal{B}) \right\} , \tag{145}$$

with the associated norm

$$\|\boldsymbol{v}\|_{H(\text{div};\mathcal{B})} = \left\{ \|\boldsymbol{v}\|^2 + |\,\text{div}\, \boldsymbol{v}|^2 \right\}^{1/2} . \tag{146}$$

References

Auricchio, F., Brezzi, F., & Lovadina, C. (2004). Mixed finite element methods. In E. Stein, R. de Borst, & T. J. R. Hughes (Eds.), *Encyclopedia of computational mechanics* (Chap. 9, pp. 238–277). Wiley and Sons.

Bathe, K.-J. (1996). *Finite element procedures*. New Jersey: Prentice Hall.

Becker, E. B., Carey, G. F. & Oden, J. T. (1981). *Finite elements, an introduction: Volume I*. Prentice-Hall.

Berdichevsky, V. L. (2009). *Variational principles of continuum mechanics*. Springer.

Boffi, D., Brezzi, F., & Fortin, M. (2009). Reduced symmetry elements in linear elasticity. *Communications on Pure and Applied Analysis*, *8*, 95–121.

Boffi, D., Brezzi, F., & Fortin, M. (2013). *Mixed finite element methods and applications*. Heidelberg: Springer.

Braess, D. (1997). *Finite elemente* (2nd ed.). Berlin: Springer.
Brenner, S. C. & Scott, L. R. (2002). The mathematical theory of finite element methods. In *Texts in applied mathematics* (Vol. 15, 2nd edition). New York: Springer.
Brezzi F. (1974). On the existence, uniqueness and approximation of saddle-point problems arising from lagrangian multipliers. *Revue française d'automatique, informatique, recherche opérationnelle. Analyse numérique, 8*(2), 129–151.
Chapelle, D., & Bathe, K.-J. (1993). The inf-sup test. *Computers and Structures*, *47*, 537–545.
Crouzeix, M., & Raviart, P.-A. (1973). Conforming and nonconforming finite element methods for solving the stationary stokes equations i. *Revue francaise d'automatique, informatique, recherche operationnelle*, *7*(3), 33–75.
Devendran, D., May, S., & Corona, E. (2009). Computational fluid dynamics reading group: Finite element methods for stokes and the infamous inf-sup condition.
Ern, A., & Guermond, J.-L. (2013). *Theory and practice of finite elements* (Vol. 159). Springer Science & Business Media.
Gockenbach, M. S. (2006). *Understanding and implementing the finite element method*. SIAM.
Goebbels, S. (2015). An inequality for negative norms with application to errors of finite element methods.
Hood, P., & Taylor, C. (1974). Navier-stokes equations using mixed interpolation. In J. T. Oden, O. C. Zienkiewicz, R. H. Gallagher, & C. Taylor (Eds.), *Finite element methods in flow problems* (pp. 121–132). UAH Press.
Hughes, T. J. R. (1987). *The finite element method*. Englewood Cliffs, New Jersey: Prentice Hall.
Kendall, R. P., & Wheeler, M. F. (1976). A Crank-Nicolson H^{-1}-Galerkin procedure for parabolic problems in a single-space variable. *SIAM Journal on Numerical Analysis*, *13*, 861–876.
Oden, J. T., & Carey, G. F. (1983). *Finite elements. Mathematical aspects. Volume IV*. Prentice-Hall.
Oden, J. T., & Reddy, J. N. (1976). *An introduction to the mathematical theory of finite elements*. Wiley.
Rachford, H. H., Wheeler, J. R., & Wheeler, M. F. (1974). An H^{-1}-Galerkin procedure for the two-point boundary value problem. In C. de Boor (Ed.), *Mathematical Aspects of Finite Elements in Partial Differential Equations, Proceedings of Symposium, Conducted by the Mathematics Research Center, The University of Wisconsin-Madison*. Academic Press.
Royden, H. L. (1968). *Real analysis*. New York: Macmillan.
Thomée, V. (1980). Negative norm estimates and superconvergence in Galerkin methods for parabolic problems. *Mathematics of Computation*, *34*(149), 93–113.
Thomée, V. (2006). *Galerkin finite element methods for parabolic problems*. Berlin, Heidelberg: Springer.
Wriggers, P. (2008). *Nonlinear finite element methods*. Springer.

Sensitivity Analysis Based Automation of Computational Problems

Jože Korelc and Teja Melink

Abstract The paper describes automation of primal and sensitivity analysis of computational models formulated and solved by the finite element method. Based on the symbolic system *AceGen* (http://symech.fgg.uni-lj.si/), fast and reliable code can be created with minimum effort and immediately tested and verified by using the associated finite element program *AceFEM* . Automation of first- and second-order sensitivity analysis with respect to an arbitrary parameter is presented. In an example, it is shown how sensitivity analysis has become an indispensable part of modern computational algorithms.

1 Introduction

Contemporary finite element software is mostly handwritten and based on formulations that were derived by scientists and software engineers. The related process is slow and can take more than several weeks to derive for a new finite element. Derivations of complex tensor fields to obtain residuals and tangent matrices are also prone to errors. To reduce the effort of developing the related new source code, symbolic code generation has been developed over the past decade. It is in a stage where the automatically generated source code is as small as the handwritten code, it is efficient and reliable. In this paper, a general approach is described that can be applied to many different applications in engineering and science. The main advantage of using symbolic code development is that the development time, especially for complex materials or elements, reduces by orders of magnitude. The paper will mainly focus on solid and structural mechanics problems. However, the general potential of the automatic code generation goes far beyond these engineering applications.

Modern finite element simulations are often coupled with optimization procedures that require additionally to the solution of primal problem also the solution of

J. Korelc (✉) · T. Melink
Faculty of Civil and Geodetic Engineering, University of Ljubljana,
Jamova 2, 1000 Ljubljana, Slovenia
e-mail: jkorelc@fgg.uni-lj.si

J. Schröder and P. de Mattos Pimenta (eds.), *Novel Finite Element Technologies for Solids and Structures*, CISM International Centre for Mechanical Sciences 597,
https://doi.org/10.1007/978-3-030-33520-5_2

sensitivity problem. The aim of the sensitivity analysis is to calculate derivatives of an arbitrary response functional with respect to chosen parameters (see e.g., Kleiber et al. (1997), Keulen et al. (2005), Choi and Kim (2005a), or Choi and Kim (2005b)). Thus, any proposed method of automation should address automation of primal as well as sensitivity analysis. The response functional can depend on arbitrary analysis model inputs (material constants, load intensity and distribution, shape parameters, etc.) as well as on arbitrary intermediate or final results of the analysis (solution vectors, derived quantities such as stress tensor, integrated quantities such as damage, etc.). The complete automation of the sensitivity analysis is thus possible only if the automatic differentiation technology is applied on the complete simulation code. This is not possible for general finite element environments. Thus, a finite difference approximation of sensitivities is used for practical applications. However, a large variety of practical problems can still be solved by the classical finite element procedure, where all problem-dependent quantities are evaluated on the individual element level and then assembled on the global level. The established algorithm is then applied on the global level to obtain the derivatives of the response functional. A comprehensive overview of the possible approaches can be found in Keulen et al. (2005). In this case we can, with the use of methods of automation, obtain analytically exact sensitivities. The use of analytically exact sensitivity analysis can significantly improve optimization procedures Choi and Kim (2005b), Kristanic and Korelc (2008), multi-scale algorithms Solinc and Korelc (2015), Korelc and Zupan (2018) and implementation of nonlinear material models Korelc and Stupkiewicz (2014), Hudobivnik and Korelc (2016).

The paper will follow the automation procedure of an analytically exact first- and second-order sensitivity analysis. In the first chapter, the necessary tools will be described that can be used to automatically derive problem-dependent quantities at the individual element level. In the second chapter, the global sensitivity problem will be formulated and solved. The third chapter introduces a set of examples that demonstrate how sensitivity analysis can be used to improve modern computational algorithms.

2 Automatic Code Generation with AceGen

The problem of automation of computational methods has been explored by researches from the fields of mathematics, computer science, and computational mechanics, resulting in a variety of approaches (e.g., the hybrid object-oriented approach by Eyheramendy and Zimmermann (2000), Logg et al. (2012) and the hybrid symbolic-numeric approach by Korelc and Wriggers (2016)) and available software tools (e.g., computer algebra systems, AD tools by Griewank (2000), problem-solving environments, and numerical libraries). Automation can address all steps of the finite element solution procedure from the strong form of a boundary-value problem to the visualization of results, or it can be applied only to the automation of the selected steps of the whole procedure.

2.1 *Hybrid Symbolic-Numerical System AceGen*

Automation of primal and sensitivity analysis is *AceGen* (http://symech.fgg.uni-lj.si/) achieved through the hybrid symbolic-numeric approach to automation of finite element method that combines symbolic and algebraic capabilities of a general computer algebra system, e.g., *Mathematica* (www.wolfram.com), an automatic differentiation technique (AD) and an automatic code generation with the general-purpose finite element environment. The structure of the hybrid symbolic-numerical system *AceGen* for multi-language and multi-environment code generation introduced by Korelc (2002) is presented in Fig. 1.

General characteristics of *AceGen* code generator are the following:

- simultaneous optimization of expressions immediately after they have been derived,
- automatic differentiation technique,
- automatic selection of the appropriate intermediate variables,
- the whole program structure can be generated,

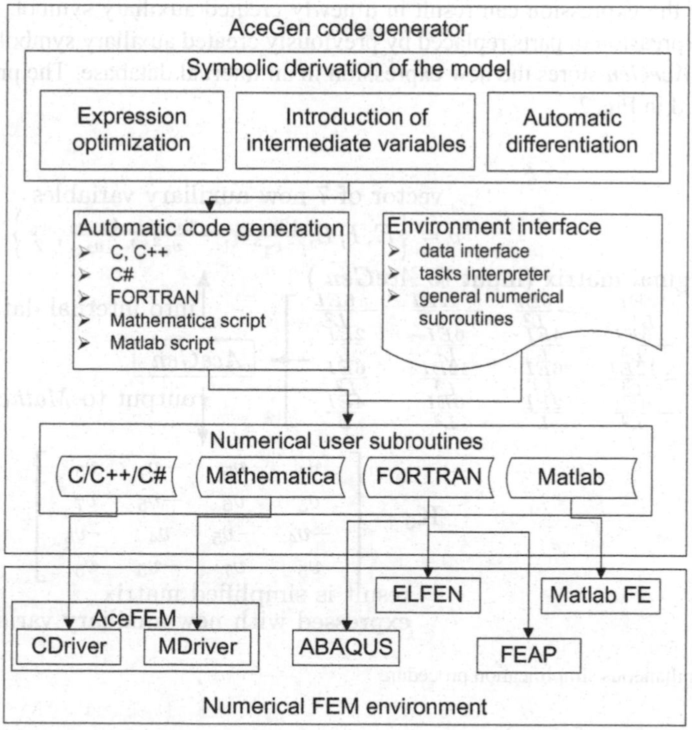

Fig. 1 Hybrid symbolic-numeric approach to automation of finite element method

- appropriate for large problems where also intermediate expressions can be subjected to uncontrolled swell,
- global expression optimization procedures with stochastic evaluation of expressions,
- differentiation with respect to indexed variables,
- automatic interface to other numerical environments,
- multi-language code generation (*Fortran/Fortran90*, C/C++, *Mathematica* language, *Matlab* language),
- advanced methods for exploring and debugging generated formulae.

The *AceGen* system is written in the symbolic language of *Mathematica*. A detailed description of the system can be found in Korelc and Wriggers (2016).

2.2 Simultaneous Simplification Procedure

Typical *AceGen* function takes the expression provided by the user, either interactively or in file, and returns an optimized version of the expression. Optimized version of the expression can result in a newly created auxiliary symbol, or in an original expression in parts replaced by previously created auxiliary symbols. In the first case, *AceGen* stores the new expression in an internal database. The procedure is presented in Fig. 2.

vector of 7 new auxiliary variables

$$v = \left\{ E, I, L, \frac{12v_1v_2}{v_3{}^3}, -\frac{6v_1v_2}{v_3{}^2}, \frac{4v_1v_2}{v_3}, \frac{v_6}{2} \right\}$$

original matrix (input to *AceGen*)

$$\mathbf{K}_0 = \begin{bmatrix} \frac{12EI}{L^3} & -\frac{6EI}{L^2} & -\frac{12EI}{L^3} & -\frac{6EI}{L^2} \\ -\frac{6EI}{L^2} & \frac{4EI}{L} & \frac{6EI}{L^2} & \frac{2EI}{L} \\ -\frac{12EI}{L^3} & \frac{6EI}{L^2} & \frac{12EI}{L^3} & \frac{6EI}{L^2} \\ -\frac{6EI}{L^2} & \frac{2EI}{L} & \frac{6EI}{L^2} & \frac{4EI}{L} \end{bmatrix}$$

into internal data base

$\boxed{AceGen}$

output to *Mathematica*

$$\mathbf{K}_0 = \begin{bmatrix} v_4 & v_5 & -v_4 & v_5 \\ v_5 & v_6 & -v_5 & v_7 \\ -v_4 & -v_5 & v_4 & -v_5 \\ v_5 & v_7 & -v_5 & v_6 \end{bmatrix}$$

result is simplified matrix
expressed with new auxiliary variables

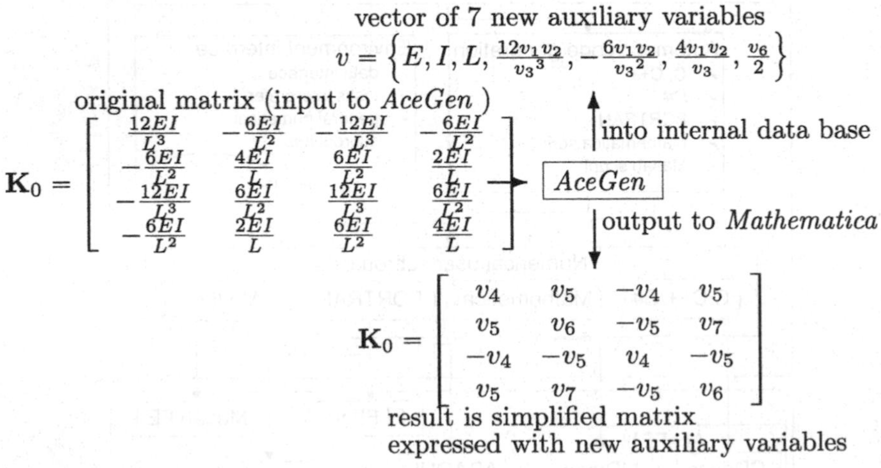

Fig. 2 Simultaneous simplification procedure

```
<< AceGen`;
SMSInitialize["DetJ", "Language" -> "C"];
SMSModule["DetJ", Real[X$$[2, 4], k$$, e$$, J$$]];
{ξ, η} ⊢ SMSReal[{k$$, e$$}];
{Xc, Yc} ⊢ SMSReal[Array[X$$, {2, 4}]];
Nh ⊨ {(1-ξ) (1-η), (1+ξ) (1-η), (1+ξ) (1+η), (1-ξ) (1+η)} / 4;
Jg ⊨ SMSD[{Nh.Xc, Nh.Yc}, {ξ, η}];
SMSExport[Det[Jg], J$$];
SMSWrite[];
```

Fig. 3 Typical *AceGen* input

2.3 Typical Example of Automatic Code Generation with AceGen

To illustrate the standard *AceGen* procedure, a simple example is considered. A typical numerical subprogram that returns a determinant of the JACOBI matrix of nonlinear transformation from the reference to initial configuration for quadrilateral element topology is derived. The syntax of the *AceGen* script language is the same as the syntax of the *Mathematica* script language with some additional functions. The input for *AceGen* is presented in Fig. 3. It can be divided into six characteristic steps:

- At the beginning of the session, the `SMSInitialize` function initializes the system.
- The `SMSModule` function defines the input and output parameters of the subroutine `DetJ`.
- The `SMSReal` function assigns the input parameters `X$$` and `k$$` and `e$$` of the subroutine to the standard *Mathematica* symbols. Double `$` character indicates that the symbol is an input or output parameter of the generated subroutine.
- During the description of the problem, special operators (⊢, ⊣, ⊨) are used to perform the simultaneous optimization of expressions and the creation of new intermediate variables. The `SMSD` function performs an automatic differentiation of one or several expressions with respect to the arbitrary variable or the vector of variables by simultaneously enhancing the already derived code.
- The results of the derivation are assigned to the output parameter `J$$` of the subroutine by the `SMSExport` function.
- At the end of the session, the `SMSWrite` function writes the contents of the vector of the generated formulae to the file in a prescribed language format.

The generated subroutine in C language is presented in Fig. 4 and in FORTRAN language in Fig. 5.

```
/****************** S U B R O U T I N E *******************/
void DetJ(double v[5001],double X[2][4],double (*k),double (*e),double (*J))
{
v[20]=(-1e0+(*k))/4e0;
v[21]=(-1e0-(*k))/4e0;
v[22]=(1e0+(*e))/4e0;
v[19]=(-1e0+(*e))/4e0;
(*J)=(v[19]*(X[0][0]-X[0][1])+v[22]*(X[0][2]-X[0][3]))*
(v[21]*(X[1][1]-X[1][2])+v[20]*(X[1][0]
 -X[1][3]))-(v[21]*(X[0][1]-X[0][2])+v[20]*(X[0][0]-X[0][3]))*
 (v[19]*(X[1][0]-X[1][1])+v[22]*
 (X[1][2]-X[1][3]));
};
```

Fig. 4 Typical automatically generated subroutine in C language

```
      SUBROUTINE DetJ(v,X,k,e,J)
      IMPLICIT NONE
      include 'sms.h'
      DOUBLE PRECISION v(5001),X(2,4),k,e,J
      v(20)=((-1d0)+k)/4d0
      v(21)=((-1d0)-k)/4d0
      v(22)=(1d0+e)/4d0
      v(19)=((-1d0)+e)/4d0
      J=(v(19)*(X(1,1)-X(1,2))+v(22)*(X(1,3)-X(1,4)))*(v(21)*(X(2,2)
     &-X(2,3))+v(20)*(X(2,1)-X(2,4)))-(v(21)*(X(1,2)-X(1,3))+v(20)*(X
     &(1,1)-X(1,4)))*(v(19)*(X(2,1)-X(2,2))+v(22)*(X(2,3)-X(2,4)))
      END
```

Fig. 5 Typical automatically generated subroutine in FORTRAN language

2.4 Automatic Differentiation

Differentiation is the most important symbolic operation needed within the algorithmic treatment of the solution process for the nonlinear boundary-value problems. This is, for example, the case for finite element methods, where NEWTON- - RAPHSON algorithms are employed to solve the nonlinear algebraic equation systems. The automatic differentiation (AD) method is used in *AceGen* for the evaluation of the exact derivatives of any arbitrary complex function via chain rule and represents an alternative solution to the numerical differentiation and symbolic differentiation. Automatic differentiation techniques are based on the fact that every computer program executes a sequence of elementary operations with known derivatives, thus allowing the evaluation of exact derivatives via the chain rule for an arbitrary complex formulation. If one has a computer code, which allows to evaluate a function f and needs to compute the gradient ∇f of f with respect to arbitrary variables, then the automatic differentiation tools, see e.g., Griewank (2000), can be applied to generate the appropriate program code.

There are two approaches for the automatic differentiation of a computer program, often recalled as the forward and the backward mode of automatic differentiation. The procedure is illustrated on a simple example of function f defined by

$$f = b\,c \quad \text{with} \quad b = \sum_{i=1}^{n} a_i^2 \quad \text{and} \quad c = \text{Sin}(b) \tag{1}$$

where $a_1, a_2, \ldots, a_n$ are n independent variables. The forward mode accumulates the derivatives of intermediate variables with respect to the independent variables as follows:

$$\begin{aligned} \nabla b &= \left\{\frac{db}{da_i}\right\} = \{2\,a_i\} && i = 1, 2, \ldots, n \\ \nabla c &= \left\{\frac{dc}{da_i}\right\} = \{\text{Cos}(b)\,\nabla b_i\} && i = 1, 2, \ldots, n \\ \nabla f &= \left\{\frac{df}{da_i}\right\} = \{\,\nabla b_i\,c + b\,\nabla c_i\,\} && i = 1, 2, \ldots, n \end{aligned} \tag{2}$$

In contrast to the forward mode, the backward mode propagates adjoin $\bar{x} = \frac{\partial f}{\partial x}$, which are the derivatives of the final values, with respect to intermediate variables:

$$\begin{aligned} \bar{f} &= \frac{df}{df} = 1 && 1 \\ \bar{c} &= \frac{df}{dc} = \frac{\partial f}{\partial c}\,\bar{f} = b\,\bar{f} && 1 \\ \bar{b} &= \frac{df}{db} = \frac{\partial f}{\partial b}\,\bar{f} + \frac{\partial c}{\partial b}\bar{c} = c\,\bar{f} + \text{Cos}(b)\,\bar{c} && 1 \\ \nabla f &= \{\bar{a}_i\} = \left\{\frac{\partial b}{\partial a_i}\,\bar{b}\right\} = \{2\,a_i\,\bar{b}\} && i = 1, 2, \ldots, n. \end{aligned} \tag{3}$$

Although obviously numerically superior when the number of functions is small, the backward mode requires potential storage of a large amount of intermediate data during the evaluation of the function that can be as high as the number of numerical operations performed. Additionally, a complete reversal of the program flow is required. This is because the intermediate variables are used in reverse order when related to their computation. For the efficient automation of the FE method, it is desirable that both approaches are available and that the software tool used for the automation can automatically select the most efficient approach for a given task. There exist many strategies how the AD procedure can be implemented, see e.g., Bischof et al. (2002). The simplest approach is to use operator overloading and during the evaluation of function f create a trace of all numerical operations and their arguments, later used to evaluate gradient in forward or backward mode. More efficient is source-to-source transformation strategy that transforms the source code for computing a function into the source code for computing the derivatives of the function.

The result of the AD procedure is called "computational derivative" and is written as $\frac{\hat{\delta} f(\mathbf{a})}{\hat{\delta}\mathbf{a}}$. The AD operator $\frac{\hat{\delta} f(\mathbf{a})}{\hat{\delta}\mathbf{a}}$ represents partial differentiation of a function $f(\mathbf{a})$ with respect to variables $\mathbf{a}$. If, for example, alternative or additional dependencies for a set of intermediate variables $\mathbf{b}$ have to be considered for differentiation, then the AD exception is indicated by the following formalism:

$$\left.\frac{\hat{\delta} f(\mathbf{a}, \mathbf{b})}{\hat{\delta} \mathbf{a}}\right|_{\frac{D\mathbf{b}}{D\mathbf{a}}=\mathbf{M}}, \tag{4}$$

which indicates that during the AD procedure, the total derivatives of variables **b** with respect to variables **a** are set to be equal to matrix **M**. The automatic differentiation exceptions are the basis for the automatic differentiation or ADB formulation of computational problem. The ADB notation can be directly translated to the AceFEM code and is part of numerically efficient code automation. Details of the method and of the corresponding software AceGen can be found in Korelc (1997), Korelc (2009) and Korelc (2018).

2.5 Automatic Differentiation and Finite Element Method

Large finite element environment usually employs a large variety of finite elements, solution procedures, and they commonly use commercial numerical libraries for which the source codes are not readily available. In such a case, it would be difficult to directly apply the AD tools to get, for example, the global stiffness matrix of a large-scale problem. However, the AD technology can still be used for the evaluation of specific quantities that appear as a part of FE simulation. For example, one can use AD at the individual element level to evaluate element-specific quantities such as

- strain and stress tensors,
- nonlinear coordinate transformations,
- consistent tangent stiffness matrix,
- residual vector and
- sensitivity pseudo-load vectors.

3 Sensitivity Analysis

The procedures for the formulation and solution of primal and sensitivity problem for an arbitrary coupled path-dependent problem are presented in detail in Korelc (2009). Here, a summary of the primal and sensitivity analysis of hyper-elastic problems is given. Let us define a primal problem with the residual equation $\mathbf{R}(\mathbf{p}) = \mathbf{0}$, where $\mathbf{p}$ represents a set of nodal unknowns of the problem. The primal problem is solved by the standard Newton–Raphson iterative procedure. For sensitivity analysis, we define the residual and the vector of unknowns as a function of a vector of design parameters $\boldsymbol{\phi} = \{\phi_1, \ldots, \phi_n\}$ as

$$\mathbf{R}(\mathbf{p}(\boldsymbol{\phi}), \boldsymbol{\phi}) = \mathbf{0}. \tag{5}$$

The sensitivity problem can be obtained from the primal problem by differentiating (5) with respect to design parameter ϕ_I. Equation (6) represents a system of linear equations for the unknown sensitivities of the primal unknowns of the problem $\frac{D\mathbf{p}}{D\phi_I}$ (8). The right-hand side (7) is called "first-order sensitivity pseudo- load vector".

$$\frac{\partial \mathbf{R}}{\partial \mathbf{p}}\frac{D\mathbf{p}}{D\phi_I}+\frac{\partial \mathbf{R}}{\partial \phi_I}=\mathbf{0} \tag{6}$$

$${}^I\tilde{\mathbf{R}}=-\frac{\partial \mathbf{R}}{\partial \phi_I} \tag{7}$$

$$\mathbf{K}\frac{D\mathbf{p}}{D\phi_I}=-{}^I\tilde{\mathbf{R}} \tag{8}$$

The sensitivity problem that is solved after the convergence of the primal problem has been reached. The second-order sensitivity problem is obtained from the first-order problem by differentiating (6) with respect to design parameter ϕ_J. It results in

$$\frac{\partial^2 \mathbf{R}}{\partial \mathbf{p}^2}\frac{D\mathbf{p}}{D\phi_I}\frac{D\mathbf{p}}{D\phi_J}+\frac{\partial^2 \mathbf{R}}{\partial \mathbf{p}\partial \phi_J}\frac{D\mathbf{p}}{D\phi_I}+\frac{\partial^2 \mathbf{R}}{\partial \mathbf{p}\partial \phi_I}\frac{D\mathbf{p}}{D\phi_J}+\frac{\partial \mathbf{R}}{\partial \mathbf{p}}\frac{D^2\mathbf{p}}{D\phi_I D\phi_J}+\frac{\partial^2 \mathbf{R}}{\partial \phi_I\partial \phi_J}=\mathbf{0} \tag{9}$$

$$\mathbf{K}\frac{D^2\mathbf{p}}{D\phi_I D\phi_J}=-{}^{IJ}\tilde{\mathbf{R}} \tag{10}$$

where $\frac{D^2\mathbf{p}}{D\phi_I D\phi_J}$ are second-order sensitivities and ${}^{IJ}\tilde{\mathbf{R}}$ represents the "second- order sensitivity pseudo-load vector" (11).

$${}^{IJ}\tilde{\mathbf{R}}=\frac{\partial^2 \mathbf{R}}{\partial \mathbf{p}^2}\frac{D\mathbf{p}}{D\phi_I}\frac{D\mathbf{p}}{D\phi_J}+\frac{\partial^2 \mathbf{R}}{\partial \mathbf{p}\partial \phi_J}\frac{D\mathbf{p}}{D\phi_I}+\frac{\partial^2 \mathbf{R}}{\partial \mathbf{p}\partial \phi_I}\frac{D\mathbf{p}}{D\phi_J}+\frac{\partial^2 \mathbf{R}}{\partial \phi_I\partial \phi_J} \tag{11}$$

The global pseudo-load vectors ${}^I\tilde{\mathbf{R}}$ and ${}^{IJ}\tilde{\mathbf{R}}$ are obtained by the standard integration over the element domain and the standard finite element assembly procedure of element contributions to global vectors

$${}^I\tilde{\mathbf{R}}=\mathop{\mathsf{A}}_{e=1}^{n_e}\sum_{g=1}^{n_g} w_g\,{}^I\tilde{\mathbf{R}}_g,\quad {}^{IJ}\tilde{\mathbf{R}}=\mathop{\mathsf{A}}_{e=1}^{n_e}\sum_{g=1}^{n_g} w_g\,{}^{IJ}\tilde{\mathbf{R}}_g \tag{12}$$

where ${}^I\tilde{\mathbf{R}}_g$ and ${}^{IJ}\tilde{\mathbf{R}}_g$ represent integration point contributions to the element pseudo-load vectors and consequently to the global pseudo-load vectors and w_g is an integration point weight. The only part of the whole procedure that depends on specific element formulation is the evaluation of the integration point pseudo-load vectors. Consequently, for the automation of the complete sensitivity analysis procedure we only need a method for automatic derivation of integration point pseudo-load vectors ${}^I\tilde{\mathbf{R}}_g$ and ${}^{IJ}\tilde{\mathbf{R}}_g$. For an arbitrary finite element formulation, this can be

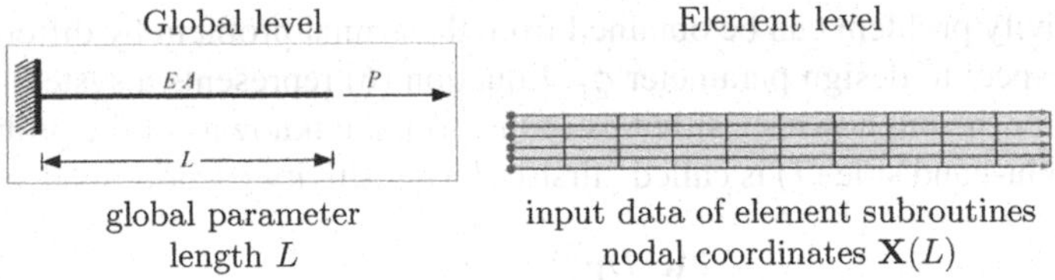

Fig. 6 Parametrization of input data of continuum and discretized problem

achieved with automatic differentiation and code optimization as described in Korelc (2009).

The obvious problem in obtaining the right-hand sides ${}^{I}\tilde{\mathbf{R}}$ and ${}^{IJ}\tilde{\mathbf{R}}$ is that an arbitrary sensitivity parameter (e.g., length of the beam in Fig. 6) does not appear explicitly as an input parameter of the finite element solution procedure, either at the global level or at the level of user subroutines. The missing dependency between an arbitrary sensitivity parameter and the finite element code is defined by "design velocity field" (Korelc and Wriggers (2016)).

3.1 Design Velocity Field

For example, let us consider shape parameter L of the beam depicted in Fig. 6 as sensitivity parameter. The relation between shape parameter L and the coordinates of an arbitrary node $\mathbf{X}^J(L)$ can be an arbitrary complex function that, in general, cannot be input data of the finite element analysis. However, it is not the relation $\mathbf{X}^J(L)$ itself that is needed within the sensitivity analysis to obtain ${}^{I}\tilde{\mathbf{R}}$ and ${}^{IJ}\tilde{\mathbf{R}}$, but its first and second derivatives. The input data for the sensitivity analysis are thus the rate of change of nodal coordinates with the change of sensitivity parameter L. The rate of change of X_1 coordinate in all nodes represents the nodal values of a scalar field $\frac{DX_1}{DL}$. The $\frac{DX_1}{DL}$ field is traditionally called the design velocity field. The discretized design velocity field $\frac{D\mathbf{X}_1}{DL}$ is evaluated for the numeric values of the design sensitivity parameter L in all nodes and is the appropriate input data for sensitivity analysis related finite element subroutines.

Evaluation of sensitivity pseudo-load vectors ${}^{I}\tilde{\mathbf{R}}$ and ${}^{IJ}\tilde{\mathbf{R}}$ for the first- and second-order sensitivity analysis of the above example then follows as

$$
{}^{I}\tilde{\mathbf{R}}_g = \frac{\partial \mathbf{R}_g}{\partial \mathbf{X}_1^e} \frac{D\mathbf{X}_1^e}{DL} \tag{13}
$$

$$
{}^{IJ}\tilde{\mathbf{R}}_g = \frac{\partial^2 \mathbf{R}_g}{\partial {\mathbf{p}_e}^2} \left(\frac{D\mathbf{p}_e}{DL} \right)^2 + 2 \frac{\partial^2 \mathbf{R}_g}{\partial \mathbf{p}_e \partial \mathbf{X}_1^e} \frac{D\mathbf{X}_1^e}{DL} \frac{D\mathbf{p}_e}{DL} + \frac{\partial^2 \mathbf{R}_g}{\partial \mathbf{X}_1^{e2}} \frac{D^2 \mathbf{X}_1^e}{DL^2} \tag{14}
$$

where $\mathbf{X}_1^e$ is a vector of X_1 coordinates of element nodes. For the automation, we also need automatic differentiation based version of formulas (13) and (14) or the ADB notation (see Korelc (2009)). For the ADB notation, the partial derivatives are replaced with computational derivatives and the AD exceptions are added for the indirect dependencies $\mathbf{X}_1(L)$, leading to

$$ {}^{I}\tilde{\mathbf{R}}_g = \left.\frac{\hat{\delta}\mathbf{R}_g}{\hat{\delta}L}\right|_{\frac{D\mathbf{X}_1^e}{DL}=\mathbf{V}_L} \tag{15} $$

$$ {}^{IJ}\tilde{\mathbf{R}}_g = \left.\frac{\hat{\delta}}{\hat{\delta}L}\left(\left.\frac{\hat{\delta}\mathbf{R}_g}{\hat{\delta}L}\right|_{\frac{D\mathbf{X}_1^e}{DL}=\mathbf{V}_L,\ \frac{D\mathbf{p}_e}{DL}=\mathbf{S}_L}\right)\right|_{\frac{D\mathbf{X}_1^e}{DL}=\mathbf{V}_L,\ \frac{D\mathbf{V}_L}{DL}=\mathbf{V}_{LL},\ \frac{D\mathbf{p}_e}{DL}=\mathbf{S}_L} \tag{16} $$

where matrices $\mathbf{V}_L = \frac{D\mathbf{X}_1^e}{DL}$ and $\mathbf{V}_{LL} = \frac{D^2\mathbf{X}_1^e}{DL^2}$ are simulation input data that represent the first- and second-order velocity fields. Components of matrix $\mathbf{S}_L = \frac{D\mathbf{p}_e}{DL}$ are zero for the DOF's with prescribed essential boundary conditions and are set to already calculated first-order sensitivities for the true DOF's. Consequently, all the first-order sensitivities have to be calculated first in order to be able to calculate the second-order sensitivities.

Shape sensitivity parameters (shape sensitivity analysis) Symbol L in (15) and (16) is a global quantity. Thus, it does not actually appear explicitly as a part of GAUSS point residual $\mathbf{R}_g$. Consequently, in formulas (15) and (16), symbol L has no meaning and it can be replaced by any symbol. Let ϕ_I and ϕ_J be an arbitrary shape parameters and $\mathbf{X}_e$ nodal wise ordered nested set of all coordinates of all element nodes ($\mathbf{X}_e = \mathbf{X}_e(\phi_I, \phi_J)$). A general ADB notation of the first- and second-order shape sensitivity analysis then follows as

$$ {}^{I}\tilde{\mathbf{R}}_g = \left.\frac{\hat{\delta}\mathbf{R}_g}{\hat{\delta}\phi_I}\right|_{\frac{D\mathbf{X}_e}{D\phi_I}=\mathbf{V}_I} \tag{17} $$

$$ {}^{IJ}\tilde{\mathbf{R}}_g = \left.\frac{\hat{\delta}}{\hat{\delta}\phi_J}\left(\left.\frac{\hat{\delta}\mathbf{R}_g}{\hat{\delta}\phi_I}\right|_{\frac{D\mathbf{X}_e}{D\phi_I}=\mathbf{V}_I,\ \frac{D\mathbf{p}_e}{D\phi_I}=\mathbf{S}_I}\right)\right|_{\frac{D\mathbf{X}_e}{D\phi_J}=\mathbf{V}_J,\ \frac{D\mathbf{V}_I}{D\phi_J}=\mathbf{V}_{IJ},\ \frac{D\mathbf{p}_e}{D\phi_J}=\mathbf{S}_J} . \tag{18} $$

The sensitivity-dependent analysis input data in (17) and (18) are matrices $\mathbf{V}_I = D\mathbf{X}_e/D\phi_I$, $\mathbf{V}_J = D\mathbf{X}_e/D\phi_I$ and $\mathbf{V}_{IJ} = D^2\mathbf{X}_e/D\phi_I D\phi_J$ that represent the first- and second- order shape design velocity fields, and $\mathbf{S}_I = \frac{D\mathbf{p}_e}{D\phi_I}$, $\mathbf{S}_J = \frac{D\mathbf{p}_e}{D\phi_J}$ are already calculated first-order sensitivities of element DOF's.

3.2 Arbitrary Sensitivity Parameters

The formulation can be extended to arbitrary sensitivity parameters. In Fig. 7, the parametrization of a general continuum problem to be solved using the finite element model is presented. Additionally to the nodal coordinates, the input data of the typical finite element procedures are material parameters and boundary conditions. The goal of automation is to preserve the standard finite element technology paradigm, where all the physical problem dependent quantities are calculated at the individual finite element level and then assembled at the global level. For the purpose of automation, each analysis input data is considered as a field defined over the domain of the problem that depends on specific sensitivity parameters, as depicted in Fig. 7. Fields and the corresponding design velocity fields are classified according to their actual appearance (or lack of it) in the formulation of the finite element problem. FE analysis input data can be, for the purpose of automation of sensitivity analysis, classified into several classes:

1. parameter (material) input data with corresponding parameter sensitivity analysis and parameter design velocity fields (e.g., E^J and $\frac{DE^J}{DE_\sigma}$),
2. nodal spatial coordinates with corresponding shape sensitivity analysis and shape design velocity fields (e.g., X_1^J and $\frac{DX_1^J}{DL}$),

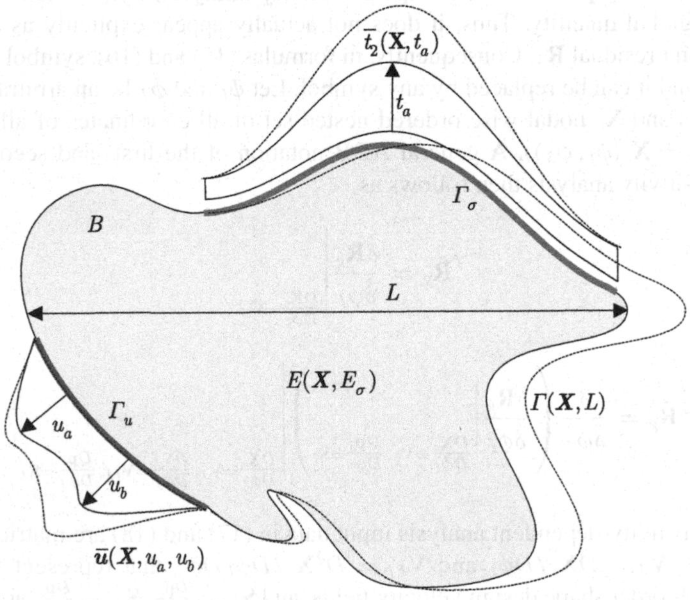

Fig. 7 Parametrization of a general continuum problem to be solved using the finite element model

3. nodal essential boundary conditions with corresponding essential boundary condition sensitivity analysis and essential boundary condition velocity fields (e.g., $\bar{u}_1^J$ and $\frac{D\bar{u}_1^J}{Du_a}, \frac{D\bar{u}_1^J}{Du_b}$),
4. nodal natural boundary conditions with corresponding natural boundary condition sensitivity analysis and natural boundary condition velocity fields (e.g., P_2^J and $\frac{DP_2^J}{Dt_a}, \frac{DP_2^J}{DL}$).

Shape sensitivity analysis Shape sensitivity analysis is described in Sect. 3.1.

Essential boundary condition sensitivity analysis Essential boundary condition sensitivity parameters are used to parametrize the distribution of the essential boundary conditions at the boundary of the problem domain (e.g. u_a and u_b are used to parametrize $\bar{\mathbf{u}}$ in Fig. 7). Let ϕ_I and ϕ_J be arbitrary essential boundary condition sensitivity parameters and $\bar{\mathbf{p}}_e$ a set of element DOF with prescribed essential boundary condition, thus $\bar{\mathbf{p}}_e \subset \mathbf{p}_e$. The $\mathbf{p}_e$ set includes both degrees of freedom with prescribed essential boundary condition and true degrees of freedom, because they are at the element-level indistinguishable. The corresponding first- and second-order essential boundary condition velocity fields are defined by

$$\mathbf{V}_I = \begin{cases} \frac{D\bar{p}_{eJ}}{D\phi_I} & \text{if } p_{eJ} \in \bar{\mathbf{p}}_e \\ 0 & \text{if } p_{eJ} \in \mathbf{p}_e \backslash \bar{\mathbf{p}}_e \end{cases} : J = 1, \ldots, n_p, \tag{19}$$

$$\mathbf{V}_{IJ} = \begin{cases} \frac{D^2\bar{p}_{eJ}}{D\phi_I D\phi_J} & \text{if } p_{eJ} \in \bar{\mathbf{p}}_e \\ 0 & \text{if } p_{eJ} \in \mathbf{p}_e \backslash \bar{\mathbf{p}}_e \end{cases} : J = 1, \ldots, n_p \tag{20}$$

where n_p is the total number of element nodal DOF. Velocity field is zero for the true degrees of freedom. Thus, proper definition of velocity fields is sufficient to make the difference between the degrees of freedom with prescribed essential boundary condition and true degrees of freedom. A general ADB notation of GAUSS point contribution to the first- and second-order essential boundary condition pseudo-load vectors then follows as

$$^I\tilde{\mathbf{R}}_g = \left.\frac{\hat{\delta}\mathbf{R}_g}{\hat{\delta}\phi_I}\right|_{\frac{D\mathbf{p}_e}{D\phi_I}=\mathbf{V}_I} \tag{21}$$

$$^{IJ}\tilde{\mathbf{R}}_g = \left.\frac{\hat{\delta}}{\hat{\delta}\phi_J}\left(\left.\frac{\hat{\delta}\mathbf{R}_g}{\hat{\delta}\phi_I}\right|_{\frac{D\mathbf{p}_e}{D\phi_I}=\mathbf{S}_I}\right)\right|_{\frac{D\mathbf{S}_I}{D\phi_J}=\mathbf{V}_{IJ}, \frac{D\mathbf{p}_e}{D\phi_J}=\mathbf{S}_J} \tag{22}$$

The sensitivity-dependent analysis input data in (21) and (22) are matrices $\mathbf{V}_I$ and $\mathbf{V}_{IJ}$. Matrices $\mathbf{S}_I$ and $\mathbf{S}_J$ are composed of the components of velocity fields for the DOF's with prescribed essential boundary conditions and already calculated first-order sensitivities for the true DOFs (23).

$$\mathbf{S}_I = \begin{cases} \frac{D\bar{p}_{eJ}}{D\phi_I} & \text{if } p_{eJ} \in \bar{\mathbf{p}}_e \\ \frac{Dp_{eJ}}{D\phi_I} & \text{if } p_{eJ} \in \mathbf{p}_e \backslash \bar{\mathbf{p}}_e \end{cases} \quad : J = 1, \ldots, n_p \tag{23}$$

Material sensitivity parameters (parameter sensitivity analysis) Input parameters of the finite element procedures can be scalars (e.g., elastic modulus E), discretized scalar fields (e.g., nodal temperatures), and discretized vector fields (e.g., nodal spatial coordinates $\mathbf{X}^J$). Without losing the generality of the formulation, a scalar can be considered as a constant scalar field discretized by its constant nodal values and a vector field can be considered component-wise. Most of the input data of the finite element procedures are associated with nodes. However some quantities, such as material constants ($E_g = E_g(E_\sigma)$ and $\nu_g = \nu_0$), are associated with integration points. Again, the integration point based quantities can be obtained from the appropriate nodal-based quantities using standard finite element interpolation techniques. Consequently, integration point based quantities are also represented as a discretized scalar field unifying all sensitivity parameters within the same framework. Let $\boldsymbol{\psi}_e$ be a set of parameters on which element residual explicitly depends ($\mathbf{R}_g = \mathbf{R}_g(\boldsymbol{\psi}_e)$). A general ADB notation of GAUSS point contribution to the first- and second-order parameter pseudo-load vectors then follows as

$$^I\tilde{\mathbf{R}}_g = \left.\frac{\hat{\delta}\mathbf{R}_g}{\hat{\delta}\phi_I}\right|_{\frac{D\boldsymbol{\psi}_e}{D\phi_I}=\mathbf{V}_I} \tag{24}$$

$$^{IJ}\tilde{\mathbf{R}}_g = \left.\frac{\hat{\delta}}{\hat{\delta}\phi_J}\left(\left.\frac{\hat{\delta}\mathbf{R}_g}{\hat{\delta}\phi_I}\right|_{\frac{D\boldsymbol{\psi}_e}{D\phi_I}=\mathbf{V}_I,\,\frac{D\mathbf{p}_e}{D\phi_I}=\mathbf{S}_I}\right)\right|_{\frac{D\boldsymbol{\psi}_e}{D\phi_J}=\mathbf{V}_J,\,\frac{D\mathbf{V}_I}{D\phi_J}=\mathbf{V}_{IJ},\,\frac{D\mathbf{p}_e}{D\phi_J}=\mathbf{S}_J} \quad . \tag{25}$$

The sensitivity-dependent analysis input data in (24) and (25) are matrices $\mathbf{V}_I = D\boldsymbol{\psi}_e/D\phi_I$, $\mathbf{V}_J = D\boldsymbol{\psi}_e/D\phi_J$ and $\mathbf{V}_{IJ} = D^2\boldsymbol{\psi}_e/D\phi_I D\phi_J$ that represent the first- and second-order parameter design velocity fields. $\mathbf{S}_I = \frac{D\mathbf{p}_e}{D\phi_I}$ and $\mathbf{S}_J = \frac{D\mathbf{p}_e}{D\phi_J}$ are the already calculated first-order sensitivities of element DOFs.

Natural boundary condition sensitivity parameters (natural boundary condition sensitivity analysis) Problems in solid mechanics and nonlinear structural mechanics, subjected to quasi-static proportional load, are frequently formulated as

$$\mathbf{R} = \mathbf{R}^{\text{int}} - \lambda\,\mathbf{R}^{\text{ref}} = \mathbf{0} \tag{26}$$

where $\mathbf{R}^{\text{int}}$ denotes the contribution of the internal forces to the global residual vector. Vector $\mathbf{R}^{\text{ref}}$ is the reference load vector associated with the pattern of the applied nodal forces (natural boundary condition input data) and λ is the loading parameter. Load vector $\lambda\,\mathbf{R}^{\text{ref}}$ is subtracted from the internal force vector and thus does not affect directly the residual vectors of the finite elements at local element level. Consequently, the contribution of variation of natural boundary conditions has to be formulated within the global solution algorithm and it does not follow the standard sensitivity analysis procedures as described in previous sections. If the contribution of the natural boundary conditions to the global residual **R** is accounted for by a special generalized finite elements then the natural boundary condition input data can be considered as a part of general input parameters $\boldsymbol{\psi}_e$ and treated accordingly.

The general equation (26) leads for an arbitrary time-dependent problem and for an arbitrary sensitivity parameter ϕ_I, ϕ_J to

$$\mathbf{R}^{\text{int}}(\mathbf{p}(\phi_I, \phi_J)) - \lambda\, \mathbf{R}^{\text{ref}}(\phi_I, \phi_J) = \mathbf{0}. \tag{27}$$

Direct differentiation of (27) with respect to ϕ_I yields the first-order pseudo-load vector and sensitivity of the response $\frac{D\mathbf{p}}{D\phi_I}$ by the solution of the linear equation systems (28).

$${}^{I}\tilde{\mathbf{R}} = -\lambda \frac{D\mathbf{R}^{\text{ref}}}{D\phi_I}, \quad \mathbf{K}\frac{D\mathbf{p}}{D\phi_I} = -{}^{I}\tilde{\mathbf{R}} \tag{28}$$

Second derivative of (27) yields

$${}^{IJ}\tilde{\mathbf{R}} = \frac{\partial^2 \mathbf{R}}{\partial \mathbf{p}^2}\frac{D\mathbf{p}}{D\phi_I}\frac{D\mathbf{p}}{D\phi_J} + \frac{\partial^2 \mathbf{R}}{\partial \mathbf{p}\partial \phi_J}\frac{D\mathbf{p}}{D\phi_I} + \frac{\partial^2 \mathbf{R}}{\partial \mathbf{p}\partial \phi_I}\frac{D\mathbf{p}}{D\phi_J} - \lambda \frac{D^2\mathbf{R}^{\text{ref}}}{D\phi_I D\phi_J}. \tag{29}$$

Equation (29) has parts that depend on internal forces and a part that depends on reference load vector. Consequently, it has to be split into parts, one that is formed globally ${}^{IJ}\tilde{\mathbf{R}}^{\text{ref}}$ (30) and one that is formed by an element-based assembly procedure ${}^{IJ}\tilde{\mathbf{R}}^{\text{int}}$ (31).

$${}^{IJ}\tilde{\mathbf{R}}^{\text{ref}} = -\lambda \frac{D^2\mathbf{R}^{\text{ref}}}{D\phi_I D\phi_J} \tag{30}$$

$${}^{IJ}\tilde{\mathbf{R}}^{\text{int}} = \frac{\partial^2 \mathbf{R}}{\partial \mathbf{p}^2}\frac{D\mathbf{p}}{D\phi_I}\frac{D\mathbf{p}}{D\phi_J} + \frac{\partial^2 \mathbf{R}}{\partial \mathbf{p}\partial \phi_J}\frac{D\mathbf{p}}{D\phi_I} + \frac{\partial^2 \mathbf{R}}{\partial \mathbf{p}\partial \phi_I}\frac{D\mathbf{p}}{D\phi_J}. \tag{31}$$

A general ADB notation of Gauss point contribution to the ${}^{IJ}\tilde{\mathbf{R}}^{\text{int}}$ pseudo-load vector then follows as

$${}^{IJ}\tilde{\mathbf{R}}_g^{\text{int}} = \frac{\hat{\delta}}{\hat{\delta}\phi_J}\left(\left.\frac{\hat{\delta}\mathbf{R}_g}{\hat{\delta}\phi_I}\right|_{\frac{D\mathbf{p}_e}{D\phi_I}=\mathbf{S}_I} \right)\Bigg|_{\frac{D\mathbf{p}_e}{D\phi_J}=\mathbf{S}_J}. \tag{32}$$

At the end, the second-order sensitivity of the response $\frac{D^2\mathbf{p}}{D\phi_I D\phi_J}$ leads from the solution of the linear equation systems

$$\mathbf{K}\frac{D^2\mathbf{p}}{D\phi_I D\phi_J} = -({}^{IJ}\tilde{\mathbf{R}}^{\text{ref}} + {}^{IJ}\tilde{\mathbf{R}}^{\text{int}}). \tag{33}$$

Table 1 Comparison of code size and AceGen evaluation time

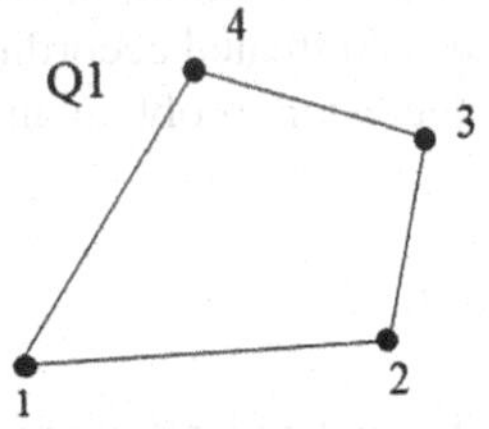

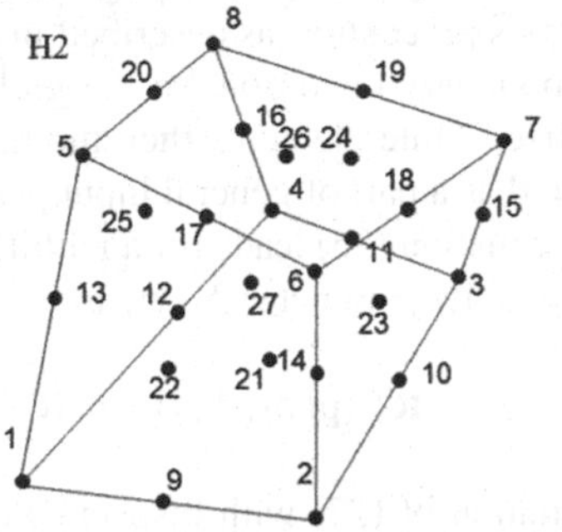

Q1 - linear elastic

Task	Code	*AceGen*
$\mathbf{K}_e$&$\mathbf{R}_e$	5 kB	4 s
$+{}^I\tilde{\mathbf{R}}_e$	8 kB	5 s
$+{}^{IJ}\tilde{\mathbf{R}}_e$	9 kB	7 s

H2 - finite strain elasto-plastic

Task	Code	*AceGen*
$\mathbf{K}_e$&$\mathbf{R}_e$	127 kB	57 s
$+{}^I\tilde{\mathbf{R}}_e$	270 kB	197 s
$+{}^{IJ}\tilde{\mathbf{R}}_e$	376 kB	377 s

3.3 Sensitivity Analysis—Code Complexity of AceGen Codes

The concept of design sensitivity velocity fields can be extended to general input parameters (e.g., nodal coordinates, material parameters, essential boundary conditions, and natural boundary conditions). For details see Korelc and Wriggers (2016). Any approach to automation is feasible only when the physical size of the generated codes stays within reasonable limits allowed by compilers and when the time to generate the code also stays within reasonable limits.

In Table 1, the code size and the *AceGen* evaluation time are compared for different finite element formulations. Two extreme cases are compared: simple two-dimensional linear elastic element and three-dimensional, finite strain, elastoplastic, 27-node brick element. For each required quantity (tangent and residual, the first-order sensitivity pseudo-load vector and the second-order sensitivity pseudo-load vector), the actual size of the code generated and the time used to generate the code are presented. We can see that also for the most complicated element the size of the code and the time to generate the code remain moderate.

4 Applications of Sensitivity Analysis

It is common for all applications of sensitivity analysis that once the element code that supports primal and sensitivity analysis for all input parameters of the individual finite elements is generated, then the only unanswered question remains "WHAT IS

THE VELOCITY FIELD OF THE PROBLEM?". In this chapter, several examples are presented and the corresponding velocity fields are identified.

4.1 Sensitivity Analysis Based Stochastic Analysis

When an input parameter of the problem is random and it also randomly varies over the domain, it can be modeled as stochastic field. A stochastic field is defined with probability density function and covariance function. Probability density function specifies the probability of the random variable falling within a particular range of values. Covariance function describes how much a variable changes along the domain. In mechanical problems, most often used is exponential covariance function $C(\boldsymbol{X}_1, \boldsymbol{X}_2) = \sigma^2 e^{-\frac{||\boldsymbol{X}_2 - \boldsymbol{X}_1||}{l_c}}$, where $\boldsymbol{X}_i$ is a position vector over the physical domain, σ is standard deviation and l_c is correlation length. The bigger l_c is, the higher correlated is stochastic field (see e.g. Ghanem and Spanos (2003)).

The representation of the Gaussian stochastic field can be done with Karhunen–Loeve expansion, which is truncated after first M terms as

$$w(\boldsymbol{X}, \theta) = \bar{w}(\boldsymbol{X}) + \sum_{k=1}^{M} \sqrt{\lambda_k} f_k(\boldsymbol{X}) \, \xi_k(\theta) \tag{34}$$

where $\boldsymbol{X}$ is a position vector over the physical domain, θ is an event of the space of random events, $\bar{w}(\boldsymbol{X})$ is expected value of the stochastic field and $\xi_k(\theta)$ are normalized uncorrelated Gaussian random variables with zero mean and unit variance. λ_k and $f_k(\boldsymbol{X})$ are the eigenvalues and eigenvectors, respectively, obtained as the solution of the homogeneus Fredholm integral equation ($\int_D C(\boldsymbol{X}_1, \boldsymbol{X}_2) f_k(\boldsymbol{X}_1)\, d\boldsymbol{X}_1 = \lambda_k f_k(\boldsymbol{X}_2)$) of the second kind with covariance function $C(\boldsymbol{X}_1, \boldsymbol{X}_2)$ as kernel. Galerkin procedure can be used to solve this equation numerically (see e.g., Melink and Korelc (2014)). The result is an approximated and discretized stochastic field according to (34).

When at least one of the input parameters is random, the response of the system is also random. The final goal of stochastic analysis is to calculate statistics (e.g., expected value and standard deviation) of the response. The response of the system is a function of a set of uncorrelated Gaussian random variables $\xi_k(\theta)$. In general, Monte Carlo method can be used to get statistics of the response for an arbitrary problem. However, Monte Carlo method requires a large number of direct simulations to be performed. An alternative approach is to use the second-order sensitivity analysis to the get second-order approximation of the response. In this case, only one direct simulation is needed.

In the presented stochastic approach, the response of the problem is approximated with a finite number of its Taylor series around the expected values of random variables (${}^0\boldsymbol{\xi} = \{{}^0\xi_1, {}^0\xi_2, \dots {}^0\xi_M\}$), which resembles higher order sensitivity analysis.

In case of Gaussian stochastic field (${}^0\boldsymbol{\xi} = \{0, 0, 0, \dots\}$) and second-order sensitivity analysis, we get

$$\mathbf{p}(\xi_1, \xi_2, \dots \xi_M) = \mathbf{p}(0, 0, 0, \dots) + \sum_{i=1}^{M} \frac{\partial \mathbf{p}}{\partial \xi_i} \xi_i + \frac{1}{2} \sum_{i=1}^{M} \sum_{j=1}^{M} \frac{\partial^2 \mathbf{p}}{\partial \xi_i \partial \xi_j} \tag{35}$$

where $\mathbf{p}$ is solution vector (in mechanics, $\mathbf{p}$ is usually vector of displacements). Derivatives of solution vector $\mathbf{p}$ with respect to random variables are calculated with sensitivity analysis. Thus, a set of sensitivity parameters of the problem is $\boldsymbol{\phi} = \boldsymbol{\xi}$. The approximation of the response is now closed-form polynomial formula. Thus, the statistics of the response (expected value and standard deviation) can be cheaply obtained either analytically or with the use of standard statistical functions in *Mathematica* . All we need to complete the derivation is the design velocity field of the problem.

A numerical example of bended clamped sinusoidal double skin cladding is chosen (see Fig. 8) to demonstrate the use of the above-described automation of the stochastic finite element method. The cladding is modeled by two-dimensional, four-node, finite strain elements. The shape of the cladding is sinusoidal with n wavelengths and constant thickness of the skin and foam. The amplitude of waves h is presumed to change stochastically along the X axis. Therefore, one-dimensional stochastic field $h(X, \boldsymbol{\xi})$ of the wave amplitude is considered. The Y coordinate of the central line nodes is then given by

$$Y(X, \boldsymbol{\xi}) = h(X, \boldsymbol{\xi}) \sin \frac{n\pi X}{L}, \tag{36}$$

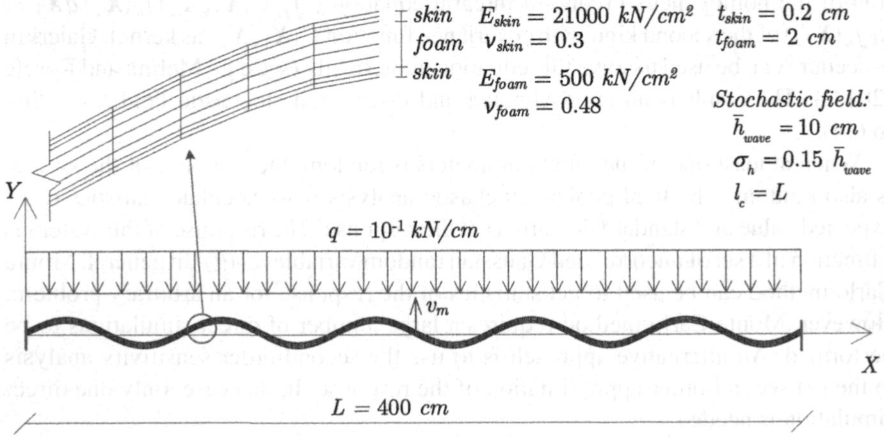

Fig. 8 Sinusoidal double skin cladding

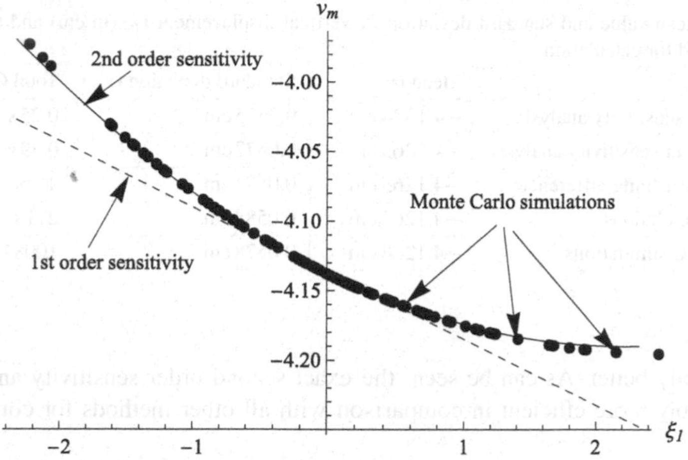

Fig. 9 Deflection in the middle of cladding, obtained with sensitivity analysis of different orders and MC simulations

$$h(X,\boldsymbol{\xi}) = \bar{h}_{wave} + \sum_{k=1}^{M} \sqrt{\lambda_k} f_k(X)\, \xi_k \tag{37}$$

The corresponding first- and second-order shape design velocity fields are then

$$\frac{\partial Y}{\partial \xi_k} = \sqrt{\lambda_k} f_k(X) \sin\frac{n\pi X}{L}, \quad \frac{\partial^2 Y}{\partial \xi_k \partial \xi_l} = 0 \tag{38}$$

where $\bar{h}_{wave}$ is the expected value of amplitude.

Stochastic field of wave amplitude change is represented via the first four terms of K-L expansion, thus $\boldsymbol{\xi} = \{\xi_1, \xi_2, \xi_3, \xi_4\}$. In Fig. 9, the vertical displacement v_m in the middle of the cladding is calculated in dependence of ξ_1, while other random variables are taken at their mean value ($\xi_2 = \xi_3 = \xi_4 = 0$). The results of the first- and second-order sensitivity analysis are compared with those obtained by 100 Monte Carlo (MC) simulations. It can be seen that the second-order sensitivity analysis suits almost exactly the direct evaluation of the response, for approximately two standard deviations from the mean value.

In Table 2, the calculated mean value, standard deviation and CPU time are compared for statistics of the response obtained by the first-order sensitivity analysis, the second-order sensitivity analysis, finite difference approximation of the second-order sensitivities, and Monte Carlo simulations. In this comparison, all four random variables ($\xi_1, \xi_2, \xi_3, \xi_4$) are considered. In MC simulations, the range of random variables was limited to the interval between 0.001 and 0.999 quantile, due to physically acceptable results. The results justify the use of the second-order sensitivity analysis instead of the analysis of the first order, since the second-order results fit the results

Table 2 Mean value and standard deviation of vertical displacement v_m (in cm) and total CPU time, needed for calculation

	Mean (v_m)	Standard deviation (v_m)	Total CPU time
First-order sensitivity analysis	−4.1374 cm	0.0555 cm	0.25 s
Second-order sensitivity analysis	−4.1263 cm	0.0577 cm	0.39 s
Second-order finite difference	−4.1263 cm	0.0577 cm	46.58 s
10^3 MC simulations	−4.1262 cm	0.0588 cm	271 s
4×10^4 MC simulations	−4.1264 cm	0.0578 cm	10084 s

considerably better. As can be seen, the exact second-order sensitivity analysis is considerably more efficient in comparison with all other methods for comparable results.

4.2 Asymptotic Numerical Methods

At present, in solid mechanics and nonlinear structural mechanics there exists no iterative method that can be applied to all different problem areas in an efficient and robust way. Additionally, for highly nonlinear problems the solution of time-independent problems cannot, in general, be achieved in one step. More efficient procedures can be derived when the resulting system of equations can be naturally parametrized in a way that for some given value of parameter the solution is trivial. The system of equations $\mathbf{R}(\mathbf{p}) = \mathbf{0}$ will be parametrized for the following considerations in the form:

$$\mathbf{R}(\mathbf{p}, \lambda) = \mathbf{0}, \tag{39}$$

where λ is parameter, and solved using the standard Newton–Raphson method. With the introduction of parameter λ, the final solution is achieved in n_{step} incremental steps with associated solution vectors $\mathbf{p}_0, \ldots \mathbf{p}_{n_{\text{step}}}$. As an example, problems in solid mechanics and nonlinear structural mechanics subjected to quasi-static proportional load are frequently parametrized by introducing the loading parameter λ as follows:

$$\mathbf{R}(\mathbf{p}, \lambda) = \mathbf{R}^{\text{int}}(\mathbf{p}) - \mathbf{F}(\lambda) = 0, \quad \mathbf{F} = \lambda\, \mathbf{F}^{\text{ref}} \tag{40}$$

where $\mathbf{R}^{\text{int}}$ denotes the contribution of internal forces to the nodal force vector and $\mathbf{F}^{\text{ref}}$ is the reference load vector associated with the pattern of the applied nodal forces.

Within the asymptotic numerical method approach (see e.g., Nezamabadi et al. (2011)), a more efficient load stepping scheme is derived by expansion of the response with respect to parameter of the problem (load level λ). Thus, sensitivity parameter of the problem is $\boldsymbol{\phi} = \{\lambda\}$ and the response is approximated as

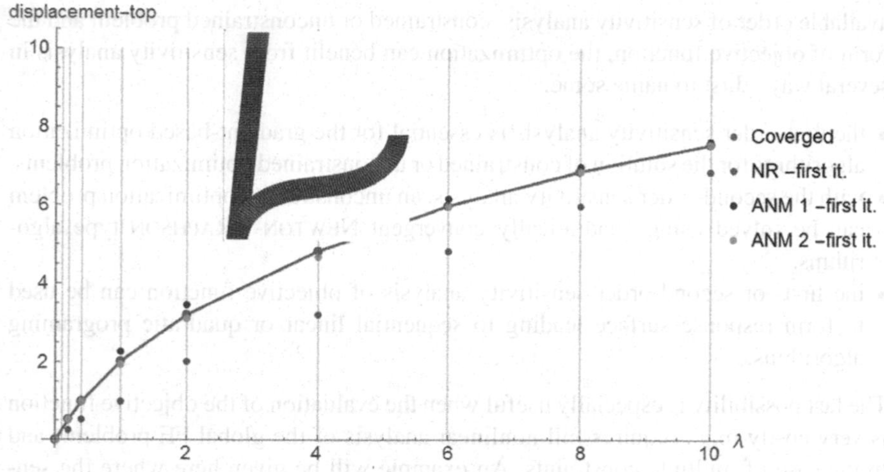

Fig. 10 Bending of column using asymptotic numerical methods

$$\mathbf{p}(\lambda) = \mathbf{p}_0 + \frac{\partial \mathbf{p}}{\partial \lambda}\delta\lambda + \frac{1}{2}\frac{\partial^2 \mathbf{p}}{\partial \lambda^2}\delta\lambda^2 \ldots \tag{41}$$

The corresponding first- and second-order natural boundary condition design velocity fields are then

$$\frac{\partial \mathbf{F}}{\partial \lambda} = \mathbf{F}^{\text{ref}}, \quad \frac{\partial^2 \mathbf{F}}{\partial \lambda^2} = \mathbf{0}. \tag{42}$$

Due to the fact that within the asymptotic numerical methods, we deal with only one sensitivity parameter, also the sensitivities of the order higher than two can be obtained in a reasonable computational time (see e.g. Nezamabadi et al. (2011)).

A numerical example of bending of column modeled by two-dimensional finite strain elements is presented in Fig. 10. The final load is achieved in 8 load steps. For each load step, the converged solution is depicted together with the converged solution from the previous load step (the usual initial guess for the standard Newton–Raphson method), the first-order ANM approximation and the second-order ANM approximation. It can be seen that in this case second-order ANM approximation gives almost an exact solution . By using even higher orders one can skip Newton iterations altogether (see e.g., Nezamabadi et al. (2011)). However, this can also change dramatically, for example, with more dense meshes and non-monotonic response.

4.3 Optimization

Optimization problems were one of the first problems, where sensitivity analysis was used to improve numerical efficiency of optimization algorithms. Depending on the available order of sensitivity analysis, constrained or unconstrained problem and the form of objective function, the optimization can benefit from sensitivity analysis in several ways. Just to name some:

- the first-order sensitivity analysis is essential for the gradient-based optimization algorithms for the solution of constrained or unconstrained optimization problems,
- with the second-order sensitivity analysis, an unconstrained optimization problem can be solved using quadratically convergent NEWTON- - RAPHSON type algorithms,
- the first- or second-order sensitivity analysis of objective function can be used to form response surface leading to sequential linear or quadratic programing algorithms.

The last possibility is especially useful when the evaluation of the objective function is very costly (e.g., requires full nonlinear analysis of the global FE problem) and in the case of multiple constraints. An example will be given here where the sensitivity analysis is used to solve the problem of worst imperfection of structures in means of ultimate limit states(Kristanic and Korelc (2008)) using sequential linear programming approach. It is well known that geometrical, structural, material, and load imperfections play a crucial role in the load-carrying behavior, especially of thin-walled structures. The idea to find such a combination of imperfections that will cause the structure to fail at the lowest possible load is as old as the ascertainment of the crucial role of imperfections itself. The review of different approaches accompanied with an impact on modern design procedures of engineering structures can be found in Schmidt (2000). When analyzing structures discretized with finite elements, it turns out that the choice of the shape and size of initial imperfections have a major influence on the response of the structure and its limit state.

With the use of direct and sensitivity analysis combined with optimization, it is possible to determine the most unfavorable combination of chosen shapes representing the initial imperfection, which leads to the least possible ultimate load. Within the optimization algorithm, the objective function is constructed by means of a fully nonlinear direct and first-order sensitivity analysis. The method is not limited to small imperfections or a linear fundamental path based on Koiters asymptotical theory (Koiter (1945)) and also allows the imposition of technological constraints on the shape of the imperfection, thus making it possible to avoid unrealistically low ultimate loads. When carefully constructed, the objective function and constraints remain linear, enabling the use of numerically efficient and readily available sequential linear programming algorithms.

Let $\mathbf{X}_p$ be a coordinate of the nodes of the perfect geometry, $\mathbf{X} = \mathbf{X}_p + \bar{\mathbf{X}}$ coordinates of the imperfect geometry, where imperfection $\bar{\mathbf{X}}$ is approximated as linear combination of M base shapes $\mathbf{\Gamma}_i$ and corresponding weights α_i (43).

$$\mathbf{X} = \mathbf{X}_p + \bar{\mathbf{X}} = \mathbf{X}_p + \sum_{i=1}^{M} \alpha_i \mathbf{\Gamma}_i \tag{43}$$

Base shapes can be chosen arbitrarily. The most convenient set of shapes is the set of buckling modes that can be extended by eigenshapes of tangent matrix, empirically known as worst shapes or deformation shapes. The response of the imperfect, materially and geometrically nonlinear structure is defined by its response curve $\mathbf{u}(\lambda)$, where λ is the load level as defined for proportional loading by (40). Let λ_l be the ultimate load factor. A limit state of a structure is generally defined with the limit point of the equilibrium path. In real, imperfect structures, this criterion proves unreliable because of the possible exceeding of permissible tolerances of displacements or deformations before reaching the limit point. The goal is to determine such coefficients α_i that the ultimate limit load factor λ_l of the structure would be minimal. Therefore, a minimization problem (44) for the limit load factor can be defined, where the imposition of technological constraints requires that the maximal amplitude of the imperfection has to be equal to or smaller than the amplitude of the prescribed equivalent geometrical imperfections e_0.

$$\begin{aligned} &\min_{\alpha_i} \lambda_l \\ &||\bar{\mathbf{X}}||_\infty \leqslant e_0 \end{aligned} \tag{44}$$

Solution of the nonlinear optimization problem (44) requires full nonlinear analysis (direct and, depending on optimization algorithm, also sensitivity analysis) of the structure at every iteration of optimization algorithm. Because of the enormous computational time required, this approach is not feasible at this time. The fully nonlinear problem (44) is simplified by expansion of the limit state load factor of the imperfect structure to a Taylor series around the imperfect geometry. The limit load factor $\lambda_l(\bar{\mathbf{X}}(\alpha_i))$ is then for k^{th} global iteration of the sequential nonlinear optimization algorithm written as

$$\lambda_l \approx \lambda_l^{k-1} + \sum_{i=1}^{M} \left.\frac{\partial \lambda_l}{\partial \alpha_i}\right|_{\alpha_i^{k-1}} \Delta\alpha_i^k \tag{45}$$

where coefficients of the series expansion $\partial\lambda_l/\partial\alpha_i|_{\alpha_i^{k-1}}$ are obtained by the first-order shape sensitivity analysis. Sensitivity parameters of the problem are weights α_i and the the corresponding shape design velocity field is obtained by the differentiation of (43) with respect to sensitivity parameters

$$\frac{\partial \mathbf{X}}{\partial \alpha_i} = \mathbf{\Gamma}_i. \tag{46}$$

Function (45) is a linear function. However, the constraint in (44) is a highly nonlinear function. A set of linear constraints for the maximal amplitude of the total imperfection vector $|\bar{X}_{l,m}| = |\sum_i \alpha_i \Gamma_{l,m}| \leqslant e_0; \forall l, m$, where $\bar{X}_{l,m}$ and $\Gamma_{l,m}$ are the mth component of the imperfection and base shape vector in l^{th} node, can be defined instead. The result is numerically highly efficient sequential linear programming problem. For each global iteration of sequential linear programming algorithm only one fully nonlinear limit state analysis together with the shape sensitivity analysis

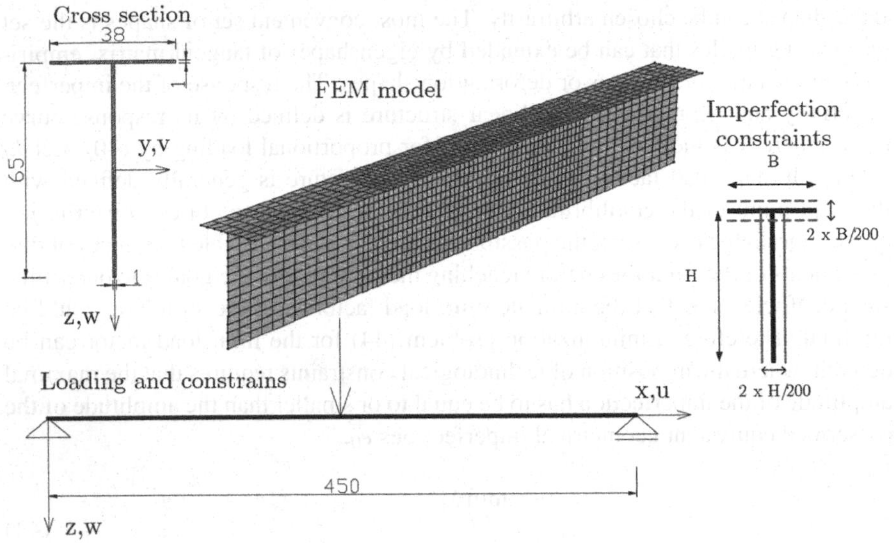

Fig. 11 Example of a T cross-sectional thin-walled beam example

has to be performed. Computational cost for the solution of the corresponding linear programming problem is in fact negligible.

The example presented refers to the ultimate load calculation of a simply supported thin-walled beam with a T cross section, loaded with a concentrate force at the mid-length. The geometrical details and loads are presented in Fig. 11. The thin-walled girders in this section were modeled by elastoplastic four node shell elements based on finite rotations, six-parameter shell theory combined with assumed natural strain formulation and two enhanced strain modes for improved performance. Within the optimization problem, it was necessary to define 3150 constraint equations for the maximal initial imperfection amplitude perpendicular to the web and 2025 constraint equations for the maximal imperfection amplitude perpendicular to the flange. The structure is analyzed considering the shape base consisting of buckling modes. In Fig. 12, the calculated limit load of the T-beam with increasing number of base shapes is shown. The results show a clear convergence of the calculated limit load.

Convergence of the global iterative optimization process of finding the most unfavorable imperfection by considering 52 base shapes ($M = 52$) is presented in Fig. 13. The most unfavorable initial imperfection is achieved within engineering tolerances in the 4th global iteration of the sequential linear programming algorithm.

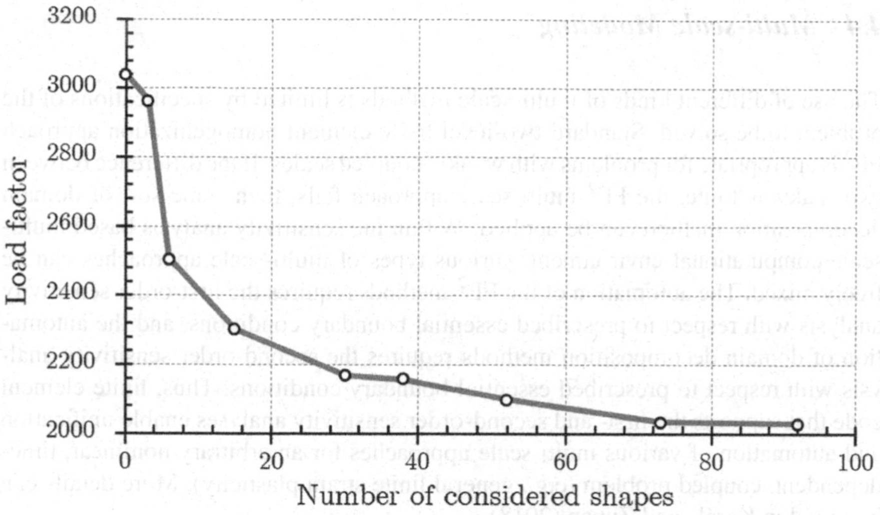

Fig. 12 Convergence of the ultimate limit load with the number of base shapes

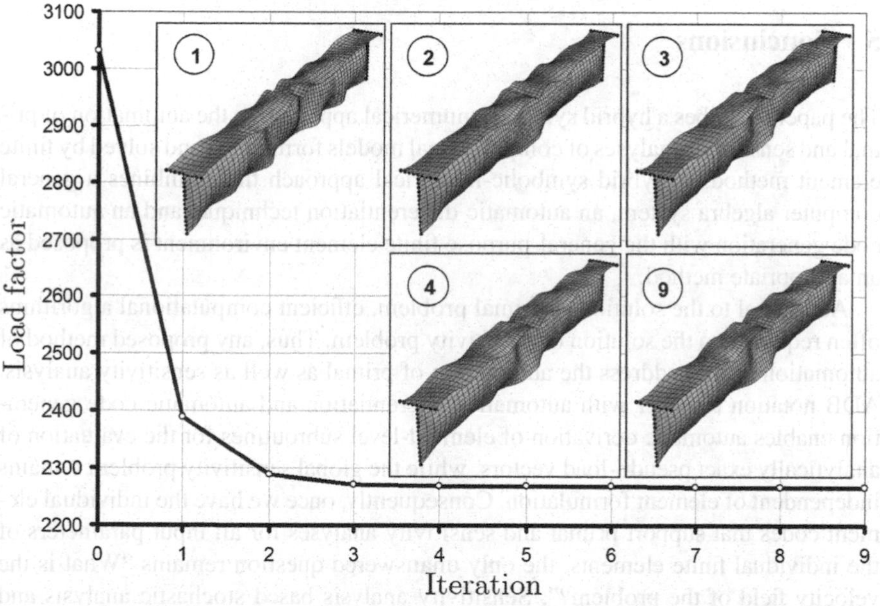

Fig. 13 Convergence of the global iterative optimization process of finding the most unfavorable imperfection shape of a T cross section

4.4 *Multi-scale Modeling*

The use of different kinds of multi-scale methods is limited by specifications of the problem to be solved. Standard two-level finite element homogenization approach FE^2 is appropriate for problems with weakly coupled scales. If the difference between two scales is finite, the FE^2 multi-scale approach fails, then some sort of domain decomposition method can be applied. Within the sensitivity analysis based multi-scale computational environment, various types of multi-scale approaches can be freely mixed. The automation of the FE^2 methods requires the first-order sensitivity analysis with respect to prescribed essential boundary conditions, and the automation of domain decomposition methods requires the second-order sensitivity analysis with respect to prescribed essential boundary conditions. Thus, finite element code that supports the first- and second-order sensitivity analyses enable unification and automation of various multi-scale approaches for an arbitrary nonlinear, time-dependent, coupled problem (e.g., general finite strain plasticity). More details can be found in Korelc and Zupan (2018).

5 Conclusions

The paper describes a hybrid symbolic-numerical approach to the automation of primal and sensitivity analyses of computational models formulated and solved by finite element method. A hybrid symbolic-numerical approach that combines a general computer algebra system, an automatic differentiation technique, and an automatic code generation with the general-purpose finite element environment is proposed as an appropriate method.

Additional to the solution of primal problem, efficient computational algorithms often require also the solution of sensitivity problem. Thus, any proposed method of automation should address the automation of primal as well as sensitivity analysis. ADB notation together with automatic differentiation and automatic code generation enables automatic derivation of element-level subroutines for the evaluation of analytically exact pseudo-load vectors, while the global sensitivity problem remains independent of element formulation. Consequently, once we have the individual element codes that support primal and sensitivity analyses for all input parameters of the individual finite elements, the only unanswered question remains "What is the velocity field of the problem?". Sensitivity analysis based stochastic analysis and asymptotic numerical methods were given as examples of identification and definition of design velocity fields. It is important to notice that no additional functionality or coding is needed for the implementation of these examples, apart from knowing the design velocity field of the problem.

Acknowledgements The financial support for this work was obtained from the Slovenian Research Agency within the research group P2-0158.

References

Bischof, Ch., Hovland, P., & Norris, B. (2002). Implementation of automatic differentiation tools. In C. Norris, & Jr. J. B. Fenwick (Eds.), *Proceedings of the ACM SIGPLAN Workshop on Partial Evaluation and Semantics-Based Program Manipulation*. New York: ACM Press.

Choi, K. K., & Kim, N. H. (2005a). *Structural sensitivity analysis and optimization 1*. New York: Linear systems. Springer Science+Business Media.

Choi, K. K., & Kim, N. H. (2005b). *Structural sensitivity analysis and optimization 2*. New York: Nonlinear systems and applications. Springer Science+Business Media.

Eyheramendy, D., & Zimmermann, Th. (2000). Object-oriented symbolic derivation and automatic programming of finite elements in mechanics. *Engineering with Computers, 15*(1), 12–36.

Ghanem, R., & Spanos, P. D. (2003). *Stochastic finite elements: A spectral approach*. New York: Dover.

Griewank, A. (2000). *Evaluating derivatives: Principles and techniques of algorithmic differentiation*. Philadelphia: SIAM.

Hudobivnik, B., & Korelc, J. (2016). Closed-form representation of matrix functions in the formulation of nonlinear material models. *Finite Elements in Analysis and Design, 111*, 19–32.

Keulen, F., Haftka, R. T., & Kim, N. H. (2005). Review of options for structural design sensitivity analysis. part 1: Linear systems. *Computer Methods in Applied Mechanics and Engineering, 194*, 3213–3243.

Kleiber, M., Antunez H., Hien, T. H., & Kowalczyk, P. (1997). *Parameter sensitivity in nonlinear mechanics*. John Wiley and Sons.

Koiter, W. T. (1945). *The stability of elastic equilibrium*. Stanford University.

Korelc, J. (1997). Automatic generation of finite-element code by simultaneous optimization of expressions. *Theoretical Computer Science, 187*, 231–248.

Korelc, J. (2002). Multi-language and multi-environment generation of nonlinear finite element codes. *Engineering with Computers, 18*, 312–327.

Korelc, J. (2009). Automation of primal and sensitivity analysis of transient coupled problems. *Computational Mechanics, 44*, 631–649.

Korelc, J. (2018). *AceFEM and AceGen user manuals* (6.902 edition). http://symech.fgg.uni-lj.si/.

Korelc, J., & Stupkiewicz, S. (2014). Closed-form matrix exponential and its application in finite-strain plasticity. *International Journal for Numerical Methods in Engineering, 98*, 960–987. https://doi.org/10.1002/nme.4653.

Korelc, J., & Wriggers, P. (2016). *Automation of finite element methods*. Switzerland: Springer.

Korelc, J.,& Zupan, N. (2018). Unified approach to sensitivity analysis based automation of multiscale modelling. In J. SORIC, P. WRIGGERS, & O. ALLIX (Eds.), *Multiscale modeling of heterogeneous structures, (Lecture notes in applied and computational mechanics)*. Berlin: Springer International Publishing AG.

Kristanic, N., & Korelc, J. (2008). Optimization method for the determination of the most unfavorable imperfection of structures. *Computational Mechanics, 42*, 859–872.

Logg, A., Mardal, K. A., & Wells, G. N. (2012). *Automated solution of differential equations by finite element method*. Berlin Heidelberg: Springer.

Melink, T., & Korelc, J. (2014). Stability of karhunen- love expansion for the simulation of gaussian stochastic fields using galerkin scheme. *Probabilistic Engineering Mechanics, 37*, 7–15.

Nezamabadi, S., Zahrouni H., & Yvonnet, J. (2011). Solving hyperelastic material problems by asymptotic numerical method. *Computational Mechanics, 47*, 77–92.

Schmidt, H. (2000). Stability of steel shell structures - general report. *Journal of Constructional Steel Research, 55*, 159–181.

Solinc, U., & Korelc, J. (2015). A simple way to improved formulation of fe2 analysis. *Computational Mechanics, 56*, 905–915.

Equilibrated Stress Reconstruction and a Posteriori Error Estimation for Linear Elasticity

Fleurianne Bertrand, Bernhard Kober, Marcel Moldenhauer and Gerhard Starke

Abstract Based on the displacement–pressure approximation computed with a stable finite element pair, a stress equilibration procedure for linear elasticity is proposed. Our focus is on the Taylor–Hood finite element space, with emphasis on the behavior for (nearly) incompressible materials. From a combination of displacement in the standard continuous finite element spaces of polynomial degrees k+1 and pressure in the standard continuous finite element spaces of polynomial degrees k, we construct an H(div)-conforming, weakly symmetric stress reconstruction. Explicit formulas are first given for a flux reconstruction and then for the stress reconstruction.

1 Introduction

The accurate resolution of displacements associated with numerical simulations is of great importance and practical interest in solid mechanics. Finite element methods for these problems have been widely used and analyzed under a general framework in many works, e.g. in Ciarlet (1988) and the references therein. In addition to standard conforming Galerkin approximations, mixed finite elements appproximating simultaneously the displacements and a pressure-like variable are very popular since they allow the variational formulation to remain stable in the incompressible limit. Convergence and optimal a priori estimates are given in Boffi et al. (2009).

It is well known that on general domains, the displacements cannot expect to be sufficiently regular to use these a priori estimates. In order to retain the optimal convergence order, adaptive procedures based on a posteriori error estimators have

F. Bertrand (✉)
Humboldt Universität zu Berlin Institut für Mathematik,
Unter den Linden 6, 10099 Berlin, Germany
e-mail: fb@math.hu-berlin.de

B. Kober · M. Moldenhauer · G. Starke
Universität Duisburg-Essen Fakultät für Mathematik,
Thea-Leymann-Straße 9, 45127 Essen, Germany

J. Schröder and P. de Mattos Pimenta (eds.), *Novel Finite Element Technologies for Solids and Structures*, CISM International Centre for Mechanical Sciences 597,
https://doi.org/10.1007/978-3-030-33520-5_3

to be used. Several approaches have been considered to construct estimators based on the residual equation presented in Ainsworth and Oden (1993). In particular Verfürth (1999) gives a review of a posteriori error estimation techniques for elasticity problems.

For the study of the optimal convergence rates of these procedures, important progress has been made during the past decades: a crucial marking provided in Dörfler (1996), Morin et al. (2000) introduced the concept of interior node property and Stevenson (2007) provided a new overall theoretical understanding in order to realize optimal computational complexity. Cascon et al. (2008) proved the optimal cardinality of the AFEM using a decay between consecutive loops, while Nochetto et al. (2009), Nochetto and Veeser (2012), Carstensen et al. (2014) present a survey and an axiomatic presentation of the proof of optimal convergence rates for adaptive finite element methods.

An alternative approach for a posteriori error estimation using a flux reconstruction of the primal variable of the source problem usually leads to guaranteed, easily, fully, and locally computable, upper bound on the error measured in the energy norm, see Braess and Schöberl (2008), Cai and Zhang (2010), Ern and Vohralík (2015). It is based on the hypercircle identity dating back at least as far as Ladevèze and Leguillon (1983) and Prager and Synge (1947). A unified framework for a posteriori error estimation based on stress reconstruction for the Stokes system is carried out in Hannukainen et al. (2012). However, the applicability of these procedures is limited to bilinear forms involving the full gradient. This includes the Stokes system which is equivalent to incompressible linear elasticity if Dirichlet conditions are prescribed on the entire boundary, but needs to be extended for the compressible case, see Bertrand et al. (2018b). For the extension to eigenvalue problem see Bertrand et al. (2019). The purpose of this work is to give the details about the extension of this approach to linear elasticity and to show how to take the symmetry of the stress–tensor into account.

2 The Displacement–Pressure Approximations

Since the impact of curved boundaries can be treated like in Bertrand et al. (2014a, b), Bertrand and Starke (2016), the elasticity problems under our consideration are based on an open and polygonally bounded domain $\Omega \subset \mathbb{R}^d$ ($d = 2, 3$) which constitutes the reference configuration of the undeformed state. The body is submitted to body forces $\mathbf{f} \in L^2(\Omega)^d$, surface traction forces $\mathbf{g} \in L^2(\Gamma_N)^d$ on Γ_N and homogeneous displacement boundary conditions $\mathbf{u} = \mathbf{0}$ on Γ_D. For simplicity, we assume the boundary $\Gamma = \partial\Omega$ to be divided into the two disjoint and non-empty subsets Γ_N and $\Gamma_D = \Gamma \setminus \Gamma_N$. This notation is summarized in Fig. 1 for the Cook's Membrane example. Numerical example for this domain will be given in Sect. 6. In the standard displacement formulation considered further, the displacements $\mathbf{u} : \Omega \to \mathbb{R}^d$ are usually approximated in a subset of the Sobolev space

$$\mathbf{H}^1_{\Gamma_D} := \left\{ \mathbf{v} \in \left(H^1(\Omega)\right)^d : \mathbf{u} = 0 \text{ on } \Gamma_D \right\}.$$

Fig. 1 Cook's Membrane

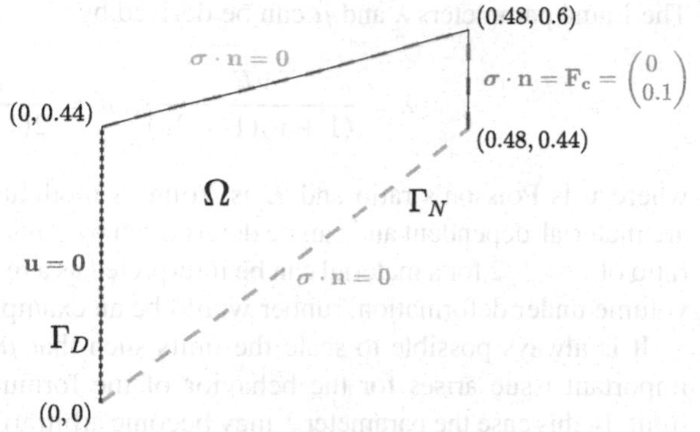

Since we assume Γ_D of positive length, Korn's inequality is valid for functions $\mathbf{v} \in \mathbf{H}^1_{\Gamma_D}$ with some constant $C_K > 0$

$$\|\boldsymbol{\varepsilon}(\mathbf{v})\|^2_{L^2(\Omega)} \geq C_K \|\nabla \mathbf{v}\|^2_{L^2(\Omega)}. \tag{1}$$

where

$$\boldsymbol{\varepsilon}_{ij}(\mathbf{v}) = \frac{1}{2}\left(\frac{\partial \mathbf{v}_i}{\partial x_j} + \frac{\partial \mathbf{v}_j}{\partial x_i}\right) \quad i, j = 1, \ldots, d$$

is the symmetric gradient, see, e.g., Braess (2013). The displacement solution $\mathbf{u} \in \mathbf{H}^1_{\Gamma_D}$ to the problem of linear elasticity is then the minimizer of

$$\mathbf{E}(\mathbf{v}) := \int_\Omega \boldsymbol{\psi}(\boldsymbol{\varepsilon}(\mathbf{v}))\,\mathrm{d}x - \int_\Omega \mathbf{f} \cdot \mathbf{v}\,\mathrm{d}x - \int_{\Gamma_N} \mathbf{g} \cdot \mathbf{v}\,\mathrm{d}x \tag{2}$$

under all $\mathbf{v} \in \mathbf{H}^1_{\Gamma_D}$, where the stored energy function is given by

$$\boldsymbol{\psi}(\boldsymbol{\varepsilon}) = \mu|\boldsymbol{\varepsilon}|^2 + \frac{\lambda}{2}(\mathrm{tr}\,\boldsymbol{\varepsilon})^2 .$$

By calculating the first necessary condition

$$\frac{\partial}{\partial t}\mathbf{E}(\mathbf{u} + t\mathbf{v})|_{t=0} = 0$$

for all $\mathbf{v} \in \mathbf{H}^1_{\Gamma_D}$ we obtain the corresponding variational formulation

$$2\mu \int_\Omega \boldsymbol{\varepsilon}(\mathbf{u}) : \boldsymbol{\varepsilon}(\mathbf{v})\,\mathrm{d}x + \lambda \int_\Omega (\mathrm{div}\,\mathbf{u})(\mathrm{div}\,\mathbf{v})\,\mathrm{d}x = \int_\Omega \mathbf{f} \cdot \mathbf{v}\,\mathrm{d}x + \int_{\Gamma_N} \mathbf{g} \cdot \mathbf{v}\,ds.$$

The Lamé parameters λ and μ can be derived by

$$\lambda = \frac{\nu E}{(1+\nu)(1-2\nu)}, \quad \mu = \frac{E}{2(1+\nu)},$$

where ν is Poisson's ratio and E is Young's modulus. Both parameters ν and E are material-dependent and can be determined by physical experiments. A Poisson's ratio of $\nu \approx 1/2$ for a material can be interpreted as a material which does not change volume under deformation, rubber would be an example.

It is always possible to scale the units such that μ is of the order of 1, but an important issue arises for the behavior of the formulation in the incompressible limit. In this case the parameter λ may become arbitrarily large and minimizing (2) it will be unstable. One idea is to introduce the new parameter p which has the physical interpretation of a pressure. Substituting the new variable $p := \lambda(\operatorname{div} \mathbf{u})$, we obtain the mixed formulation:
Find $\mathbf{u} \in \mathbf{H}^1_{\Gamma_D}$ and $p \in L^2(\Omega)$ such that

$$\begin{aligned} a(\mathbf{u}, \mathbf{v}) + b(\mathbf{v}, p) &= (\mathbf{f}, \mathbf{v}) + \langle \mathbf{g}, \mathbf{v} \rangle_{\Gamma_N} \qquad && \forall \mathbf{v} \in \mathbf{H}^1_{\Gamma_D} \\ b(\mathbf{u}, q) - c(p, q) &= 0 && \forall q \in L^2(\Omega) \end{aligned} \tag{MF}$$

where

$$a(\mathbf{u}, \mathbf{v}) := 2\mu \int_\Omega \boldsymbol{\varepsilon}(\mathbf{u}) : \boldsymbol{\varepsilon}(\mathbf{v}) \, \mathrm{d}x, \; b(\mathbf{v}, p) := \int_\Omega p \, (\operatorname{div} \mathbf{v}) \, \mathrm{d}x,$$
$$c(p, q) := \frac{1}{\lambda} \int_\Omega p \, q \, \mathrm{d}x.$$

The incompressible case reduces to the problem:
Find $\mathbf{u} \in \mathbf{H}^1_{\Gamma_D}$ and $p \in L^2(\Omega)$ such that

$$\begin{aligned} a(\mathbf{u}, \mathbf{v}) + b(\mathbf{v}, p) &= (\mathbf{f}, \mathbf{v}) + \langle \mathbf{g}, \mathbf{v} \rangle_{\Gamma_N} \qquad && \forall \mathbf{v} \in \mathbf{H}^1_{\Gamma_D} \\ b(\mathbf{u}, q) &= 0 && \forall q \in L^2(\Omega) \end{aligned} \tag{MFI}$$

An important question is the solvability and uniqueness of a solution of (**MF**) and (**MFI**). In the case of well-behaved λ one can use a fairly easy proof to show those solution properties for (**MF**), see Boffi et al. (2013).

Theorem 2.1 *The problem (MF) has a unique solution for all* $(\mathbf{f}, \mathbf{g}) \in L^2(\Omega) \times L^2(\Gamma_N)$ *if* $a(\cdot,\cdot)$, $b(\cdot,\cdot)$, $c(\cdot,\cdot)$ *are continuous bilinear forms on* $\mathbf{H}^1_{\Gamma_D} \times \mathbf{H}^1_{\Gamma_D}$, $\mathbf{H}^1_{\Gamma_D} \times L^2(\Omega)$, $L^2(\Omega) \times L^2(\Omega)$ *respectively and* $a(\cdot,\cdot)$, $c(\cdot,\cdot)$ *are coercive on* $\mathbf{H}^1_{\Gamma_D}$, $L^2(\Omega)$ *respectively and* $c(\cdot,\cdot) \not\equiv 0$. *This means there exist* $c_a, c_b, c_c, \gamma_a, \gamma_c > 0$ *such that*

$$|a(\mathbf{u}, \mathbf{v})| \le c_a \|\mathbf{u}\|_{\mathbf{H}^1_{\Gamma_D}} \|\mathbf{v}\|_{\mathbf{H}^1_{\Gamma_D}} \qquad \forall \mathbf{u}, \mathbf{v} \in \mathbf{H}^1_{\Gamma_D}, \tag{3a}$$

$$|b(\mathbf{u}, q)| \le c_b \|\mathbf{u}\|_{\mathbf{H}^1_{\Gamma_D}} \|q\|_{L^2(\Omega)} \qquad \forall \mathbf{u} \in \mathbf{H}^1_{\Gamma_D}, q \in L^2(\Omega), \tag{3b}$$

$$|c(p,q)| \leq c_c \|p\|_{L^2(\Omega)} \|q\|_{L^2(\Omega)} \qquad \forall p, q \in L^2(\Omega), \tag{3c}$$

$$a(\mathbf{u},\mathbf{u}) \geq \gamma_a \|\mathbf{u}\|^2_{\mathbf{H}^1_{\Gamma_D}} \qquad \forall \mathbf{u} \in \mathbf{H}^1_{\Gamma_D}, \tag{3d}$$

$$c(q,q) \geq \gamma_c \|q\|^2_{L^2(\Omega)} \qquad \forall q \in L^2(\Omega) \tag{3e}$$

To show the same result for (**MFI**) we introduce the Inf-Sup condition which is fulfilled if there exists a $\beta > 0$ such that

$$\beta \leq \inf_{q \in L^2(\Omega)} \sup_{\mathbf{v} \in \mathbf{H}^1_{\Gamma_D}} \frac{b(\mathbf{v},q)}{\|q\|_{L^2(\Omega)} \|\mathbf{v}\|_{\mathbf{H}^1_{\Gamma_D}}}. \tag{4}$$

Theorem 2.2 *The problem (**MFI**) has a unique solution for all* $(\mathbf{f},\mathbf{g}) \in L^2(\Omega) \times L^2(\Gamma_N)$ *if* $a(\cdot,\cdot)$, $b(\cdot,\cdot)$ *are continuous bilinear forms on* $\mathbf{H}^1_{\Gamma_D} \times \mathbf{H}^1_{\Gamma_D}$ *and* $\mathbf{H}^1_{\Gamma_D} \times L^2(\Omega)$ *respectively. The bilinear form* $a(\cdot,\cdot)$ *is coercive on* $K := \{\mathbf{v} \in \mathbf{H}^1_{\Gamma_D} : b(\mathbf{v},q) = 0 \; \forall q \in L^2(\Omega)\}$ *and* $b(\cdot,\cdot)$ *satisfies the Inf-Sup condition.*

Considering a family of shape-regular triangulations $\{\mathcal{T}_h\}$ of Ω, we denote the diameter of an element $T \in \mathcal{T}_h$ by h_T and derive a discrete system using the finite element spaces $\mathbf{V}_h$ and Q_h based on this triangulation for the approximation of $\mathbf{H}^1_{\Gamma_D}$ and $L^2(\Omega)$ respectively. The construction of the mesh will have importance for the reconstruction of the stress–tensor. In particular, the case involving Dirichlet and Neumann boundary conditions in the same nodal patch has to be considered separately. In order to avoid this particularity, it is possible to exclude this type of nodal patches. For instance two different initial triangulations of the Cook's membrane are shown in Fig. 2. The discrete counterparts to (**MF**) and (**MFI**) have the following form.

Find $\mathbf{u}_h \in \mathbf{V}_h$ and $p_h \in Q_h$ such that

$$\begin{aligned} a(\mathbf{u}_h,\mathbf{v}_h) + b(\mathbf{v}_h,p_h) &= (\mathbf{f},\mathbf{v}_h) + \langle \mathbf{g},\mathbf{v}_h \rangle_{\Gamma_N} && \forall \mathbf{v}_h \in \mathbf{V}_h \\ b(\mathbf{u}_h,q_h) - c(p_h,q_h) &= 0 && \forall q \in Q_h \end{aligned} \tag{$\mathbf{MF}_h$}$$

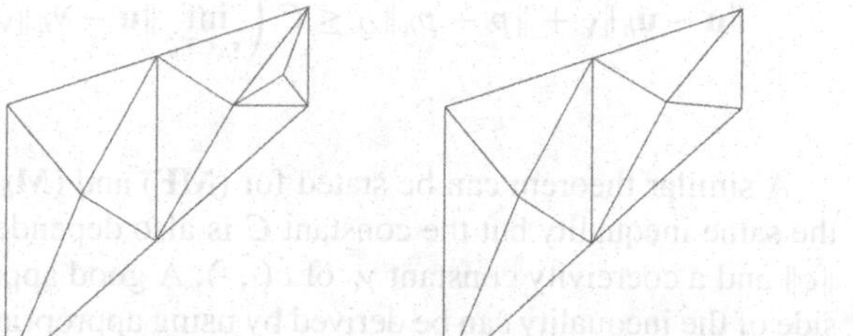

Fig. 2 Two different initial triangulations

Find $\mathbf{u}_h \in \mathbf{V}_h$ and $p_h \in Q_h$ such that

$$\begin{aligned} a(\mathbf{u}_h, \mathbf{v}_h) + b(\mathbf{v}_h, p_h) =& (\mathbf{f}, \mathbf{v}_h) + \langle \mathbf{g}, \mathbf{v}_h \rangle_{\Gamma_N} \quad && \forall \mathbf{v}_h \in \mathbf{V}_h \\ b(\mathbf{u}_h, q_h) =& 0 && \forall q \in Q_h \end{aligned} \qquad (\mathbf{MFI}_h)$$

It is important to differ between conforming finite-dimensional spaces where $\mathbf{V}_h \subset \mathbf{H}^1_{\Gamma_D}$ and $Q_h \subset L^2(\Omega)$ applies and non-conforming finite-dimensional spaces where we allow $\mathbf{V}_h \not\subset \mathbf{H}^1_{\Gamma_D}$ and $Q_h \not\subset L^2(\Omega)$. In case of conforming spaces a lot of properties carry over from the continuous cases (**MF**) and (**MFI**). For well behaved λ Theorem 2.1 does not change at all because $\mathbf{V}_h$ and Q_h are subsets of $\mathbf{H}^1_{\Gamma_D}$ and $L^2(\Omega)$. To apply Theorem 2.2 in the conforming finite-dimensional setting the Inf-Sup condition for $\mathbf{V}_h$ and Q_h remains to be stated, i.e., that there exists a scalar $\beta > 0$ such that

$$\beta \leq \inf_{q \in Q_h} \sup_{\mathbf{v} \in \mathbf{V}_h} \frac{b(\mathbf{v}_h, q_h)}{\|q_h\|_{Q_h} \|\mathbf{v}_h\|_{\mathbf{V}_h}}. \tag{5}$$

In case of non-conforming spaces more work has to be done, for example Korn's inequality is usually not an implication of the continuous case and has to be proven for the specific non-conforming space.

The system (**MF**) and the discrete counterpart ($\mathbf{MF}_h$) can be interpreted as perturbed Stokes problems, where the incompressible cases (**MFI**) and ($\mathbf{MFI}_H$) are equivalent to the Stokes problem. As a conclusion, all Stokes stable elements are also suitable in the linear elastic setting, which is easy to comprehend because the Inf-Sup condition (5) is the same for the Stokes problem.

In this section we recall some convergence results for the approximation of (**MFI**) and ($\mathbf{MEI}_h$) using conforming finite element spaces. First we derive general error bounds, which we then use in an example for the Taylor–Hood element.

Theorem 2.3 *Let* $(\mathbf{u}, p) \in \mathbf{V} \times Q$ *and* $(\mathbf{u}_h, p_h) \in \mathbf{V}_h \times Q_h$ *solutions of* (***MFI***) *and* ($\boldsymbol{MEI}_h$) *respectively. The spaces* $\mathbf{V}_h$ *and* Q_h *satisfy the discrete Inf-Sup condition with a constant* β*. The bilinear forms* $a(\cdot, \cdot)$ *and* $b(\cdot, \cdot)$ *are continuous with constants* $\|a\|$ *and* $\|b\|$ *respectively. The bilinear form* $a(\cdot, \cdot)$ *is also coercive on* $K_h := \{\mathbf{v}_h \in \mathbf{V}_h : b(\mathbf{v}_h, q_h) = 0 \; \forall q_h \in Q_h\}$ *with a constant* α*. Then there exists a constant* C *depending only on* α, β, $\|a\|$, $\|b\|$ *and independent of h, such that*

$$\|\mathbf{u} - \mathbf{u}_h\|_{\mathbf{V}} + \|p - p_h\|_Q \leq C \left(\inf_{\mathbf{v}_h \in \mathbf{V}_h} \|\mathbf{u} - \mathbf{v}_h\|_{\mathbf{V}} + \inf_{q_h \in Q_h} \|p - q_h\|_Q \right).$$

A similar theorem can be stated for (**MF**) and ($\mathbf{M}_h$). In this case one can derive the same inequality but the constant C is also dependent on the continuity constant $\|c\|$ and a coercivity constant γ_c of $c(\cdot, \cdot)$. A good approximation for the right-hand side of the inequality can be derived by using appropriate interpolations for the finite element spaces $\mathbf{V}_h$ and Q_h.

The Taylor–Hood Element A popular choice for $\mathbf{V}_h \times Q_h$ is the Taylor–Hood element, defined with $\mathbf{V}_h = (\mathcal{P}_k(\mathcal{T}_h))^d \cap H^1_{\Gamma_D}$ and $Q_h = \mathcal{P}_{k-1}(\mathcal{T}_h)$ for $k \geq 1$, where $\mathcal{P}_k(\mathcal{T}_h)$ denotes the space of continuous piecewise polynomials of order k with respect to $\mathcal{T}_h$. Since $\mathbf{V}_h \subset H^1_{\Gamma_D}$ and $Q_h \subset L^2(\Omega)$ the Taylor–Hood element is a conforming finite element for our formulation. The Korn inequality (1) is also valid for functions $\mathbf{v}_h \in \mathbf{V}_h$ and many results of the continuous cases are also valid in the discrete cases. One major question is the validation of the discrete Inf-Sup condition for the Taylor–Hood element.

Theorem 2.4 *If for every triangle in 2D, or tetrahedron in 3D, $T \in \mathcal{T}_h$ of Ω at least one corner is inside Ω then the Taylor–Hood elements satisfy the discrete Inf-Sup condition for $k \geq 1$.*

By ensuring Inf-Sup stability we can use Theorem 2.3 and suitable interpolation operators for the Taylor–Hood elements and derive the following convergence result.

Theorem 2.5 *For a solution $(\mathbf{u}, p)$ of (**MF**) or (**MF1**) which is sufficiently smooth and a solution $(\mathbf{u}_h, p_h)$ of (**MF**$_h$) or (**MFI**$_h$) in the Taylor–Hood space, respectively, it holds*

$$\|\mathbf{u} - \mathbf{u}_h\|_{H^1(\Omega)} + \|p - p_h\|_{L^2(\Omega)} \leq Ch^k \left(|\mathbf{u}|_{H^{k+1}(\Omega)} + |p|_{H^k(\Omega)} \right),$$

where $C > 0$ is a constant independent of h.

3 Stress Approximations

Since large stress components may cause plastic behavior or damage, an accurate resolution of the stress–tensor is needed. In particular, surface traction forces are of interest where the elastic body is clamped due to the risk of failure of the material. For linear elasticity, the stress–tensor is given by the strain–stress relation $\mathcal{C}$

$$\sigma(\mathbf{u}) = \mathcal{C}\varepsilon(\mathbf{u}) = 2\mu\epsilon(\mathbf{u}) + \lambda \mathrm{tr}(\varepsilon(\mathbf{u}))\mathbf{I}\cdot \tag{6}$$

Note that with this strain–stress relation, the linear elasticity model may then be written as the first-order system

$$\begin{aligned} \operatorname{div} \sigma + \mathbf{f} &= \mathbf{0} \\ \sigma - \mathcal{C}\varepsilon(\mathbf{u}) &= \mathbf{0}\,. \end{aligned} \tag{7}$$

As mentioned in the introduction, an important issue is the behavior in the incompressible limit, i.e., when the Lamé parameter λ tends to infinity. Because of the second term $\lambda \mathrm{tr}(\varepsilon(\mathbf{u}))\mathbf{I}$, the operator $\mathcal{C}$ can not remain well-defined in the incompressible limit. Fortunately, the following calculation shows that the inverse $\mathcal{C}^{-1}$ remains well-defined in the incompressible limit:

$$\begin{aligned} \mathcal{C}^{-1}\boldsymbol{\sigma} &= \frac{1}{2\mu}\left(\boldsymbol{\sigma} - \frac{\lambda}{2\mu + d\lambda}\mathrm{tr}(\boldsymbol{\sigma})\mathbf{I}\right) \\ &\overset{\lambda\to\infty}{\to} \frac{1}{2\mu}\left(\boldsymbol{\sigma} - \frac{1}{d}\mathrm{tr}(\boldsymbol{\sigma})\mathbf{I}\right) = \frac{1}{2\mu}\,\mathbf{dev}\,\boldsymbol{\sigma}, \end{aligned}$$

In particular, in the incompressible limit, $\mathcal{C}^{-1}$ constitutes the orthogonal projection onto the trace-free matrices. In order to allow this first-order system to remain well-defined in the incompressible limit we will use the inverse $\mathcal{C}^{-1}$ instead of $\mathcal{C}$ and we write $\mathcal{A}$ instead of $\mathcal{C}^{-1}$ to avoid confusion, since $\mathcal{C}^{-1}$ itself is not invertible any more. The first-order system (7) now reads

$$\begin{aligned} \operatorname{div}\boldsymbol{\sigma} + \mathbf{f} &= \mathbf{0} \\ \mathcal{A}\boldsymbol{\sigma} - \boldsymbol{\varepsilon}(\mathbf{u}) &= \mathbf{0}\,. \end{aligned} \tag{8}$$

Note that for an arbitrary approximation $(\mathbf{u}_h)$ and a computed stress $\boldsymbol{\sigma}(\mathbf{u}_h) = 2\mu\boldsymbol{\varepsilon}(\mathbf{u}_h) + \lambda\mathrm{tr}(\boldsymbol{\varepsilon}(\mathbf{u}_h))\mathbf{I}$ the strain–stress relation $\mathcal{A}\boldsymbol{\sigma}(\mathbf{u}_h) = \boldsymbol{\varepsilon}(\mathbf{u}_h)$ does not hold in general. Similarly, if we want to use the pressure inserted in the last section, the stress–strain relation reads

$$\boldsymbol{\sigma}(\mathbf{u}, p) = \mathcal{C}\boldsymbol{\varepsilon}(\mathbf{u}) = 2\mu\boldsymbol{\varepsilon}(\mathbf{u}) + p\mathbf{I} \tag{9}$$

and the relation $\mathcal{A}\boldsymbol{\sigma}(\mathbf{u}_h, p_h) = \boldsymbol{\varepsilon}(\mathbf{u}_h)$ does not hold for an arbitrary approximation $(\mathbf{u}_h, p_h)$ computed with an inf-sup stable discretization of the displacement–pressure formulation. In fact, the definition of the stress leads to

$$\begin{aligned} \mathrm{tr}\,\boldsymbol{\sigma} &= 2\mu\operatorname{div}\mathbf{u} + dp = \left(\frac{2\mu}{\lambda} + d\right)p\,, \\ \mathrm{tr}\,\boldsymbol{\sigma}_h &= 2\mu\operatorname{div}\mathbf{u}_h + dp_h = \left(\frac{2\mu}{\lambda} + d\right)p_h + 2\mu\left(\operatorname{div}\mathbf{u}_h - \frac{1}{\lambda}p_h\right), \end{aligned} \tag{10}$$

which implies

$$\begin{aligned} \boldsymbol{\varepsilon}(\mathbf{u}_h) &= \frac{1}{2\mu}\left(\boldsymbol{\sigma}_h - p_h\mathbf{I}\right) \\ &= \frac{1}{2\mu}\left(\boldsymbol{\sigma}_h - \frac{\lambda}{2\mu + d\lambda}(\mathrm{tr}\,\boldsymbol{\sigma}_h)\mathbf{I}\right) + \frac{\lambda}{2\mu + d\lambda}\left(\operatorname{div}\mathbf{u}_h - \frac{1}{\lambda}p_h\right)\mathbf{I} \\ &= \mathcal{A}\boldsymbol{\sigma}_h + \frac{\lambda}{2\mu + d\lambda}\left(\operatorname{div}\mathbf{u}_h - \frac{1}{\lambda}p_h\right)\mathbf{I}\,. \end{aligned} \tag{11}$$

In order to be able to evaluate the surface traction forces, i.e., the normal component of the stress–tensor, the stress–tensor needs to be in $H(\operatorname{div}, \Omega)^d$. Unfortunately, considering a finite element approximation $(\mathbf{u}_h, p_h)$ from the displacement–pressure formulation, the associated stress–tensor $\boldsymbol{\sigma}(\mathbf{u}_h, p_h)$ is contained in $L^2(\Omega)^{d\times d}$ but can not be assumed to belong to $H(\mathrm{div}, \Omega)^d$. In order to evaluate the normal component of the boundary traces, a reconstruction of an $H(\mathrm{div}, \Omega)^d$-conforming stress–tensor is therefore crucial and will be the topic of the next section. However, there exist variational principles involving stresses directly like saddle point formulations of Hellinger–Reissner-type. Since their background will be important for the proof of the existence of the stress reconstruction, we should introduce this formulation in this section.

The Hellinger–Reissner Formulation

In order to obtain a saddle point formulation, the stress–strain relation $\boldsymbol{\varepsilon}(\mathbf{u}) = \mathcal{A}\boldsymbol{\sigma}$ will be tested with test functions $\tau \in H_{\Gamma_N}(\mathrm{div}, \Omega)^d$, a subspace that consider the Neumann boundary conditions:

$$H_{\Gamma_N}(\mathrm{div}, \Omega)^d = \{\boldsymbol{\tau} \in H(\mathrm{div}, \Omega)^d : \boldsymbol{\tau} \cdot \mathbf{n} = \mathbf{0} \text{ on } \Gamma_N\} .$$

Integrating the by parts and using the definition asymmetric part $\mathbf{as}\ \boldsymbol{\tau} = (\boldsymbol{\tau} - \boldsymbol{\tau}^T)/2$ leads then to

$$(\mathcal{A}\boldsymbol{\sigma}, \boldsymbol{\tau}) + (\mathbf{u}, \mathrm{div}\, \boldsymbol{\tau}) + (\mathbf{as}\, \nabla \mathbf{u}, \mathbf{as}\, \boldsymbol{\tau}) = 0 , \tag{12}$$

for all $\boldsymbol{\tau} \in H_{\Gamma_N}(\mathrm{div}, \Omega)^d$. Note that the last two terms correspond to the momentum balance and symmetry constraints

$$\begin{aligned} (\mathrm{div}\, \boldsymbol{\sigma} + \mathbf{f}, \mathbf{v}) &= 0 \text{ for all } \mathbf{v} \in L^2(\Omega)^d , \\ (\mathbf{as}\ \boldsymbol{\sigma}, \boldsymbol{\theta}) &= 0 \text{ for all } \boldsymbol{\theta} \in L^2(\Omega)^{d\times d,\mathrm{as}} , \end{aligned} \tag{13}$$

where $L^2(\Omega)^{d\times d,\mathrm{as}}$ is the subspace of $L^2(\Omega)^{d\times d}$ with vanishing symmetric part. This means that introducing a new variable $\boldsymbol{\gamma}$ for $\mathbf{as}\ \nabla\mathbf{u}$ correspond to the minimization of the energy $(\mathcal{A}\boldsymbol{\sigma}, \boldsymbol{\sigma})$ subject to the constraints (13). Therefore, the mixed variational formulation of Hellinger–Reissner consists in finding $(\boldsymbol{\sigma}, \mathbf{u}, \boldsymbol{\gamma}) \in (\boldsymbol{\sigma}^N + H_{\Gamma_N}(\mathrm{div}, \Omega)^d) \times L^2(\Omega)^d \times L^2(\Omega)^{d\times d,\mathrm{as}}$ such that

$$\begin{aligned} (\mathcal{A}\boldsymbol{\sigma}, \boldsymbol{\tau}) + (\mathbf{u}, \mathrm{div}\, \boldsymbol{\tau}) + (\boldsymbol{\gamma}, \mathbf{as}\ \boldsymbol{\tau}) &= 0 \text{ for all } \boldsymbol{\tau} \in H_{\Gamma_N}(\mathrm{div}, \Omega)^d , \\ (\mathrm{div}\, \boldsymbol{\sigma} + \mathbf{f}, \mathbf{v}) &= 0 \text{ for all } \mathbf{v} \in L^2(\Omega)^d , \\ (\mathbf{as}\ \boldsymbol{\sigma}, \boldsymbol{\theta}) &= 0 \text{ for all } \boldsymbol{\theta} \in L^2(\Omega)^{d\times d,\mathrm{as}} , \end{aligned} \tag{14}$$

hold.

Applying the theory described in the previous section, the well-posedness of the system (14) follows from the coercivity of $a_{HR}(\boldsymbol{\sigma}, \boldsymbol{\tau}) = (\mathcal{A}\boldsymbol{\sigma}, \boldsymbol{\tau})$ on the kernel

$$H^0_{\Gamma_N}(\mathrm{div}, \Omega)^d = \{\boldsymbol{\tau} \in H_{\Gamma_N}(\mathrm{div}, \Omega)^d : \mathrm{div}\, \boldsymbol{\tau} = \mathbf{0}\} , \tag{15}$$

i.e.,

$$(\mathcal{A}\boldsymbol{\tau}, \boldsymbol{\tau}) \gtrsim \|\boldsymbol{\tau}\|^2 \text{ for all } \boldsymbol{\tau} \in H^0_{\Gamma_N}(\mathrm{div}, \Omega)^d . \tag{16}$$

and the inf-sup condition, see Boffi et al. (2013, Prop. 9.3.2). Since

$$(\mathcal{A}\boldsymbol{\tau}, \boldsymbol{\tau}) \geq \frac{1}{2\mu} \| \, \mathbf{dev} \;\; \boldsymbol{\tau} \|^2 \tag{17}$$

holds, the coercivity of a_{HR} on $H^0_{\Gamma_N}(\text{div}, \Omega)^d$ follows from the fact that, a function can be controlled by its deviatoric part, as stated in the following theorem:

Theorem 3.1 *Assume that $\Gamma_N \subseteq \partial\Omega$ consists of a finite number of connected components each of which has positive $(d-1)$-dimensional measure. Then,*

$$\|\boldsymbol{\tau}\| \lesssim \| \, \boldsymbol{dev} \;\; \boldsymbol{\tau} \| \tag{18}$$

holds for all $\boldsymbol{\tau} \in H^0_{\Gamma_N}(div, \Omega)^d$.

For the discretization of the Hellinger–Reissner formulation, as well as for the reconstruction of the $H(\text{div}, \Omega)^d$-conforming stress–tensor, the discretization of $H(\text{div}, \Omega)^d$ is crucial. The smallest polynomial $H(\text{div}, \Omega)^d$-conforming space such that the divergence maps onto the polynomial of degree k is given by the Raviart–Thomas space $RT_k(\mathcal{T}_h)$. This space, that we will introduce below, can be combined with the discontinuous polynomial of order k and the continuous asymmetric polynomial tensors of order k to form a stable inf-sup tuple

$$RT_k(\mathcal{T}_h)^d \times DP_k(\mathcal{T}_h)^d \times P_k(\mathcal{T}_h)^{d\times d,\text{as}} \tag{19}$$

for the discretization of (14), see Boffi et al. (2009).

The next subsection recalls the major properties of this space.

The Raviart–Thomas Elements

The Raviart–Thomas space is a vector-valued space involving on each element $T \in \mathcal{T}_h$ the space $RT_k(T) \subset (\mathcal{P}_{k+1}(T))^2$ of the Raviart–Thomas functions. For $T \in \mathcal{T}_h$, let ∂T denote its boundary, $\mathcal{E}(T)$ the set of its edges, $\mathbf{n}_T$ the outward oriented normal and

$$R_k(\partial T) = \{\phi \in L^2(\partial T) : \phi|_e \in \mathcal{P}_k(e) \;\; \forall\, e \in \mathcal{E}(T)\} \tag{20}$$

the polynomial space on the edges. When there is no ambiguity, the subscript T in $\mathbf{n}_T$ will be dropped. Then, the definition of the Raviart–Thomas space reads

$$RT_k(\mathcal{T}_h) = \{\mathbf{v}_h \in H_{\text{div}}(\Omega) : \mathbf{v}_h|_T \in RT_k(T) \; \forall T \in \mathcal{T}_h\} \tag{21}$$

with $RT_k(T) = (\mathcal{P}_k(T))^2 + \mathbf{x}\mathcal{P}_k(T)$. Note that for $\mathbf{v}_h \in RT_k(T)$, it holds

$$\text{div}\,\mathbf{v}_h \in \mathcal{P}_k(T)\,, \tag{22a}$$

$$\mathbf{v}_h \cdot \mathbf{n} \in R_k(\partial T) \tag{22b}$$

$$\int_{\partial T} (\mathbf{n} \cdot \mathbf{v}_h) p_k \, ds, \quad p_k \in R_k(\partial T), \tag{22c}$$

$$\int_T \mathbf{v}_h \cdot \mathbf{p}_{k-1} \, d\mathbf{x}, \quad \mathbf{p}_{k-1} \in (P_{k-1}(T))^2, \tag{22d}$$

(see Ern and Guermond (2004) as well). The proof of the unisolvence (see Boffi et al. (2013)) ensures that the last two equations can be used to define

- $k+1$ degrees of freedom as the fluxes across edges of the mesh on each edge of the triangulation $\mathcal{T}_h$,
- $\frac{1}{2}k(k+1)$ degrees of freedom as the value at some interior points in each element of the triangulation $\mathcal{T}_h$.

Note that for $k \geq 1$, the interior points can be chosen as the interior points of the Lagrange element of type $k+2$, each point represents two degrees of freedom (see Figs. 3, 4), such that

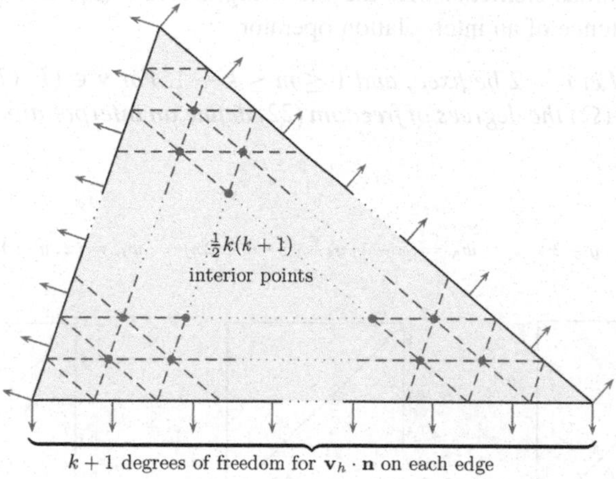

Fig. 3 Degrees of freedom for the Raviart–Thomas element

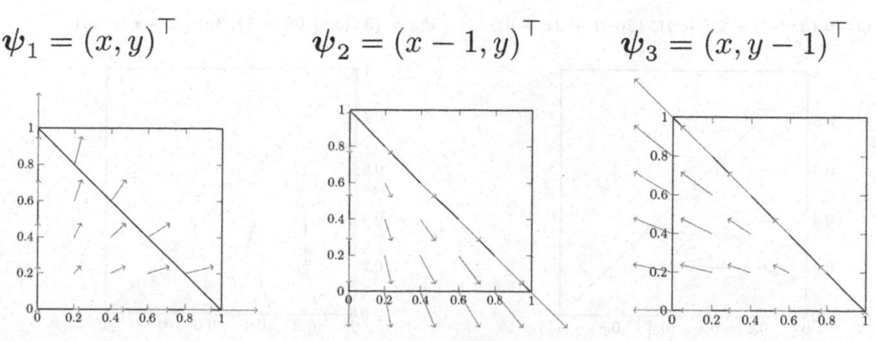

Fig. 4 Ansatz functions on reference triangle for RT_0

$$\dim RT_k(K) = (k+3)(k+1). \tag{23}$$

Note that in the literature, the choice of the degrees of freedom may differ. Bahriawati and Carstensen (2005) for instance chose to evaluate $(\mathbf{v}_h \cdot \mathbf{n})|_E$ instead of $\langle \mathbf{n} \cdot \mathbf{v}_h, 1\rangle$ for the definition of the degrees of freedom of RT_0. Since $\mathbf{v}_h \cdot \mathbf{n}$ is constant on each edge, this is in fact just a different scaling of the basis functions. In particular, in order to give a hierarchical procedure for the flux reconstruction, we will use the following degrees of freedom for a function $\psi \in RT_1$:

$$\begin{cases} \int_{E_i} \boldsymbol{\psi} \cdot \mathbf{n}\, \mathrm{ds} & i = 1, \ldots, 3 \ E_i \text{ facet of } T \\ \int_{E_i} \boldsymbol{\psi} \cdot \mathbf{n}\, q_{E_i}\, \mathrm{ds} & i = 1, \ldots, 3 \ E_i \text{ facet of } T \\ \int_T (\operatorname{div} \boldsymbol{\psi})(\mathbf{x} - \mathbf{x}_m)\mathbf{e}_j\, \mathrm{d}\mathbf{x} & j = 1, 2 \end{cases} \tag{24}$$

The impact of this choice is illustrated in Figs. 5 and 6. We close the introduction of the Raviart–Thomas elements with the following theorem (see Boffi et al. (2013)) stating the existence of an interpolation operator.

Theorem 3.2 *Let $r > 2$ be fixed, and $1 \le m \le k+1$. For $\mathbf{v} \in (L^r(T))^2 \cap H^m(\Omega)$ with div $\mathbf{v} \in L^2(\Omega)$ the degrees of freedom (22) define an interpolation operator $\hat{\mathcal{R}}_h$,*

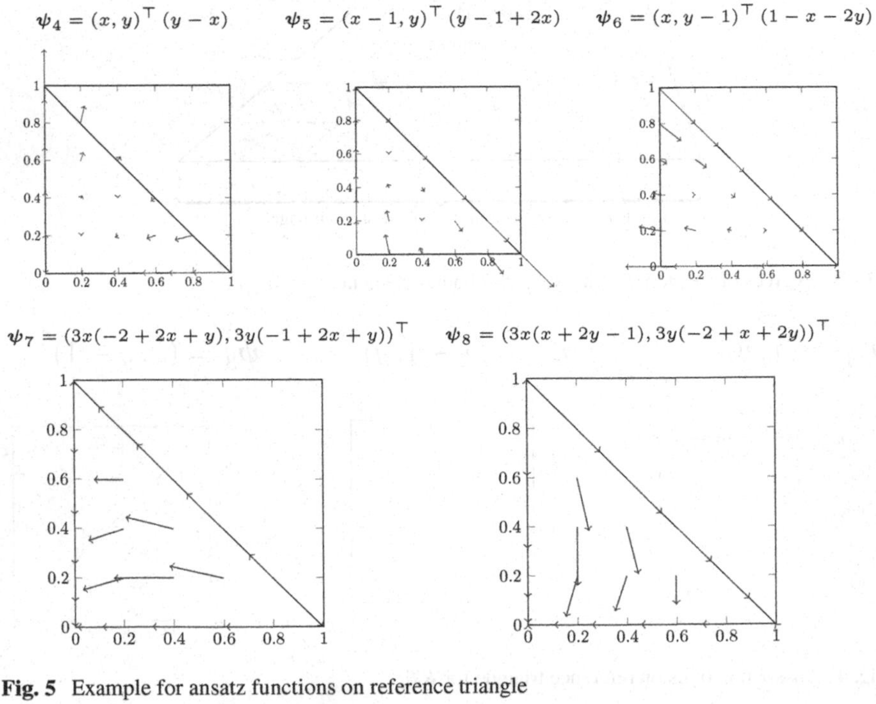

Fig. 5 Example for ansatz functions on reference triangle

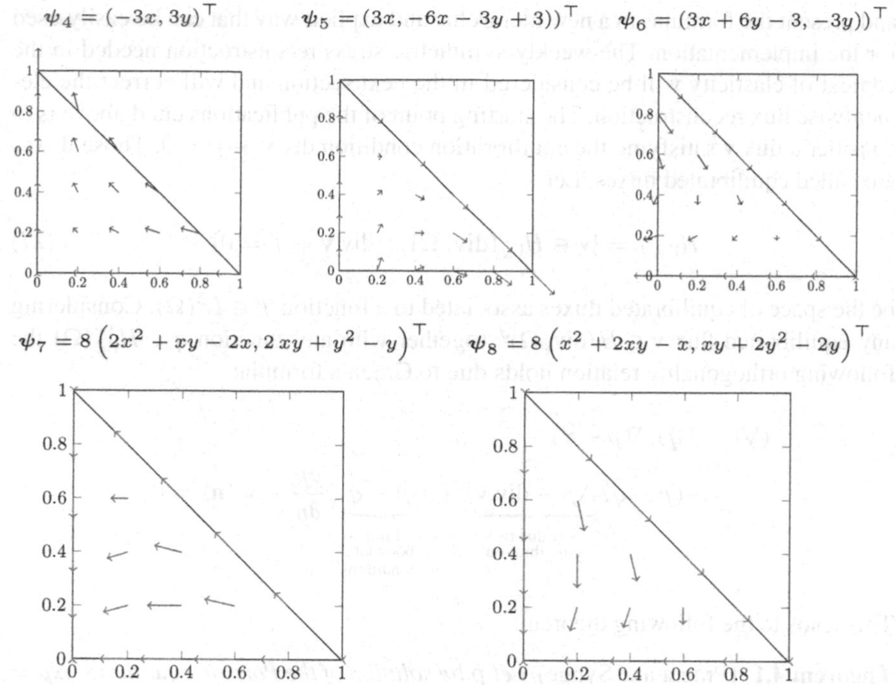

Fig. 6 Ansatz functions on reference triangle corresponding to (24)

and it holds for $s \leq k+1$:

$$\|\mathbf{v} - \hat{\mathcal{R}}_h \mathbf{v}\|_{0,\Omega} \lesssim h^m |\mathbf{v}|_{m,\Omega} \tag{25a}$$

$$\|div(\hat{\mathcal{R}}_h \mathbf{v})\|_{0,\Omega} \lesssim h^s |div\ \mathbf{v}|_{s,\Omega}\ . \tag{25b}$$

4 Flux Equilibration

The principal idea of stress reconstruction can be traced back to Prager and Synge (1947). Although the original publication is concerned with elasticity problems, the idea is often presented using the solution p of the Poisson equation

$$-\Delta p = f \quad \text{in } \Omega \tag{26a}$$

$$p = 0 \quad \text{on } \Gamma_D \tag{26b}$$

$$\nabla p \cdot \mathbf{n} = 0 \quad \text{on } \Gamma_N \tag{26c}$$

(see also Braess (2013), Braess and Schöberl (2008)). In fact, we will see in the next section, that our stress reconstruction procedure can use an elementwise flux reconstruction procedure. In this section, we therefore recall the idea of flux reconstruction and present the formulas in a new hierarchic and explicit way that can be easily used for the implementation. The weekly symmetric stress reconstruction needed in the context of elasticity will be considered in the next section and will correct the elementwise flux reconstruction. The starting point of the publications cited above is to consider a flux $\mathbf{v}$ satisfying the equilibration condition div $\mathbf{v} + f = 0$. Those fluxes are called equilibrated fluxes. Let

$$H_{\Gamma_N,f} = \{\mathbf{v} \in H_{\Gamma_N}(\text{div}, \Omega) \; : \; \text{div}\, \mathbf{v} + f = 0\} \tag{27}$$

be the space of equilibrated fluxes associated to a function $f \in L^2(\Omega)$. Considering any equilibrated flux $\mathbf{v} \in H(\text{div}, \Omega)^d$ together with any function $q \in H^1_{\Gamma_D}(\Omega)$ the following orthogonality relation holds due to Green's formula:

$$\begin{aligned}(\nabla(p-q), \nabla p - \mathbf{v}) \\ = -(p-q, \underbrace{\Delta p - \text{div}\, \mathbf{v}}_{\substack{=0 \text{ due to} \\ \text{equilibration}}}) + \langle \underbrace{p-q}_{\substack{=0 \text{ due to} \\ \text{boundary} \\ \text{conditions}}}, \frac{\partial p}{\partial n} - \mathbf{v}\cdot\mathbf{n}\rangle = 0\,.\end{aligned}$$

This leads to the following theorem:

Theorem 4.1 (Prager and Synge) *Let p be solution of the Poisson equation $-\Delta p = f$ and $(\mathbf{v}, q) \in H_{\Gamma_N,f}(div, \Omega) \times H^1_{\Gamma_D}(\Omega)$ be any approximation. Then it holds*

$$\|\nabla p - \nabla q\|^2 + \|\nabla p - \mathbf{v}\|^2 = \|\nabla q - \mathbf{v}\|^2\,. \tag{28}$$

Let p_h be any conforming approximation of p. Applying the Prager and Synge theorem leads directly to the following a posteriori estimate

$$\|\nabla p - \nabla p_h\| \le \|\nabla p_h - \mathbf{v}\| \quad \forall \mathbf{v} \in H_{\Gamma_N,f}(\text{div}, \Omega)\,. \tag{29}$$

This motivates the construction of an equilibrated flux $\mathbf{v}_h$. Solving a global problem would lead to such an equilibrated flux, but would be too expensive for the computation of a posteriori estimates. Moreover, we would like to use the approximated flux $\mathbf{u}_h = \nabla p_h$, computed from the approximation p_h. Unfortunately, $\mathbf{u}_h$ does not belong to $H(\text{div}, \Omega)^d$. We are therefore looking for a local procedure, usually called equilibration, to reconstruct a flux from ∇p_h. Usually, a correction $\mathbf{v}_h^\Delta$ is computed in a discontinuous finite element space, such that

$$\left(\mathbf{v}_h^\Delta + \nabla p_h\right) \in H_{\Gamma_N,f}\,.$$

Let $\mathcal{E}_{h,I}$ being the set of the interior edges in the triangulations. We will further denote $\mathcal{E}_{h,N}$ and $\mathcal{E}_{h,D}$ the sets of the edges in the triangulation intersecting with Γ_N and Γ_D, respectively. Intuitively, $\mathcal{E}_h$ denotes the union $\mathcal{E}_{h,D} \cup \mathcal{E}_{h,N} \cup \mathcal{E}_{h,I}$. Denoting the jump of a function q define piecewise by $q|_T = v^T$ over a facet E by

$$[\![q(\mathbf{x})]\!]_E = \begin{cases} q^1(\mathbf{x})|_E - q^2(\mathbf{x})|_E & \text{if } E = T_1 \cap T_2 \text{and } \mathbf{n}_E = \mathbf{n}_{T_1} \\ q(\mathbf{x})|_E & \text{if } E \in \mathcal{E}_{h,D} \cup \mathcal{E}_{h,N} \end{cases},$$

the above consideration means that the equilibration procedure is equivalent to find $\mathbf{u}_h^\Delta \in H_{\Gamma_N}(\text{div}, \Omega)^d$ such that

$$\text{div}\, \mathbf{u}_h^\Delta = -f \tag{30a}$$

$$[\![\mathbf{u}_h^\Delta \cdot \mathbf{n}]\!]_E = -[\![\nabla p_h \cdot \mathbf{n}]\!]_E \qquad \forall E \in \mathcal{E}_{h,I}\,. \tag{30b}$$

Lower Order Case Stress Equilibration

Braess and Schöberl (2008) proved that if p_h belongs to $\mathcal{P}^1(\mathcal{T}_h)$ and $f \in \mathcal{P}^0(\mathcal{T}_h)$, it is possible to reconstruct an equilibrated flux $(\mathbf{u}_h^{\Delta,0} + \nabla p_h) \in RT_0$ in a stable way, with $\mathbf{u}_h^{\Delta,0} \in DRT_0$. Note that if f is not piecewiese constant, it can be replaced by its projection $\Pi^0 f$ on the piecewise constants, and the term $\|f - \Pi^0 f\|$ is proven to be an oscillation term.

The equilibration procedure involves the partition of unity

$$1 \equiv \sum_{z \in \mathcal{V}_h} \phi_z \text{ on } \Omega\,. \tag{31}$$

where $\mathcal{V}_h$ denotes the set of vertices of the triangulation and ϕ_z is the continuous, piecewise linear Lagrange function with $\phi_z(z) = 1$ and whose support is the vertex patch

$$\omega_z := \bigcup \{T \in \mathcal{T}_h : z \text{ is a vertex of } T\}\,. \tag{32}$$

The correction $\mathbf{u}_h^{\Delta,0}$ can therefore be constructed from functions

$$\mathbf{u}_h^{\Delta,0} = \sum_{z \in \mathcal{V}_h} \phi_z \mathbf{u}_h^{\Delta,0} = \sum_{z \in \mathcal{V}_h} \mathbf{u}_{h,z}^{\Delta,0}\,, \tag{33}$$

where

$$\begin{aligned} \mathbf{u}_{h,z}^{\Delta,0} \in DRT_{0,z} = \{\mathbf{u}_h \in L^2(\omega_z)^d : \mathbf{u}_h \cdot \mathbf{n} = \mathbf{0} \text{ on } \partial\omega_z \backslash \partial\Omega\,, \\ \mathbf{u}_h|_T \in RT_0(T)\ \forall\, T \in \omega_z\,, \\ \mathbf{u}_h \equiv \mathbf{0} \text{ on } \Omega \backslash \omega_z\}\,. \end{aligned}$$

Since each two elements share one facet, the conditions (30a) and (30b) means that the function $\mathbf{u}_{h,z}^{\Delta,0}$ has to fulfill

$$\text{div}\, \mathbf{u}_{h,z}^{\Delta,0} = -\frac{1}{|T|}(f, \phi_z) \quad \forall T \in \omega_z \tag{34a}$$

$$[\![\mathbf{u}_{h,z}^{\Delta,0} \cdot \mathbf{n}]\!]_E = -\frac{1}{2}[\![\nabla p_h \cdot \mathbf{n}]\!]_E \quad \forall E \in \mathcal{E}_{h,I}(\omega_z)\backslash\mathcal{E}_{h,D} \tag{34b}$$

where $E_{h,I}(\omega_z)$ denotes the interior edges of the patch ω_z.

Let n_z denote the number of elements in ω_z. Then assume that the elements of $\omega_z = \{T_1, \ldots, T_{n_z}\}$ are numbered such that T_i and T_{i+1} share a facet, denoted by $E_{i,z}$ for $i = 1, \ldots, n_z - 1$. Moreover, if z is an interior node, T_1 and T_{n_z} share the facet $E_{n_z} = E_0$. If z lies on the Neumann boundary, T_1 is assumed to be an element with a facet E_0 on Γ_N. More generally, if z lies on the boundary, T_1 is assumed to be an element with a facet E_0 on $\partial\Omega$ and E_{n_z} is the facet of T_{n_z} intersecting with $\partial\Omega$, see Figs. 7 and 8. Then, let $P_{i,z}$ be the vertex of the element T_i that does not belong to the facet E_i.

For the purpose of clarity, we now consider a two-dimensional triangulation. Then, note that the set of the functions

$$\left\{\mathbf{v}_{P_{i,z},T_i}\right\}_{i=i_1}^{n_z} \bigcup \left\{\mathbf{v}_{P_{i,z},T_{i+1}}\right\}_{i=0}^{n_z-i_z} \tag{35}$$

with

$$\mathbf{v}_{P,T}(\mathbf{x}) = \frac{1}{2|T|}(\mathbf{x} - P)$$

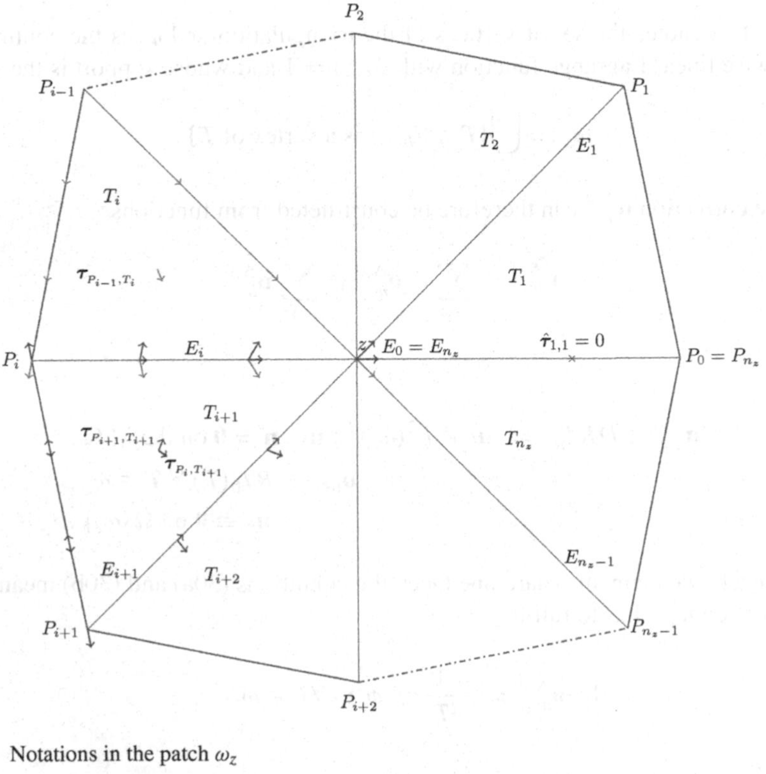

Fig. 7 Notations in the patch ω_z

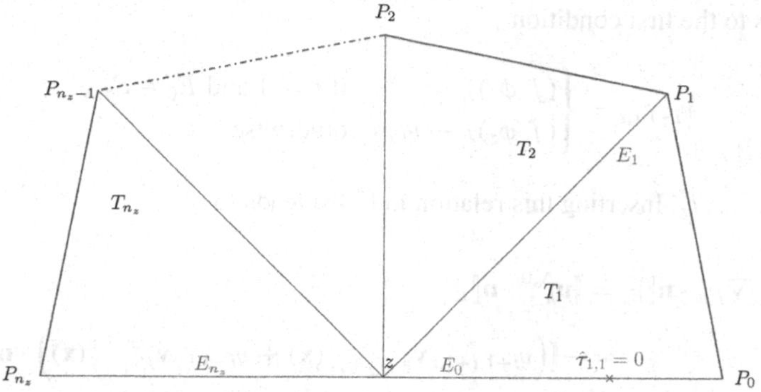

Fig. 8 Notations in the patch ω_z

and

$$i_1 = \begin{cases} 2 & \text{if } E_0 \in \mathcal{E}_{h,N} \\ 1 & \text{otherwise} \end{cases} , \quad i_z = \begin{cases} n_z - 1 & \text{if } E_0 \in \mathcal{E}_{h,N} \\ n_z & \text{otherwise} \end{cases}$$

and the convention P_{n_z} is P_0, constitutes a basis for $DRT_{0,z}$. Let $v_{i,j,z}$ denote the coefficient associated to the basis function $\mathbf{v}_{P_{i,z},T_j}$ such that

$$\mathbf{u}_{h,z}^{\Delta,0}(\mathbf{x}) = \sum_{i=i_1}^{n_z} v_{i,i,z} \mathbf{v}_{P_{i,z},T_i}(\mathbf{x}) + \sum_{i=1}^{i_z} v_{i-1,i,z} \mathbf{v}_{P_{i-1,z},T_i}(\mathbf{x}) .$$

Note that

$$\int_{E_{i-1}} \mathbf{v}_{P_{i,z},T_i}(s) \cdot \mathbf{n} \, ds = \int_{E_i} \mathbf{v}_{P_{i,z},T_{i+1}}(s) \cdot \mathbf{n} \, ds = 1$$

This means, that the divergence condition (34a) in each element $T_i \subset \omega_z$ can be rewritten as

$$\begin{aligned} -\frac{1}{|T|}(f, \phi_z)_{T_i} &= \operatorname{div} \mathbf{u}_{h,z}^{\Delta,0}|_{T_i} \\ &= \frac{1}{|T|} \left(\operatorname{div} \mathbf{u}_{h,z}^{\Delta,0}, 1 \right)_{0,T_i} \\ &= \frac{1}{|T|} \left(v_{i,i,z} \mathbf{v}_{P_{i,z},T_i}(\mathbf{x}) + v_{i-1,i,z} \mathbf{v}_{P_{i-1,z},T_i}(\mathbf{x}), 1 \right)_{0,T_i} \\ &= -\frac{1}{|T|} \left(v_{i,i,z} \langle \mathbf{v}_{P_{i,z},T_i}(\mathbf{x}) + \cdot \mathbf{n} \rangle_{E_{i-1}} + v_{i-1,i,z} \langle \mathbf{v}_{P_{i-1,z},T_i}(\mathbf{x}) \cdot \mathbf{n} \rangle_{E_i} \right) \\ &= -\frac{1}{|T|} \left(v_{i,i,z} + v_{i-1,i,z} \right) . \end{aligned}$$

This leads to the first condition

$$v_{i-1,i,z} = \begin{cases} (f,\phi_z)_{T_i} & \text{if } i=1 \text{ and } E_0 \in \mathcal{E}_{h,N} \\ (f,\phi_z)_{T_i} - v_{i,i,z} & \text{otherwise} \end{cases} \tag{36}$$

for $i = 1, \dots, i_z$. Inserting this relation in (34b) leads to

$$\begin{aligned} -\frac{1}{2}[\![\nabla p_h \cdot \mathbf{n}]\!]_{E_i} &= [\![\mathbf{u}_{h,z}^{\Delta,0} \cdot \mathbf{n}]\!]_{E_i} \\ &= -[\![\Big(v_{i+1,i+1,z}\mathbf{v}_{P_{i+1,z},T_{i+1}}(\mathbf{x}) + v_{i-1,i,z}\mathbf{v}_{P_{i-1,z},T_i}(\mathbf{x})\Big) \cdot \mathbf{n}]\!]_{E_i} \\ &= -\frac{1}{|E_i|}\left(v_{i+1,i+1,z} + v_{i-1,i,z}\right) \\ &= \frac{1}{|E_i|}\left(v_{i,i,z} - v_{i+1,i+1,z} - (f,\phi_z)_{T_i}\right) \end{aligned}$$

for $i = i_1, \dots, n_z - 1$ if E_i does not intersect with $\partial\Omega$ and

$$\begin{aligned} -\frac{1}{2}[\![\nabla p_h \cdot \mathbf{n}]\!]_{E_{n_z}} &= [\![\mathbf{u}_{h,z}^{\Delta,0} \cdot \mathbf{n}]\!]_{E_{n_z}} \\ &= -[\![\Big(v_{1,1,z}\mathbf{v}_{P_{1,z},T_1}(\mathbf{x}) + v_{n_z-1,i}, z\mathbf{v}_{P_{n_z-1,z},T_{n_z}}(\mathbf{x})\Big) \cdot \mathbf{n}]\!]_{E_{n_z}} \\ &= -\frac{1}{|E_{n_z}|}\left(v_{1,1,z} + v_{n_z-1,n_z,z}\right) \\ &= \frac{1}{|E_{n_z}|}\left(v_{n_z,n_z,z} - v_{1,1,z} - (f,\phi_z)_{T_{n_z}}\right) \end{aligned}$$

if T_{n_z} and T_1 share the edge E_0. If $E_{n_z} \in \mathcal{E}_{h,D}$, there is no condition on E_{n_z}, such that the final condition on the coefficients reads

$$\begin{cases} v_{i+1,i+1,z} = v_{i,i,z} + \frac{E_i}{2}[\![\nabla p_h \cdot \mathbf{n}]\!]_{E_i} - (f,\phi_z)_{T_i} & \text{for } i = 1,\dots,n_z-1 \\ v_{1,1,z} = v_{n_z,n_z,z} + \frac{E_{n_z}}{2}[\![\nabla p_h \cdot \mathbf{n}]\!]_{E_{n_z}} - (f,\phi_z)_{T_{n_z}} & \text{if } T_1 \cap T_{n_z} = E_{n_z} \\ v_{0,1,z} = (f,\phi_z)_{T_1} & \text{if } E_0 \in \mathcal{E}_{h,N} \\ v_{i,i,z} = (f,\phi_z)_{T_i} - v_{i-1,i,z} & i = 1,\dots,i_z \end{cases}.$$

Choosing $v_{1,1,z} = 0$ in the case $E_0 \notin \mathcal{E}_{h,N}$, we have the explicit formulas

$$\begin{cases} v_{1,1,z} = 0 & \text{if } E_0 \notin \mathcal{E}_{h,N} \\ v_{0,1,z} = (f,\phi_z)_{T_1} & \text{if } E_0 \in \mathcal{E}_{h,N} \\ v_{i+1,i+1,z} = \sum_{k=1}^{i} \frac{E_k}{2}[\![\nabla p_h \cdot \mathbf{n}]\!]_{E_k} - (f,\phi_z)_{T_k} & \text{for } i = 1,\dots,n_z-1 \\ v_{i-1,i,z} = (f,\phi_z)_{T_i} - v_{i,i,z} & i = i_1,\dots,i_z \end{cases}$$

since integrating by parts in the case $E_0 \notin \mathcal{E}_{h,N}$ leads to

$$\sum_{k=1}^{n_z} \frac{|E_k|}{2} [\![\nabla p_h \cdot \mathbf{n}]\!]_{E_k} - (f, \phi_z)_{T_k} = \sum_{k=1}^{n_z} \langle [\![\nabla p_h \cdot \mathbf{n}]\!]_{E_k}, \phi_z \rangle - (\nabla p_h, \nabla \phi_z)_{T_i} = 0\,,$$

such that it is in fact possible to fulfill the condition (34).

Note that in order to construct a hierarchical reconstruction, we need to consider the case div $\nabla p_h \neq 0$. Then, the term $-(f, \phi_z)$ has to be replaced by $(-f + \mathrm{div}\nabla p_h, \phi_z)$ and the formula reads

$$\begin{cases} v_{1,1,z} = 0 & \text{if } E_0 \notin \mathcal{E}_{h,N} \\ v_{0,1,z} = (f - \mathrm{div}\nabla p_h, \phi_z)_{T_1} & \text{if } E_0 \in \mathcal{E}_{h,N} \\ v_{i+1,i+1,z} = \sum_{k=1}^{i} \frac{E_k}{2} [\![\nabla p_h \cdot \mathbf{n}]\!]_{E_k} + (\mathrm{div}\nabla p_h - f, \phi_z)_{T_k} & \text{for } i = 1, \ldots, n_z - 1 \\ v_{i-1,i,z} = (f - \mathrm{div}\nabla p_h, \phi_z)_{T_i} - v_{i,i,z} & i = i_1, \ldots, i_z\,. \end{cases}$$

Next-to-lower order case stress equilibration

The lower order equilibration procedure of the previous section can be extended to the higher order case, see Cai and Zhang (2012) and Braess et al. (2009). In this section, we will concentrate on the next-to-lower order case, i.e., the approximated and reconstructed flux are quadratic. Therefore, we now assume $p_h \in \mathcal{P}^2(\mathcal{T}_h)$, $f \in \mathcal{DP}^1(\mathcal{T}_h)$ and reconstruct an equilibrated flux $(\mathbf{u}_h^{\Delta,1} + \nabla p_h) \in RT_1$ with $\mathbf{u}_h^{\Delta,1} \in DRT_1$.

Using the degrees of freedom defined in (24) for $RT_1(T)$ for the Ansatz functions $\boldsymbol{\psi}$ corresponding to the decomposition

$$DRT_1(\mathcal{T}_h) = DRT_0(\mathcal{T}_h) + DRT_{1,\times}(\mathcal{T}_h) + RT_{1,\blacktriangle}(\mathcal{T}_h) \tag{37}$$

with

$$DRT_{1,\times}(\mathcal{T}_h) = \{\mathbf{v} \in DRT1(\mathcal{T}_h) \ : \ (\mathrm{div}\,\mathbf{v})|_T = 0\} \tag{38}$$

and

$$RT_{1,\blacktriangle}(\mathcal{T}_h) = \{\mathbf{v} \in RT1(\mathcal{T}_h) \ : \ \mathbf{v} \cdot \mathbf{n} = 0 \text{ on all facet in } \mathcal{T}_h\} \tag{39}$$

we note that we can separate divergence and jump condition in the corresponding hierarchical next-to-order conditions for (30). In fact, we can choose a function in $RT_{1,\blacktriangle}$ that fulfills the divergence condition and will have no impact on the jump condition. This means, we use $\mathbf{u}_h^{\Delta,0}$ from the previous section and search for a correction $\bar{\mathbf{u}}_h^{\Delta} = \mathbf{u}_h^{\Delta,1} - \mathbf{u}_h^{\Delta,0}$ in

$$DRT_{1,\Delta} = DRT_{1,\times} \cup RT_{1,\blacktriangle}$$

does not require a partition of unity or patch decomposition any more and the correction $\bar{\mathbf{u}}_h^{\Delta}$ can be given elementwise.

Note that now, $\operatorname{div} \mathbf{u}_h^{\Delta,1}$ and $[\![\mathbf{u}_h^{\Delta,1} \cdot \mathbf{n}]\!]_E$ are linear functions. Additionally, $\operatorname{div} \nabla p_h$ does not vanish anymore, such that we need the projection Π^1 onto the space of the linear polynomials defined on a triangle T by

$$(\Pi_T^1 p, q)_T = (p, q)_T \quad \forall q \in \mathcal{P}^1(T) \tag{40}$$

and on an edge E by

$$(\Pi_E^1 p, q)_E = (p, q)_E \quad \forall q \in \mathcal{P}^1(E)\,. \tag{41}$$

Recall that the conditions that have to be fulfilled pointwise for $\mathbf{u}_h^{\Delta,1}$ read

$$\operatorname{div} \mathbf{u}_h^{\Delta} = -f + \operatorname{div} \nabla p_h \tag{42a}$$

$$[\![\mathbf{u}_h^{\Delta} \cdot \mathbf{n}]\!]_E = -[\![\nabla p_h \cdot \mathbf{n}]\!]_E \forall E \in \mathcal{E}_{h,I}\,. \tag{42b}$$

and this would imply

$$\begin{aligned}\left(\operatorname{div} (\mathbf{u}_h^{\Delta,1} - \mathbf{u}_h^{\Delta,0}), 1\right) &= \left(-f + \operatorname{div} \nabla p_h - \operatorname{div} \bar{\mathbf{u}}_h^{\Delta,0}, 1\right)\\ &= (-f + \operatorname{div} \nabla p_h, 1) - \left(\operatorname{div} \bar{\mathbf{u}}_h^{\Delta,0}, 1\right)\\ &= 0\end{aligned}$$

and for all $E \in \mathcal{E}_{h,I}$

$$\langle [\![(\mathbf{u}_h^{\Delta,1} - \mathbf{u}_h^{\Delta,0}) \cdot \mathbf{n}]\!]_E, 1\rangle_E = -\langle [\![\nabla p_h \cdot \mathbf{n}]\!]_E, 1\rangle_E - \langle [\![\mathbf{u}_h^{\Delta,0} \cdot \mathbf{n}]\!]_E, 1\rangle_E = 0\,.$$

This means that it is actually possible to find a correction $\bar{\mathbf{u}}_h^{\Delta} = \mathbf{u}_h^{\Delta,1} - \mathbf{u}_h^{\Delta,0}$ in $DRT_{1,\Delta}$ satisfying

$$\operatorname{div} \bar{\mathbf{u}}_h^{\Delta} = -f + \operatorname{div} \nabla p_h - \operatorname{div} \bar{\mathbf{u}}_h^{\Delta,0} \tag{43a}$$

$$[\![\bar{\mathbf{u}}_h^{\Delta} \cdot \mathbf{n}]\!]_E = -[\![\nabla p_h \cdot \mathbf{n}]\!]_E - [\![\mathbf{u}_h^{\Delta,0} \cdot \mathbf{n}]\!]_E \qquad \forall E \in \mathcal{E}_{h,I}\,. \tag{43b}$$

Using the freedoms degree defined in (24), these equations read

$$\left(\operatorname{div} \bar{\mathbf{u}}_h^{\Delta}, x - x_m\right)_T = (-f + \operatorname{div} \nabla p_h, (x - x_m))_T \quad \forall T \in \mathcal{T}_h \tag{44a}$$

$$\left(\operatorname{div} \bar{\mathbf{u}}_h^{\Delta}, y - y_m\right)_T = (-f + \operatorname{div} \nabla p_h, (y - y_m))_T \quad \forall T \in \mathcal{T}_h \tag{44b}$$

$$\langle \bar{\mathbf{u}}_h^{\Delta} \cdot \mathbf{n}, q_E\rangle_E = -\langle [\![\nabla p_h \cdot \mathbf{n}]\!]_E, q_E\rangle_E \quad \forall E \in \mathcal{E}_{h,I} \tag{44c}$$

and decomposing

$$\bar{\mathbf{u}}_h^{\Delta} = \mathbf{u}_{h,\times}^{\Delta} + \mathbf{u}_{h,\blacktriangle}^{\Delta}$$

with $\mathbf{u}_{h,\times}^{\Delta} \in DRT_{1,\times}(\mathcal{T}_h)$ and $\mathbf{u}_{h,\blacktriangle}^{\Delta} \in DRT_{1,\blacktriangle}(\mathcal{T}_h)$ leads to

$$\left(\operatorname{div} \mathbf{u}_{h,\blacktriangle}^{\Delta}, x - x_m\right)_T = (-f + \operatorname{div} \nabla p_h, (x - x_m))_T \quad \forall T \in \mathcal{T}_h \tag{45a}$$

$$\left(\operatorname{div} \mathbf{u}_{h,\blacktriangle}^{\Delta}, y - y_m\right)_T = (-f + \operatorname{div} \nabla p_h, (y - y_m))_T \quad \forall T \in \mathcal{T}_h \tag{45b}$$

$$\langle \mathbf{u}_{h,\times}^{\Delta} \cdot \mathbf{n}, q_E \rangle_E = -\langle [\![\nabla p_h \cdot \mathbf{n}]\!]_E, q_E \rangle_E \quad \forall E \in \mathcal{E}_{h,I} \,. \tag{45c}$$

The function $\mathbf{u}_{h,\blacktriangle}^{\Delta}$ defined piecewise on each triangle T by

$$\mathbf{u}_{h,\blacktriangle}^{\Delta} = -\frac{1}{3} \sum_{i=1}^{3} r_T(\mathbf{P}_i)(\mathbf{x} - \mathbf{P}_i) \tag{46}$$

fulfills (45a) and (45b) with $r_T = -f + \operatorname{div} \nabla p_h$. For the remaining condition (45c), note that we need two ansatz functions for each edge $E \in \mathcal{E}_h$ to span the space $DRT_{1,\times}$. However, in order to find a function that fulfills (45c), only one function for each edge is needed. Therefore, for each edge $E \in \mathcal{E}_{h,I}$, fix $T^+(E)$ being the one of the two adjacent triangles of E whose outwards normal component coincides with $\mathbf{n}_E$. Therefore, define the piecewise polynomial function

$$\mathbf{u}_{h,\times,E}^{\Delta} = \begin{cases} \varrho\left(\left[\mathbf{P}_E^+ \, \mathbf{P}_E^-\right]\right) \mathbf{v}_{\mathbf{P}^+(E),T^+(E)} & \text{on } T^+ \\ 0 & \text{elsewhere} \end{cases}, \tag{47}$$

where $\mathbf{P}_E^+$ and $\mathbf{P}^-$ are the ending and starting point of E, $\mathbf{P}^+(E)$ is the vertex of T^+ that does not lie on E, $\mathbf{v}_{\mathbf{P}^+(E),T^+(E)}$ the corresponding $RT_0(T^+)$ ansatz function and

$$\varrho(A) = A \begin{pmatrix} 0 & 1 \\ 1 & 0 \end{pmatrix} A^T \begin{pmatrix} 0 & -1 \\ 1 & 0 \end{pmatrix} .$$

Then, the sum

$$\mathbf{u}_{h,\times}^{\Delta} = \sum_{E \in \mathcal{E}_{h,I}} \mathbf{u}_{h,\times,E}^{\Delta} \tag{48}$$

fulfills (45c).

5 Stress Equilibration

The aim of this section is to extend the previous flux reconstruction to the context of linear elasticity. Therefore, consider the exact solution $(\mathbf{u}, p)$ of the problem **MF**, and its approximation $(\mathbf{u}_h, p_h)$, that solves the discrete saddle point problem

$\mathbf{MF}_h$ in the Taylor–Hood finite element space. Similarly to the previous section, we first assume $\mathbf{f}$ to be piecewise linear, $\mathbf{g} = \mathbf{0}$ and consider the approximation $\boldsymbol{\sigma}_h = 2\mu\boldsymbol{\varepsilon}(\mathbf{u}_h) + p_h\mathbf{I}$ that is discontinuous and piecewise linear. Then, we aim to reconstruct an $H(\text{div})$-conforming stress–tensor $\boldsymbol{\sigma}_h^R \in H(div)^d$ and denote the difference between the reconstructed and the original stress by $\boldsymbol{\sigma}_h^\Delta \in DRT(\mathcal{T}_h)^d$. Note that the proof of the Prager–Synge orthogonality in Theorem 4.1 involves the integration by parts of the bilinear form such that a straightforward extension for bilinear forms not involving full gradients would require to perform the stress reconstruction in a symmetric space like the Arnold–Winther elements. In order to keep the stress reconstruction procedure simple, we are looking for a weakly symmetric construction. Moreover, using the displacement–pressure formulations the computation remains stable to the incompressible limit. Since in the Taylor–Hood case, $\text{div}\,\mathbf{u}_h \neq \frac{p_h}{\lambda}$, the term

$$\eta_{N,T} = \left\|\text{div}\,\mathbf{u}_h - \frac{p_h}{\lambda}\right\|_T$$

occurs additionally to the norms of the reconstructed stress $\eta_{R,T} = \|\boldsymbol{\sigma}_h^\Delta\|_{\mathcal{A},T}$ and $\eta_S = \|\mathbf{as}\ \boldsymbol{\sigma}_h^\Delta\|_T$ in the error estimator. Similarly, define

$$\eta_N = \left\|\text{div}\,\mathbf{u}_h - \frac{p_h}{\lambda}\right\|, \eta_R = \|\boldsymbol{\sigma}_h^\Delta\|_{\mathcal{A}} \text{ and } \eta_S = \|\mathbf{as}\ \boldsymbol{\sigma}_h^\Delta\|. \tag{49}$$

Using these notations, the energy norm defined by

$$|||(\mathbf{u}-\mathbf{u}_h, p-p_h)||| = \left(2\mu\|\boldsymbol{\varepsilon}(\mathbf{u}-\mathbf{u}_h)\|^2 + \frac{1}{\lambda}\|p-p_h\|^2\right)^{1/2}, \tag{50}$$

the inner product $(\,\cdot\,,\,\cdot\,)_{\mathcal{A}} := (\,\mathcal{A}(\cdot)\,,\,\cdot\,)$, the matrix

$$\mathbf{J}^2(\theta) := \begin{pmatrix} 0 & \theta \\ -\theta & 0 \end{pmatrix}, \qquad \mathbf{J}^3(\boldsymbol{\theta}) := \begin{pmatrix} 0 & \theta_3 & -\theta_2 \\ -\theta_3 & 0 & \theta_1 \\ \theta_2 & -\theta_1 & 0 \end{pmatrix} \tag{51}$$

for every $(2d-1)$-dimensional vector $\boldsymbol{\theta}$, and the abbreviation

$$\zeta = \frac{\lambda}{2\mu + d\lambda}\left(\text{div}\,\mathbf{u}_h - \frac{1}{\lambda}p_h\right) \tag{52}$$

we have

$$\begin{aligned} \eta_R^2 &= \|\boldsymbol{\sigma}_h^R - \boldsymbol{\sigma} + \boldsymbol{\sigma} - \boldsymbol{\sigma}_h\|_{\mathcal{A}}^2 \\ &= \|\boldsymbol{\sigma}_h^R - \boldsymbol{\sigma}\|_{\mathcal{A}}^2 + \left(2(\boldsymbol{\sigma}_h^R - \boldsymbol{\sigma}) + \boldsymbol{\sigma} - \boldsymbol{\sigma}_h, \boldsymbol{\sigma} - \boldsymbol{\sigma}_h\right)_{\mathcal{A}} \\ &= \left(\boldsymbol{\sigma}_h^R - \boldsymbol{\sigma}, \mathcal{A}(\boldsymbol{\sigma}_h^R - \boldsymbol{\sigma})\right) + \left(2(\boldsymbol{\sigma}_h^R - \boldsymbol{\sigma}) + \boldsymbol{\sigma} - \boldsymbol{\sigma}_h, \mathcal{A}(\boldsymbol{\sigma} - \boldsymbol{\sigma}_h)\right). \end{aligned}$$

Inserting $\mathcal{A}(\boldsymbol{\sigma}-\boldsymbol{\sigma}_h)=\boldsymbol{\varepsilon}(\mathbf{u}-\mathbf{u}_h)+\zeta(\mathbf{u}_h,p_h)\mathbf{I}$ leads to

$$\begin{aligned}
\eta_R^2 &= \|\boldsymbol{\sigma}_h^R-\boldsymbol{\sigma}\|_{\mathcal{A}}^2+2\mu\|\boldsymbol{\varepsilon}(\mathbf{u}-\mathbf{u}_h)\|^2 \\
&+2\left((\boldsymbol{\sigma}_h^R-\boldsymbol{\sigma}),\boldsymbol{\varepsilon}(\mathbf{u}-\mathbf{u}_h)\right)+((p-p_h)\mathbf{I},\boldsymbol{\varepsilon}(\mathbf{u}-\mathbf{u}_h)) \\
&+\left(2(\boldsymbol{\sigma}_h^R-\boldsymbol{\sigma})+2\mu\boldsymbol{\varepsilon}(\mathbf{u}-\mathbf{u}_h)+(p-p_h)\mathbf{I},\zeta(\mathbf{u}_h,p_h)\mathbf{I}\right) \\
&= \|\boldsymbol{\sigma}_h^R-\boldsymbol{\sigma}\|_{\mathcal{A}}^2+2\mu\|\boldsymbol{\varepsilon}(\mathbf{u}-\mathbf{u}_h)\|^2 \\
&+2\left((\boldsymbol{\sigma}_h^R-\boldsymbol{\sigma}),\nabla(\mathbf{u}-\mathbf{u}_h)\right)-2\left((\boldsymbol{\sigma}_h^R-\boldsymbol{\sigma}),\mathbf{as}\ \nabla(\mathbf{u}-\mathbf{u}_h)\right) \\
&+((p-p_h),\operatorname{div}(\mathbf{u}-\mathbf{u}_h)) \\
&+\left(2\operatorname{tr}(\boldsymbol{\sigma}_h^R-\boldsymbol{\sigma})+2\mu\operatorname{div}(\mathbf{u}-\mathbf{u}_h)+2(p-p_h),\zeta(\mathbf{u}_h,p_h)\right).
\end{aligned}$$

Inserting the additional assumption $(\boldsymbol{\sigma}_h^R,\mathbf{J}(\gamma_h))=0$ for all $\gamma_h\in P^1(\mathcal{T}_h)$, we obtain

$$\begin{aligned}
\eta_R^2 &= \frac{1}{2\mu}\left(\|\boldsymbol{\sigma}_h^R-\boldsymbol{\sigma}\|^2-\frac{\lambda}{2\mu+d\lambda}\|\operatorname{tr}(\boldsymbol{\sigma}_h^R-\boldsymbol{\sigma})\|^2\right)+2\mu\|\boldsymbol{\varepsilon}(\mathbf{u}-\mathbf{u}_h)\|^2 \\
&-2\left((\boldsymbol{\sigma}_h^R-\boldsymbol{\sigma}),\mathbf{as}\ \nabla(\mathbf{u}-\mathbf{u}_h)\right) \\
&+\left(2\operatorname{tr}(\boldsymbol{\sigma}_h^R-\boldsymbol{\sigma}),\zeta(\mathbf{u}_h,p_h)\right)+\frac{1}{\lambda}\|p-p_h\|^2-\frac{2\mu\lambda}{2\mu+d\lambda}\eta_N^2 \\
&= \frac{1}{2\mu}\|\ \mathbf{dev}\ :\boldsymbol{\sigma}_h^R-\boldsymbol{\sigma}\|^2+|||(\mathbf{u}-\mathbf{u}_h,p-p_h)|||^2 \\
&-2\left(\mathbf{as}\ (\boldsymbol{\sigma}_h^R-\boldsymbol{\sigma}),\ \nabla(\mathbf{u}-\mathbf{u}_h)\right) \\
&+\left(2\operatorname{tr}(\boldsymbol{\sigma}_h^R-\boldsymbol{\sigma}),\zeta(\mathbf{u}_h,p_h)\right)-\frac{2\mu\lambda}{2\mu+d\lambda}\eta_N^2.
\end{aligned}$$

Upper bounds for both mixed terms remaining in this equation were provided in Bertrand et al. (2018a). According to their Lemma 2, if the stress reconstruction satisfies the weak symmetry condition $(\boldsymbol{\sigma}_h^R,\mathbf{J}(\gamma_h))=0$ for all $\gamma_h\in\mathcal{P}^1(\mathcal{T}_h)$, it holds

$$\left|(\mathbf{as}\ \boldsymbol{\sigma}_h^R,\nabla(\mathbf{u}-\mathbf{u}_h))\right|\le C_K\|\mathbf{as}\ \boldsymbol{\sigma}_h^R\|\ \|\boldsymbol{\varepsilon}(\mathbf{u}-\mathbf{u}_h)\|, \tag{53}$$

$$\left|(\operatorname{tr}(\boldsymbol{\sigma}-\boldsymbol{\sigma}_h^R),\operatorname{div}\mathbf{u}_h-\frac{1}{\lambda}p_h)\right|\le C_A\|\ \mathbf{dev}\ (\boldsymbol{\sigma}-\boldsymbol{\sigma}_h^R)\|\ \|\operatorname{div}\mathbf{u}_h-\frac{1}{\lambda}p_h\|, \tag{54}$$

where the constants C_K and C_A depend only on the largest interior angle in the triangulation $\mathcal{T}_h$. Thus,

$$\begin{aligned}
\eta_R^2 &\ge \frac{1}{2\mu}\|\ \mathbf{dev}\ \boldsymbol{\sigma}_h^R-\boldsymbol{\sigma}\|^2+|||(\mathbf{u}-\mathbf{u}_h,p-p_h)|||^2-2C_K\eta_S\|\boldsymbol{\varepsilon}(\mathbf{u}-\mathbf{u}_h)\| \\
&+2C_A\alpha\|\ \mathbf{dev}\ (\boldsymbol{\sigma}_h^R-\boldsymbol{\sigma})\|\eta_N-\frac{2\mu\lambda}{2\mu+d\lambda}\eta_N^2.
\end{aligned}$$

Weighting the mixed terms correctly leads to

$$\frac{1}{2}|||(\mathbf{u}-\mathbf{u}_h, p-p_h)|||^2 \le \eta_R^2 + \frac{C_K^2}{\mu}\eta_S^2 + 2\mu\left(\frac{C_A^2\lambda^2 + 2\mu\lambda + d\lambda^2}{(2\mu+d\lambda)^2}\right)\eta_N^2 ,$$

such that the error in the energy norm can be bounded by the three error estimator terms. It remains to be able to reconstruct such a stress $\boldsymbol{\sigma}_R^\Delta$ satisfying the conditions

$$\begin{aligned} \operatorname{div} \boldsymbol{\sigma}_h^\Delta &= \mathbf{f} + \operatorname{div} \boldsymbol{\sigma}_h \\ [\![\boldsymbol{\sigma}_h^\Delta \cdot \mathbf{n}]\!]_E &= -[\![\boldsymbol{\sigma}(\mathbf{u}_h, p_h) \cdot \mathbf{n}]\!]_E \text{ for all } E \in \mathcal{E}_{h,I}, \\ (\boldsymbol{\sigma}_h^\Delta, \mathbf{J}(\boldsymbol{\gamma}_h)) &= 0 \text{ for all } \boldsymbol{\gamma}_h \in \mathcal{P}^1(\mathcal{T}_h) . \end{aligned} \tag{55}$$

The RT_0 and RT_1 flux reconstructions can be applied on $(\boldsymbol{\sigma}_h)_i$ instead of ∇p_h. Let $\boldsymbol{\sigma}_h^{\Delta,1}$ be this reconstructed stress satisfying

$$\begin{aligned} \operatorname{div} \boldsymbol{\sigma}_h^{\Delta,1} &= \mathbf{f} + \operatorname{div} \boldsymbol{\sigma}_h \\ [\![\boldsymbol{\sigma}_h^{\Delta,1} \cdot \mathbf{n}]\!]_E &= -[\![\boldsymbol{\sigma}(\mathbf{u}_h, p_h) \cdot \mathbf{n}]\!]_E \text{ for all} E \in \mathcal{E}_{h,I} \end{aligned} .$$

Then, it remains to find a divergence-free $\boldsymbol{\sigma}_h^S \in RT_1(\mathcal{T}_h)$ such that

$$(\boldsymbol{\sigma}_h^S, \mathbf{J}(\boldsymbol{\gamma}_h)) = -(\boldsymbol{\sigma}_h^{\Delta,1}, \mathbf{J}(\boldsymbol{\gamma}_h)) \text{ for all } \boldsymbol{\gamma}_h \in \mathcal{P}^1(\mathcal{T}_h) . \tag{56}$$

Note that if $z \notin \Gamma_D$, inserting $\mathbf{v}_h = (\phi_z \boldsymbol{\rho})$ in $\mathbf{MF}_h$ leads to

$$0 = (\mathbf{f}, \phi_z \boldsymbol{\rho})_{\omega_z} - (\boldsymbol{\sigma}_h, \boldsymbol{\varepsilon}(\phi_z \boldsymbol{\rho}))_{\omega_z}$$

for each rigid body mode $\boldsymbol{\rho}$ on the patch ω_z, due to the fact that $(\phi_z \boldsymbol{\rho}) \in \mathcal{P}^2(\Omega)$. Integrating by parts, we obtain

$$\begin{aligned} 0 &= (\mathbf{f}, \phi_z \boldsymbol{\rho})_{\omega_z} - (\operatorname{div} \boldsymbol{\sigma}_h, (\phi_z \boldsymbol{\rho}))_{\omega_z} + \sum_{E \in \mathcal{E}_{h,I}(\omega_z)} \langle [\![\boldsymbol{\sigma}_h \cdot \mathbf{n}]\!], \phi_z \boldsymbol{\rho} \rangle_S \\ &= ((\mathbf{f} + \operatorname{div} \boldsymbol{\sigma}_h)\phi_z, \boldsymbol{\rho})_{\omega_z} + \sum_{E \in \mathcal{E}_{h,I}(\omega_z)} \langle [\![\boldsymbol{\sigma}_h \cdot \mathbf{n}]\!], \phi_z \boldsymbol{\rho} \rangle_S \\ &= (\operatorname{div} \boldsymbol{\sigma}_h^{\Delta,1} \phi_z, \boldsymbol{\rho})_{\omega_z} + \sum_{E \in \mathcal{E}_{h,I}(\omega_z)} \langle [\![\boldsymbol{\sigma}_h^{\Delta,1} \cdot \mathbf{n}]\!], \phi_z \boldsymbol{\rho} \rangle_S \\ &= -(\boldsymbol{\sigma}_h^{\Delta,1} \phi_z, \nabla \boldsymbol{\rho})_{\omega_z} = (\boldsymbol{\sigma}_h^{\Delta,1} \phi_z, \mathbf{as}\,(\nabla \boldsymbol{\rho}))_{\omega_z} . \end{aligned}$$

Therefore, we can construct a divergence-free function $\boldsymbol{\sigma}_h^{S,0} \in RT_0$ such that

$$(\boldsymbol{\sigma}_h^{S,0} \phi_z, \mathbf{J}(\phi_z))_{\omega_z} = -(\boldsymbol{\sigma}_h^{\Delta,0} \phi_z, \mathbf{J}(\phi_z))_{\omega_z}$$

holds for all $z \in \mathcal{V}$ with $z \notin \Gamma_D$. Note that there is no divergence-free function in $RT_{1,\blacktriangle}$ to correct $\boldsymbol{\sigma}_h^{\Delta,1}$. Therefore, it remains to find $\boldsymbol{\sigma}_h^{S,1} \in RT_{1,\times}(\mathcal{T}_h)$ with

$$(\boldsymbol{\sigma}_h^{S,1}, \mathbf{J}(\gamma_h))_{\bar{T}_E} = -(\boldsymbol{\sigma}_h^{\Delta,1} + \boldsymbol{\sigma}_h^{S,0}, \mathbf{J}(\gamma_h))_{\bar{T}_E}. \tag{57}$$

for all $\gamma_h \in \mathcal{P}^2(\bar{T}_E)$. This can be done choosing similarly to the previous section, using the function

$$\mathbf{v}_{h,\times,\bar{T}_E}^{\Delta} = \begin{cases} \varrho\left(\left[\mathbf{P}_E^+ \, \mathbf{P}_E^-\right]\right) \mathbf{v}_{\mathbf{P}^+(E),T^+(E)} & \text{on } T^+ \\ \varrho\left(\left[\mathbf{P}_E^+ \, \mathbf{P}_E^-\right]\right) \mathbf{v}_{\mathbf{P}^+(E),T^-(E)} & \text{on } T^- \\ 0 & \text{elsewhere} \end{cases}, \tag{58}$$

and

$$\boldsymbol{\sigma}_h^{S,1} = \sum_{E \in \mathcal{E}_{h,I}} \sum_{z \in \mathcal{V}(E)} \phi_z \left(\mathbf{v}_{h,\times,\bar{T}_E}^{\Delta}\right)^{\top} \left(\alpha_{E,z} \; \beta_{E,z}\right) \tag{59}$$

with $\mathcal{V}_E = \{P_E^+, P_E^-, P(T^+, E), P(T^-, E)\}$, $\bar{T}_E = T_E^+ \cup T_E^-$ and coefficients $\alpha_{E,z}$ and $\beta_{E,z}$ such that

$$\begin{cases} \left(\phi_z \left(\mathbf{v}_{h,\times,\bar{T}_E}^{\Delta}\right)^{\top} \left(\alpha_{E,z} \; \beta_{E,z}\right), \mathbf{J}(\phi_{P_E^+})\right)_T = -\frac{1}{3}(\boldsymbol{\sigma}_h^{\Delta,1} + \boldsymbol{\sigma}_h^{S,0}, \phi_z \mathbf{J}(\phi_{P_E^+}))_T \\ \left(\phi_z \left(\mathbf{v}_{h,\times,\bar{T}_E}^{\Delta}\right)^{\top} \left(\alpha_{E,z} \; \beta_{E,z}\right), \mathbf{J}(\phi_{P_E^-})\right)_T = -\frac{1}{3}(\boldsymbol{\sigma}_h^{\Delta,1} + \boldsymbol{\sigma}_h^{S,0}, \phi_z \mathbf{J}(\phi_{P_E^-}))_T \end{cases}$$

if $z \notin E$, $z \in T$ and

$$\begin{cases} \left(\phi_z \left(\mathbf{v}_{h,\times,\bar{T}_E}^{\Delta}\right)^{\top} \left(\alpha_{E,z} \; \beta_{E,z}\right), \mathbf{J}(\phi_{P_E^+})\right)_{\bar{T}_E} = -\frac{1}{3}(\boldsymbol{\sigma}_h^{\Delta,1} + \boldsymbol{\sigma}_h^{S,0}, \phi_z \mathbf{J}(\phi_{P_E^+}))_{\bar{T}_E} \\ \left(\phi_z \left(\mathbf{v}_{h,\times,\bar{T}_E}^{\Delta}\right)^{\top} \left(\alpha_{E,z} \; \beta_{E,z}\right), \mathbf{J}(\phi_{P_E^-})\right)_{\bar{T}_E} = -\frac{1}{3}(\boldsymbol{\sigma}_h^{\Delta,1} + \boldsymbol{\sigma}_h^{S,0}, \phi_z \mathbf{J}(\phi_{P_E^-}))_{\bar{T}_E} \end{cases}$$

if $z \in E$ and $P = \{P_E^+, P_E^-\} \backslash \{z\}$.

6 Numerical Illustration

The numerical tests for our finite element method are performed for Cook's Membrane as well as for a more regular problem. For the adaptive refinement we applied a Dörfler marking strategy, which consists of finding the smallest set of triangles $\tilde{\mathcal{T}}_h \subset \mathcal{T}_h$ such that

$$\left(\sum_{T \in \tilde{\mathcal{T}}_h} \eta_T^2\right)^{1/2} \geq \theta \left(\sum_{T \in \mathcal{T}_h} \eta_T^2\right)^{1/2}$$

holds for a chosen parameter $\theta = 0.8$, with $\eta_T^2 := \eta_{A,T}^2 + \eta_{B,T}^2 + \eta_{C,T}^2$. All triangles in this set are then refined as well as those adjacent triangles necessary to avoid hanging nodes.

We ran our tests both for a perfectly incompressible material ($\lambda = \infty$) as well as for the compressible case ($\lambda = 2$). The second Lamé constant μ was set to 0.5 for all tests.

Cook's Membrane

The geometry and boundary conditions of the Cook's membrane problem are summarized in Fig. 1, and we set $\mathbf{x}_{\text{tip}} := (0.48, 06)$.

We first discuss the results for the incompressible case. Plots of the deformed mesh and reconstructed stress distribution on the 14th adaptive refinement level can be found in Figs. 9 and 10. Refinement is concentrated at the upper left corner where the solution of the Cook's Membrane problem is known to have the strongest singularity. This singularity is most evident in the plot of the von Mises stress defined as

$$\sigma_{VM} := \sqrt{\left(\frac{3}{2}\ \mathbf{dev}\ \sigma\right)}.$$

In Fig. 23 we summarized the values of our error estimator as well as the tip displacement on the different triangulations. As mentioned in Sect. 4 the modification of the partition of unity can be avoided if the used triangulation is such that each vertex on the

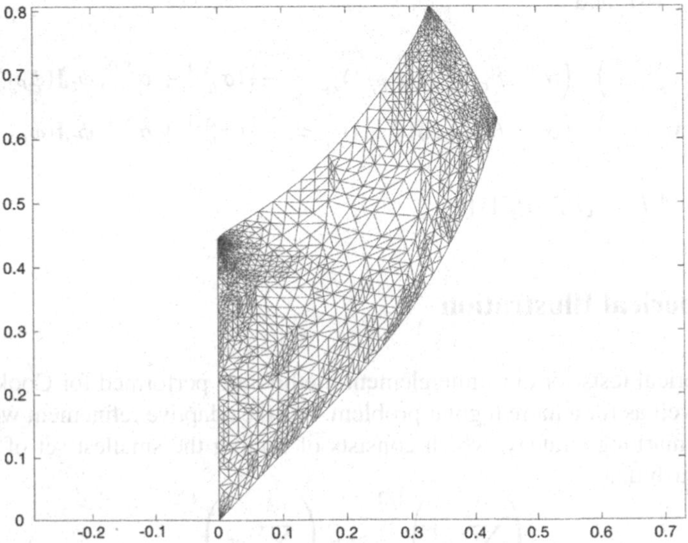

Fig. 9 Deformed configuration after 14 adaptive refinements ($\lambda = \infty$)

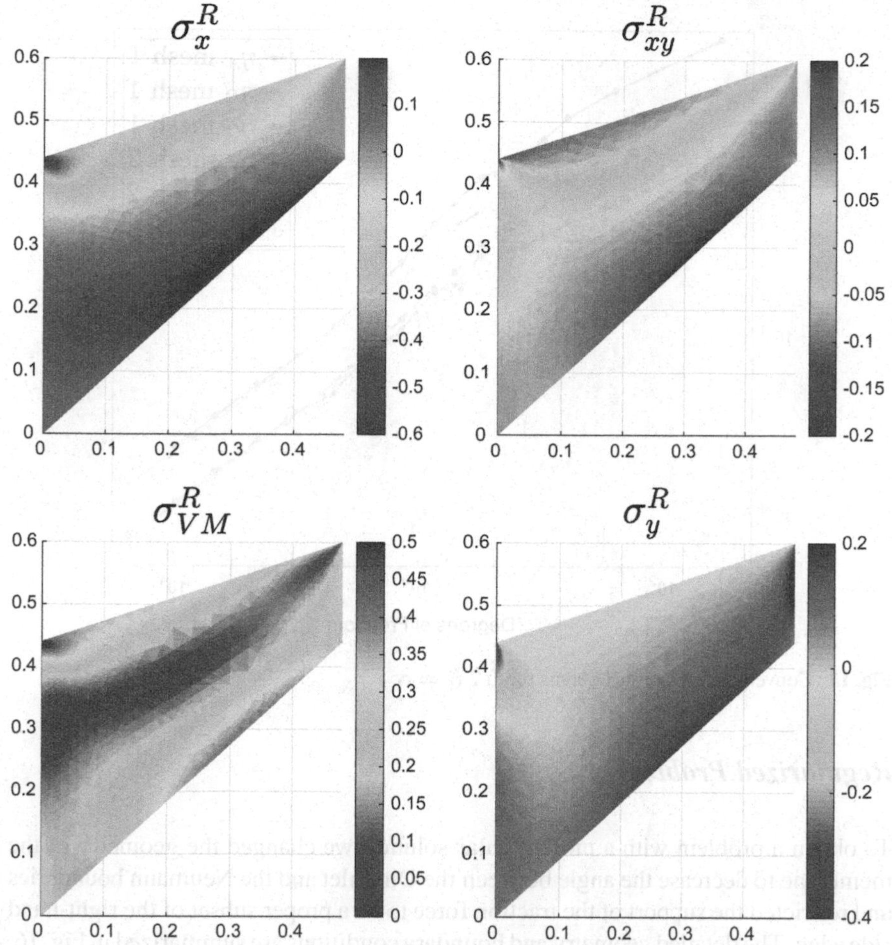

Fig. 10 Stress distribution after 14 adaptive refinements ($\lambda = \infty$)

Neumann boundary is connected to at least two vertices not on Neumann boundary. We verified this by applying our method on the two different start meshes displayed in Fig. 2, using the standard partition of unity for mesh 1 und the modified partition of unity for mesh 2 and comparing the performance in Fig. 11. The improved efficiency of our adaptive method as compared to uniform refinement is evident in Fig. 14.

For the compressible case, we obtained similar results which are illustrated in Figs. 12, 13, 15 and 24.

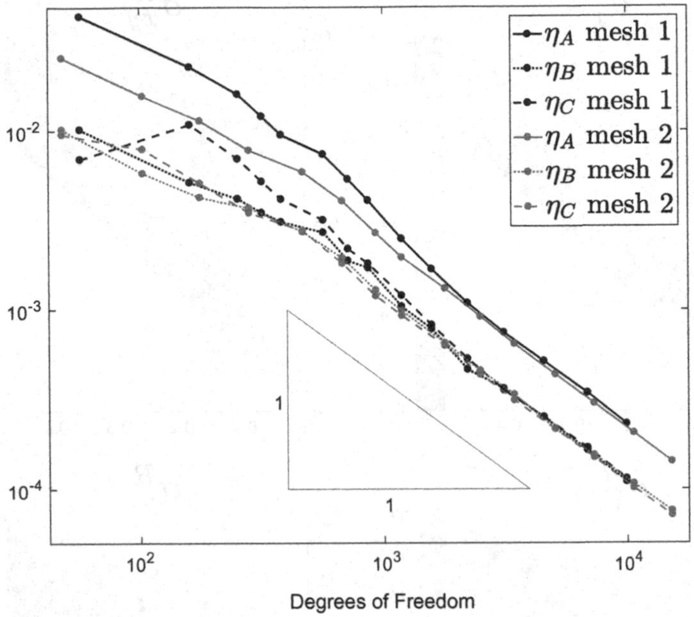

Fig. 11 Convergence of mesh 1 versus mesh 2 ($\lambda = \infty$)

Regularized Problem

To obtain a problem with a more regular solution we changed the geometry of the membrane to decrease the angle between the Dirichlet and the Neumann boundaries and restricted the support of the traction force to be a proper subset of the right-hand side edge. The detailed geometry and boundary conditions are summarized in Fig. 16.

The lack of corner singularities can be observed in Fig. 18 for the incompressible case and consequently the adaptive mesh refinement is less concentrated and the uniform refinement performance is improved as can be seen in Figs. 17 and 19. The quantitative data is summarized in Fig. 25. Again the compressible case yields similar results found in Figs. 20, 21, 22, 23, 24 and 26.

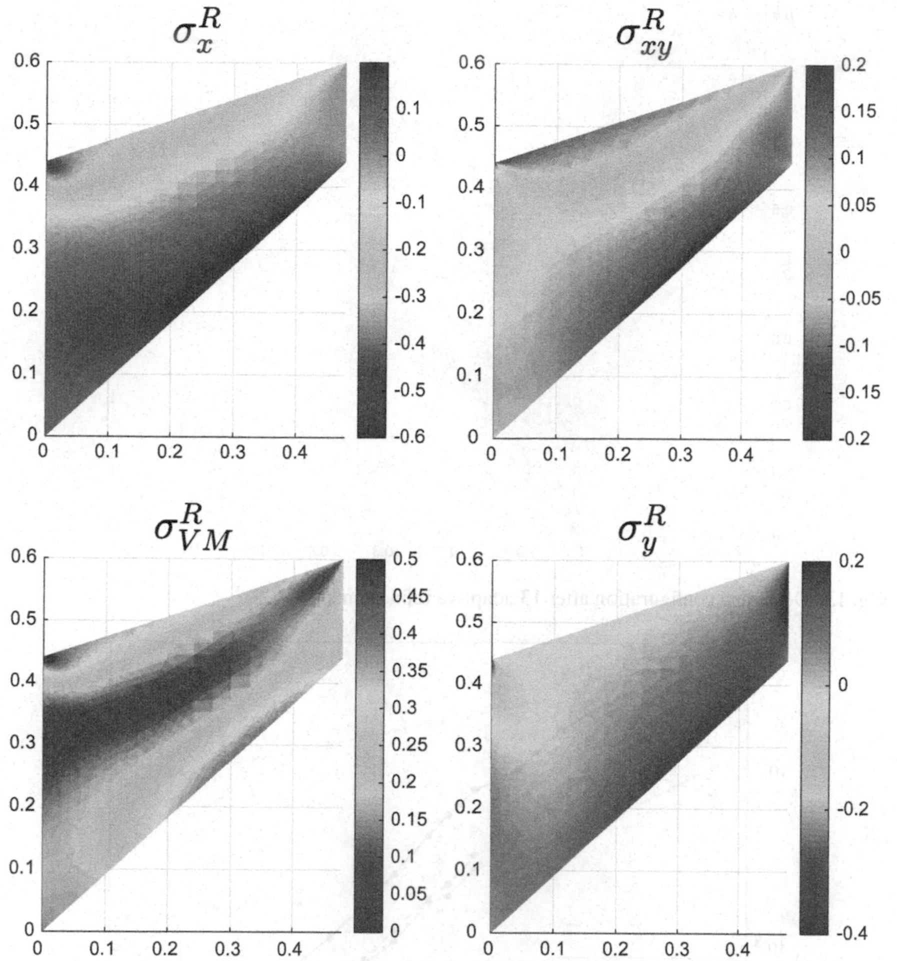

Fig. 12 Stress distribution after 13 adaptive refinements ($\lambda = 2$)

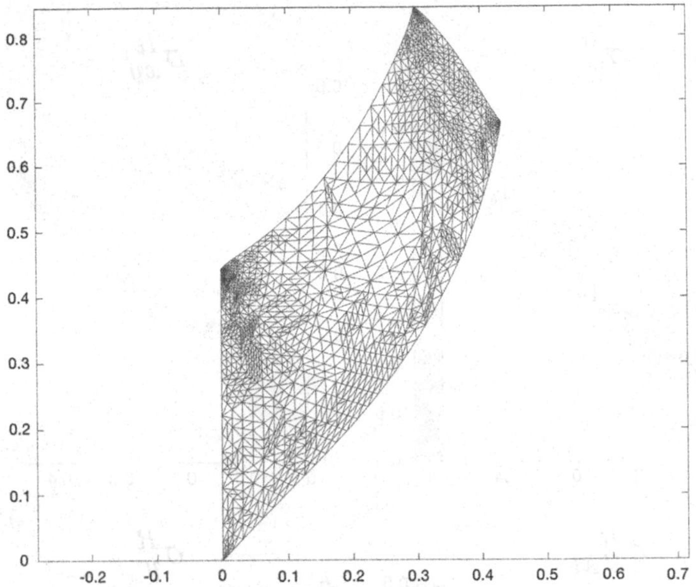

Fig. 13 Deformed configuration after 13 adaptive refinements ($\lambda = 2$)

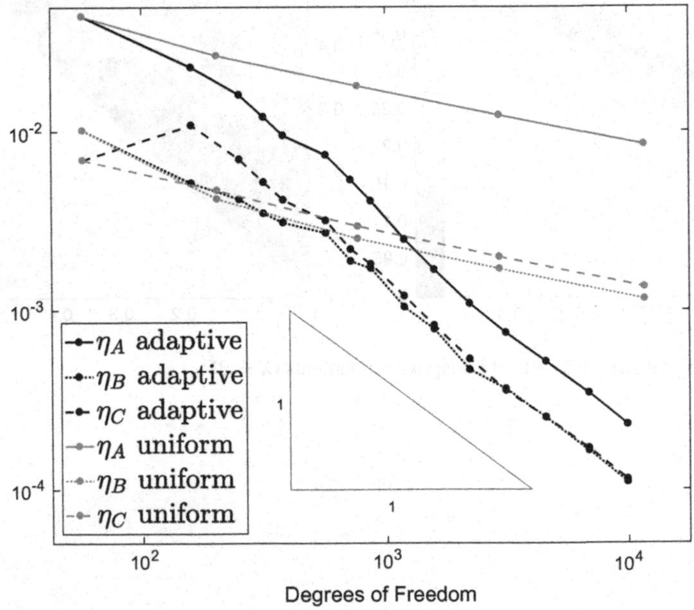

Fig. 14 Convergence of adaptive versus uniform refinement ($\lambda = \infty$)

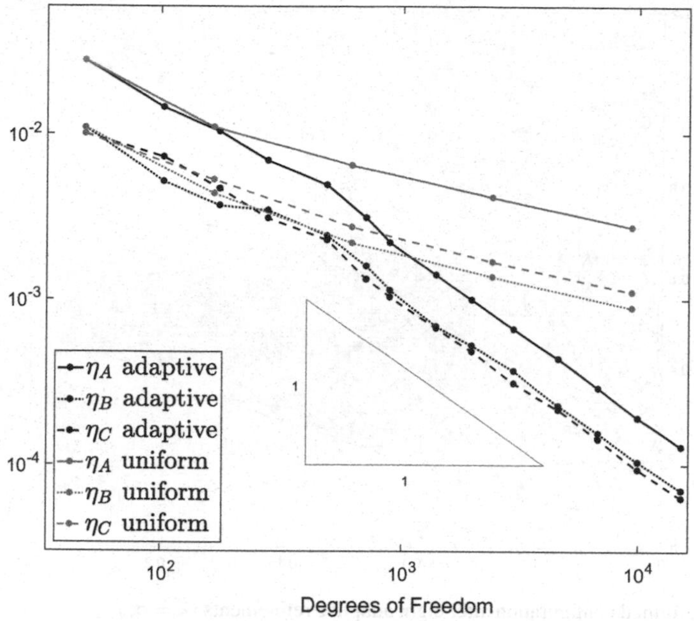

Fig. 15 Convergence of adaptive versus uniform refinement ($\lambda = 2$)

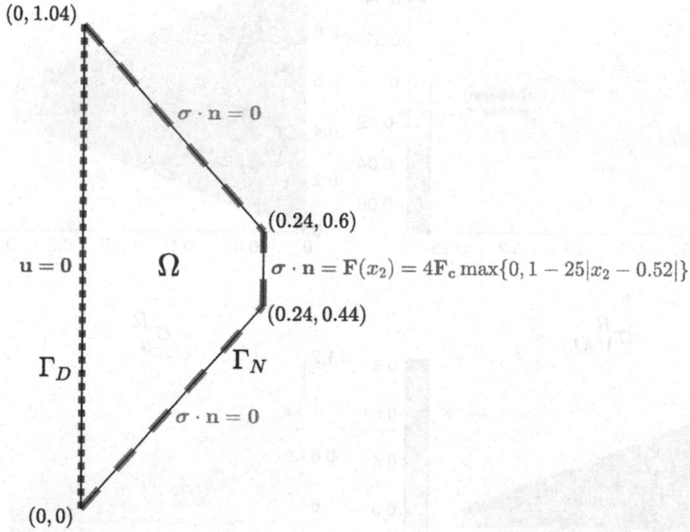

Fig. 16 Regularized problem

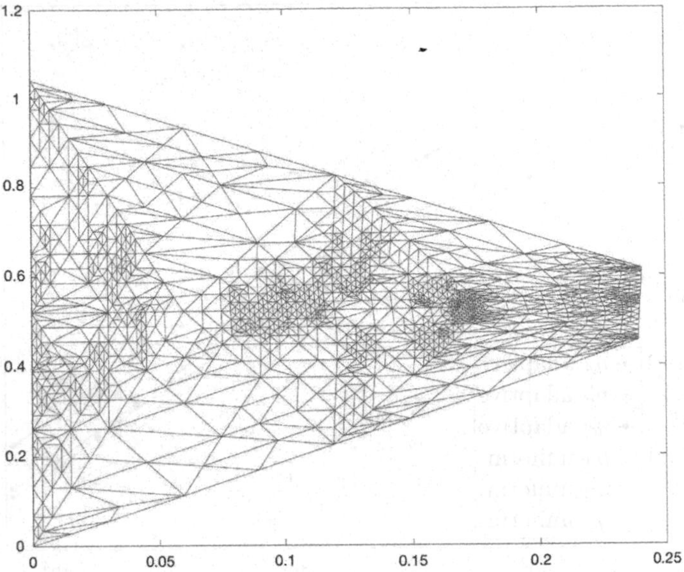

Fig. 17 Deformed configuration after eight adaptive refinements ($\lambda = \infty$)

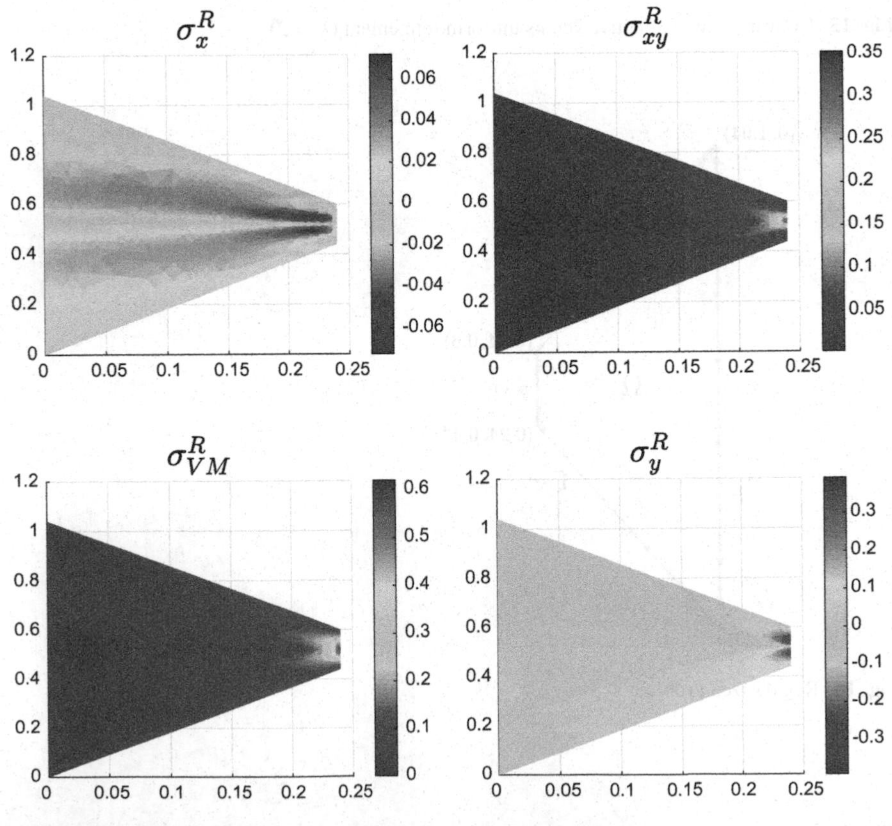

Fig. 18 Stress distribution after eight adaptive refinements ($\lambda = \infty$)

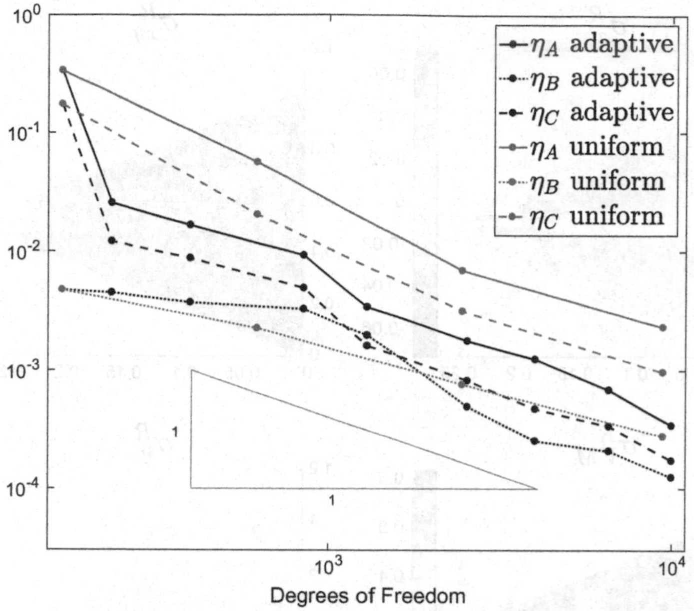

Fig. 19 Convergence of adaptive versus uniform refinement ($\lambda = \infty$)

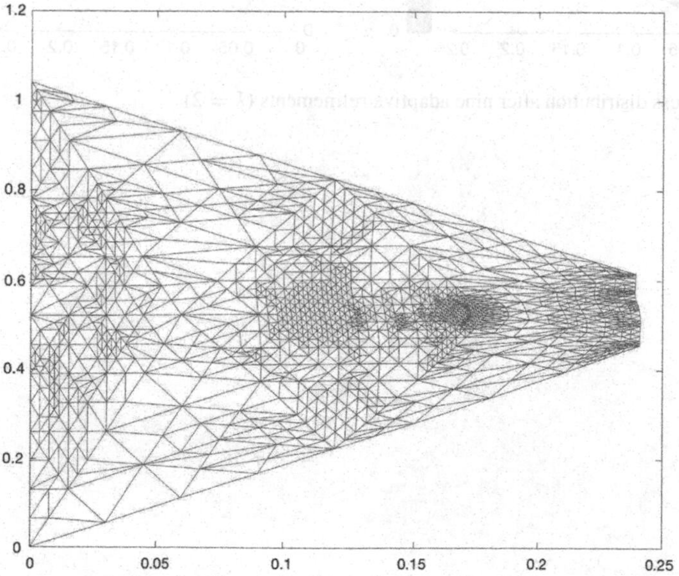

Fig. 20 Deformed configuration after nine adaptive refinements ($\lambda = 2$)

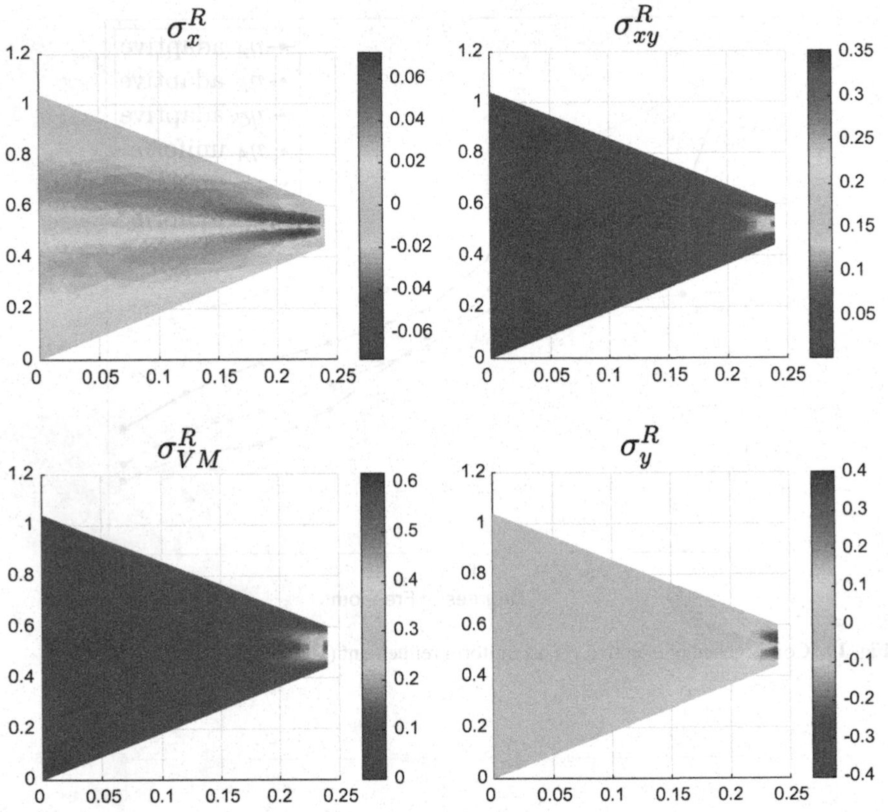

Fig. 21 Stress distribution after nine adaptive refinements ($\lambda = 2$)

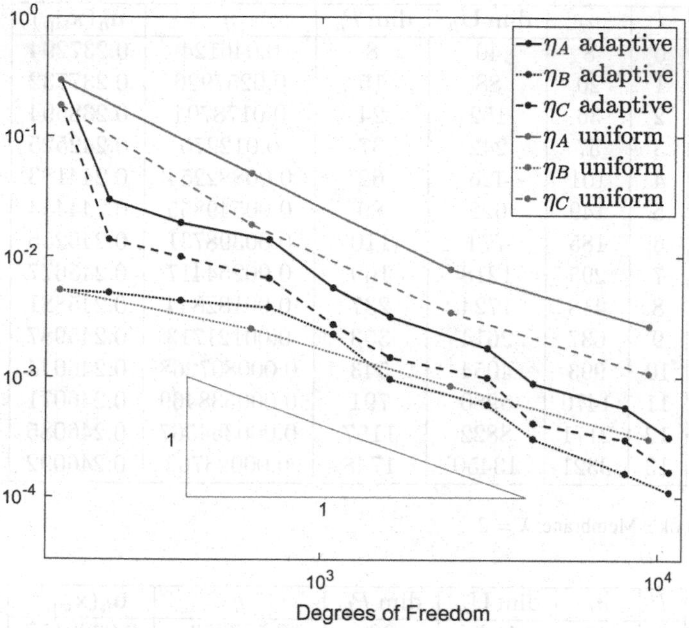

Fig. 22 Convergence of adaptive versus uniform refinement ($\lambda = 2$)

l	n_t	$\dim \mathbf{U}_h$	$\dim P_h$	η	$\mathbf{u}_h(\mathbf{x}_{\text{tip}})$
0	8	40	8	0.0431095	0.201041
1	20	88	15	0.0281644	0.199812
2	36	152	24	0.0197404	0.200857
3	57	242	37	0.0137582	0.204314
4	97	406	59	0.0104914	0.20555
5	142	592	85	0.00716875	0.205864
6	196	816	115	0.00473496	0.20656
7	251	1046	146	0.00353926	0.206869
8	382	1580	219	0.00237943	0.20703
9	539	2210	302	0.00165592	0.207123
10	749	3062	414	0.00117939	0.207133
11	1106	4512	606	0.000796602	0.207169
12	1593	6478	861	0.000552066	0.207188
13	2319	9394	1234	0.000377115	0.207199
14	3351	13554	1769	0.000262978	0.207204

Fig. 23 Cook's Membrane: $\lambda = \infty$

l	n_t	$\dim \mathbf{U}_h$	$\dim P_h$	η	$\mathbf{u}_h(\mathbf{x}_{\text{tip}})$
0	8	40	8	0.046124	0.237294
1	20	88	15	0.0257926	0.237732
2	36	152	24	0.0178704	0.238594
3	57	242	37	0.012275	0.242525
4	101	426	62	0.00882254	0.244183
5	149	622	89	0.00549855	0.244454
6	185	774	110	0.00398731	0.245255
7	295	1218	169	0.00254417	0.245677
8	418	1724	237	0.00182844	0.245881
9	637	2614	353	0.00121713	0.245987
10	993	4054	543	0.000807268	0.246044
11	1470	5980	791	0.000538469	0.246071
12	2171	8822	1157	0.000354567	0.246085
13	3321	13450	1748	0.00023762	0.246092

Fig. 24 Cook's Membrane: $\lambda = 2$

l	n_t	$\dim \mathbf{U}_h$	$\dim P_h$	η	$\mathbf{u}_h(\mathbf{x}_{\text{tip}})$
0	32	144	23	0.587656	0.0300157
1	47	202	31	0.0437496	0.0299908
2	83	346	50	0.0296502	0.0298429
3	183	742	101	0.0169036	0.0296102
4	280	1136	151	0.00629771	0.0146575
5	542	2192	286	0.00305551	0.0144975
6	859	3478	451	0.00202166	0.0144282
7	1439	5786	746	0.00120929	0.0144275
8	2199	8834	1135	0.000614436	0.0144014

Fig. 25 Regularized problem: $\lambda = \infty$

l	n_t	$\dim \mathbf{U}_h$	$\dim P_h$	η	$\mathbf{u}_h(\mathbf{x}_{\text{tip}})$
0	32	144	23	0.544967	0.0329654
1	47	202	31	0.0503809	0.0330178
2	79	334	48	0.0337426	0.0328692
3	150	616	85	0.0238685	0.0326998
4	236	960	130	0.00984681	0.0160559
5	346	1404	186	0.00545432	0.0159859
6	673	2714	353	0.00324157	0.0158735
7	920	3716	480	0.00155134	0.0158072
8	1756	7072	907	0.00101678	0.0157795
9	2386	9596	1228	0.000569833	0.0157796

Fig. 26 Regularized problem: $\lambda = 2$

References

Ainsworth, M., & Oden, J. T. (1993). A unified approach to a posteriori error estimation using element residual methods. *Numerical Mathematics*, *65*, 23–50.

Bahriawati, C., & Carstensen, C. (2005). Three matlab implementations of the lowest-order raviart-thomas mfem with a posteriori error control. *Computational Methods in Applied Mathematics*, *5*(4), 333–361.

Bertrand, F., Boffi, D., & Stenberg, R. (2019). Guaranteed a posteriori error analysis for the mixed laplace eigenvalue problem. arXiv:1812.11203.

Bertrand, F., Kober, B., Moldenhauer, M., & Starke, G. (2018a). Weakly symmetric stress equilibration and a posteriori error estimation for linear elasticity. arXiv:1808.02655.

Bertrand, F., Moldenhauer, M., & Starke, G. (2018b). A posteriori error estimation for planar linear elasticity by stress reconstruction. *Computational Methods in Applied Mathematics*.

Bertrand, F., Münzenmaier, S., & Starke, G. (2014a). First-order system least squares on curved boundaries: Higher-order Raviart-Thomas elements. *SIAM Journal on Numerical Analysis*, *52*, 3165–3180.

Bertrand, F., Münzenmaier, S., & Starke, G. (2014b). First-order system least squares on curved boundaries: Lowest-order Raviart-Thomas elements. *SIAM Journal on Numerical Analysis*, *52*, 880–894.

Bertrand, F., & Starke, G. (2016). Parametric Raviart-Thomas elements for mixed methods on domains with curved surfaces. *SIAM Journal on Numerical Analysis*, *54*, 3648–3667.

Boffi, D., Brezzi, F., & Fortin, M. (2009). Reduced symmetry elements in linear elasticity. *Communications on Pure and Applied Analysis*, *8*, 95–121.

Boffi, D., Brezzi, F., & Fortin, M. (2013). *Mixed finite element methods and applications*. Heidelberg: Springer.

Braess, D. (2013). *Finite Elemente: Theorie, schnelle Löser und Anwendungen in der Elastizitätstheorie* (p. 5). Berlin: Springer. (Auflage).

Braess, D., Pillwein, V., & Schöberl, J. (2009). Equilibrated residual error estimates are p-robust. *Computer Methods in Applied Mechanics and Engineering*, *198*, 1189–1197.

Braess, D., & Schöberl, J. (2008). Equilibrated residual error estimator for edge elements. *Mathematics of Computation*, *77*(262), 651–672.

Cai, Z., & Zhang, S. (2010). Flux recovery and a posteriori error estimators: Conforming elements for scalar elliptic equations. *SIAM Journal on Numerical Analysis*, *48*, 578–602.

Cai, Z., & Zhang, S. (2012). Robust equilibrated residual error estimator for diffusion problems: Conforming elements. *SIAM Journal on Numerical Analysis*, *50*, 151–170.

Carstensen, C., Feischl, M., Page, M., & Praetorius, D. (2014). Axioms of adaptivity. *Computers and Mathematics with Applications*, *67*(6), 1195–1253.

Cascon, J. M., Kreuzer, C., Nochetto, R. H., & Siebert, K. G. (2008). Quasi-optimal convergence rate for an adaptive finite element method. *SIAM Journal on Numerical Analysis*, *46*(5), 2524–2550.

Ciarlet, P. G. (1988). *Mathematical elasticity Volume I: Three–Dimensional elasticity*. North-Holland, Amsterdam.

Dörfler, W. (1996). A convergent adaptive algorithm for Poisson's equation. *SIAM Journal on Numerical Analysis*, *33*(3), 1106–1124. ISSN 0036-1429.

Ern, A., & Guermond, J.-L. (2004). *Theory and practice of finite elements*. New York: Springer.

Ern, A., & Vohralík, M. (2015). Polynomial-degree-robust a posteriori error estimates in a unified setting for conforming, nonconforming, discontinuous Galerkin, and mixed discretizations. *SIAM Journal on Numerical Analysis*, *53*, 1058–1081.

Hannukainen, A., Stenberg, R., & Vohralík, M. (2012). A unified framework for a posteriori error estimation for the Stokes equation. *Numerical Mathematics*, *122*, 725–769.

Ladevèze, P., & Leguillon, D. (1983). Error estimate procedure in the finite element method and applications. *SIAM Journal on Numerical Analysis*, *20*, 485–509.

Morin, P., Nochetto, R. H., & Siebert, K. G. (2000). Data oscillation and convergence of adaptive FEM. *SIAM Journal on Numerical Analysis*, *38*(2), 466–488.

Nochetto, R. H., Siebert, K. G., & Veeser, A. (2009). Theory of adaptive finite element methods: an introduction. *Multiscale* (pp. 409–542). nonlinear and adaptive approximation Berlin: Springer.

Nochetto, R. H., & Veeser A. (2012). Primer of adaptive finite element methods. In *Multiscale and adaptivity: Modeling, numerics and applications*, volume 2040 of *Lecture Notes in Mathematics* (pp. 125–225). Heidelberg: Springer.

Prager, W., & Synge, J. L. (1947). Approximations in elasticity based on the concept of function space. *Quarterly of Applied Mathematics*, *5*, 241–269.

Stevenson, R. (2007). Optimality of a standard adaptive finite element method. *Foundations of Computational Mathematics*, *7*(2), 245–269.

Verfürth, R. (1999). A review of a posteriori error estimation techniques for elasticity problems. *Computer Methods in Applied Mechanics and Engineering*, *176*, 419–440.

A Concept for the Extension of the Assumed Stress Finite Element Method to Hyperelasticity

Nils Viebahn, Jörg Schröder and Peter Wriggers

Abstract The proposed work extends the well-known assumed stress elements to the framework of hyperelasticity. In order to obtain the constitutive relationship, a nonlinear set of equations is solved implicitly on element level. A numerical verification, where two assumed stress elements are compared to classical enhanced assumed strain elements, depicts the reliability and efficiency of the proposed concept. This work is closely related to the publication of Viebahn et al. (2019)

1 Introduction

The research on the family of assumed stress finite elements goes back to the 80s and was mainly pushed by the workgroup of Pian at the Massachusetts Institute of Technology, see Pian (1964), Pian and Chen (1982), Pian and Sumihara (1984) and Pian and Tong (1986). The elements are based on the variational principle of (Prange-)Hellinger–Reissner which has been elaborated independently by Prange (1916), Hellinger (1913) and Reissner (1950). This approach treats the displacement and stresses as independent fields, whereas the constitutive equation is described via a complementary stored energy function in terms of the stresses. In this approach the displacement act as a Lagrange multiplier. Thus, the resulting finite elements can be assigned to the family of mixed finite elements and their stability in the linear range can be shown within the theory of Babuska and Brezzi (BB conditions), see Babuška (1973) and Brezzi (1974). Especially the elements of Pian and Sumihara (1984) for

The financial support by the Deutsche Forschungsgemeinschaft (DFG, German Research Foundation) is gratefully acknowledged - 255432295.

N. Viebahn (✉) · J. Schröder
Department of Civil Engineering, Faculty of Engineering, Institute of Mechanics,
University of Duisburg-Essen, Essen, Germany
e-mail: nils.viebahn@uni-due.de

P. Wriggers
Faculty of Mechanical Engineering, Institute of Continuum Mechanics,
Leibniz Universität Hannover, Hanover, Germany

J. Schröder and P. de Mattos Pimenta (eds.), *Novel Finite Element Technologies for Solids and Structures*, CISM International Centre for Mechanical Sciences 597,
https://doi.org/10.1007/978-3-030-33520-5_4

plane elasticity and Pian and Tong (1986) for 3D applications, constitute still eminently efficient formulations possessing a superior behavior in bending dominated situations but also in the framework of incompressibility or shell-like structures. Their main drawback is the necessity of an explicit complementary stored energy function which in general does not exist in the nonlinear regime, see Ogden (1984). However, a couple of approaches to extend the family of assumed stresses to more general material descriptions has been done, e.g., Atluri (1973), Simo et al. (1989), Wriggers (2009), Schröder et al. (1997) and Schröder et al. (2017). This inflexibility, of the formulation lead to the development of the enhanced assumed strain formulations, which can be mainly attributed to the pioneering works of Simo and Rifai (1990) and Pantuso and Bathe (1995). It can be shown that this approach embeds the method of incompatible modes, e.g., Wilson et al. (1973), into a variational consistent and conforming method based on a Hu-Washizu-like formulation, see Hu (1955) an Washizu (1955). In contrast to the assumed stress approach, the enhanced assumed strain method can be extended to complex nonlinear constitutive descriptions in a straight forward manner and have been successfully applied in the engineering community.

However, in linear elasticity assumed stress and assumed strain elements possess a very close relationship, which can also result in an equivalence under certain conditions. This observation has been first published by Andelfinger and Ramm (1993) and was later analyzed by Yeo and Lee (1996), Bischoff et al. (1999), and Djoko et al. (2006).

The main goal of this work is to propose a concept which allows for an extension of the assumed stress finite elements to general hyperelasticity. Therefore, in the following section, the concept of these elements will be briefly discussed in the linear elastic framework. This is followed by a short survey on continuum mechanics, which are needed in the hyperelastic setting. Then, the concept for the extension to hyperelastic is discussed in detail and the resulting elements will be verified and compared with classical and enhanced strain elements in a set of different numerical examples.

2 Assumed Stress Elements in Linear Elasticity

In the following a body $\mathcal{B} \subset \mathbb{R}^3$ is considered, where its boundary $\partial\mathcal{B}$ is subdivided into a nontrivial Dirichlet $\partial\mathcal{B}_u$ and a Neumann $\partial\mathcal{B}_\sigma$ part. The strong form of the elasticity problem is given by

$$\begin{aligned} \mathbb{C}^{-1} : \boldsymbol{\sigma} &= \boldsymbol{\varepsilon}(\boldsymbol{u}) && \text{on } \mathcal{B}\,, \\ \operatorname{div}\boldsymbol{\sigma} + \boldsymbol{f} &= \boldsymbol{0} && \text{in } \mathcal{B}\,, \\ \boldsymbol{u} &= \overline{\boldsymbol{u}} && \text{in } \partial\mathcal{B}_u \\ \boldsymbol{\sigma} \cdot \boldsymbol{n} &= \overline{\boldsymbol{t}} && \text{on } \partial\mathcal{B}_\sigma\,, \end{aligned} \tag{1}$$

where $\boldsymbol{\sigma}$ denotes the Cauchy stresses, $\boldsymbol{u}$ the displacements, $\mathbb{C}$ the fourth-order elasticity tensor, $\boldsymbol{f}$ the body forces and $\overline{\boldsymbol{u}}$, and $\overline{\boldsymbol{t}}$ the prescribed displacements and tractions on $\partial\mathcal{B}_u$ and $\partial\mathcal{B}_\sigma$, respectively. The strains are defined by $\boldsymbol{\varepsilon}(\boldsymbol{u}) = \frac{1}{2}(\nabla\boldsymbol{u} + (\nabla\boldsymbol{u})^T)$ with ∇ as the gradient operator. The solution of (1) is equivalent to the Hellinger–Reissner principle which seeks a saddle point for $\boldsymbol{\sigma}$ and $\boldsymbol{u}$ of the functional

$$\mathcal{F}^*(\boldsymbol{\sigma}, \boldsymbol{u}) = \int_{\mathcal{B}} \left(\frac{1}{2}\boldsymbol{\sigma} : \mathbb{C}^{-1} : \boldsymbol{\sigma} + (\operatorname{div}\boldsymbol{\sigma} + \boldsymbol{f}) \cdot \boldsymbol{u} \right) \mathrm{d}V\,, \tag{2}$$

including appropriate boundary conditions. With help of the divergence theorem and integration by parts we reformulate equation (2) and obtain a formulation which seeks a saddle point for $\boldsymbol{\sigma}$ and $\boldsymbol{u}$ of the functional

$$\mathcal{F}(\boldsymbol{\sigma}, \boldsymbol{u}) = \int_{\mathcal{B}} \left(-\frac{1}{2}\boldsymbol{\sigma} : \mathbb{C}^{-1} : \boldsymbol{\sigma} + \boldsymbol{\sigma} : \boldsymbol{\varepsilon}(\boldsymbol{u}) - \boldsymbol{f} \cdot \boldsymbol{u} \right) \mathrm{d}V - \int_{\partial\mathcal{B}_\sigma} \boldsymbol{u} \cdot \overline{\boldsymbol{t}}\, \mathrm{d}A\,, \tag{3}$$

including appropriate boundary conditions. The variation with respect to the displacements $\boldsymbol{u}$ and the stresses $\boldsymbol{\sigma}$ of the considered functional (3) are given by

$$\begin{aligned} \delta_\sigma \mathcal{F}(\boldsymbol{\sigma}, \boldsymbol{u}) &= \int_{\mathcal{B}} (\boldsymbol{\varepsilon}(\boldsymbol{u}) - \mathbb{C}^{-1} : \boldsymbol{\sigma}) : \delta\boldsymbol{\sigma}\, \mathrm{d}V = 0\,, \\ \delta_u \mathcal{F}(\boldsymbol{\sigma}, \boldsymbol{u}) &= \int_{\mathcal{B}} (\boldsymbol{\sigma} : \delta\boldsymbol{\varepsilon}(\boldsymbol{u}) - \boldsymbol{f} \cdot \delta\boldsymbol{u})\, \mathrm{d}V - \int_{\partial\mathcal{B}_\sigma} \overline{\boldsymbol{t}} \cdot \delta\boldsymbol{u}\, \mathrm{d}A = 0\,. \end{aligned} \tag{4}$$

Introducing a shape regular triangulation $\mathcal{T} = \bigcup_e \mathcal{T}^e$ of $\mathcal{B}$ into a finite number of elements $\mathcal{T}^e$ and considering a conforming choice for the solution spaces, which demands continuity of the displacements and allows for jumps of the stresses at element faces. The vectors, containing the nodal degrees of freedom for the displacements and stresses, are denoted by $\underline{\boldsymbol{d}}_u$ and $\underline{\boldsymbol{d}}_\sigma$. The discretized counterparts of the displacements, strains and their variations in matrix notation are obtained by

$$\begin{aligned} \underline{\boldsymbol{u}} &= \underline{\mathbb{N}}\, \underline{\boldsymbol{d}}_u\,, \quad \underline{\boldsymbol{\varepsilon}}(\boldsymbol{u}) = \underline{\mathbb{B}}\, \underline{\boldsymbol{d}}_u\,, \\ \delta\underline{\boldsymbol{u}} &= \underline{\mathbb{N}}\, \delta\underline{\boldsymbol{d}}_u\,, \; \delta\underline{\boldsymbol{\varepsilon}}(\boldsymbol{u}) = \underline{\mathbb{B}}\, \delta\underline{\boldsymbol{d}}_u\,, \end{aligned} \tag{5}$$

where $\underline{\mathbb{N}}$ and $\underline{\mathbb{B}}$ represent suitable matrices, including the trilinear Lagrangian shape functions and its spatial derivatives. The stresses and their variations are discretized on the isoparametric reference element by

$$\begin{aligned} \widehat{\underline{\boldsymbol{\sigma}}} &= (\widehat{\sigma}_{11}, \widehat{\sigma}_{22}, \widehat{\sigma}_{33}, \widehat{\sigma}_{12}, \widehat{\sigma}_{23}, \widehat{\sigma}_{13})^T = \widehat{\underline{\mathbb{L}}}\, \underline{\boldsymbol{d}}_\sigma\,, \\ \delta\widehat{\underline{\boldsymbol{\sigma}}} &= (\delta\widehat{\sigma}_{11}, \delta\widehat{\sigma}_{22}, \delta\widehat{\sigma}_{33}, \delta\widehat{\sigma}_{12}, \delta\widehat{\sigma}_{23}, \delta\widehat{\sigma}_{13})^T = \widehat{\underline{\mathbb{L}}}\, \delta\underline{\boldsymbol{d}}_\sigma\,, \end{aligned} \tag{6}$$

whereas, the matrix $\widehat{\underline{\mathbb{L}}}$ contains the interpolation functions of the stresses. In the following we will distinguish two different sets of interpolation functions. The general structure of $\widehat{\underline{\mathbb{L}}}$ follows by

$$\widehat{\mathbb{L}} = \operatorname{diag}\left(\widehat{\mathbb{L}}_{11}, \widehat{\mathbb{L}}_{22}, \widehat{\mathbb{L}}_{33}, \widehat{\mathbb{L}}_{12}, \widehat{\mathbb{L}}_{23}, \widehat{\mathbb{L}}_{13}\right) , \tag{7}$$

where $\widehat{\mathbb{L}}_{ij}$ are suitable vectors corresponding to the interpolation of the ij-th entry of $\widehat{\sigma}$. In general, the construction of a stress discretization of this type is limited by two conditions. Too few stress unknowns lead to a nonstable formulation. In this context a necessary, but not sufficient condition, is the *count condition*, see Zienkiewicz et al. (1986), which states that the number of stress unknowns per element has to be equal or larger than the number of unknowns for the displacements. A violation of the *count condition* leads to a singular tangent matrix. However, unstable formulations may be obtained although if this condition is satisfied. Thus, a sufficient condition for the stability of such elements is the satisfaction of the LBB-conditions, see Ladyzhenskaya (1969), Babuška (1973), Brezzi (1974). On the other side, too many stress unknowns introduces additional stiffness to the system, resulting in the *principle of limitation*, going back to Fraejis de Veubeke (1965). Thus, a mixed assumed stress element is equivalent to a classical displacement-based approach, if the stress interpolation of the mixed element comprises the stress space of the displacement element. In the sense of these two conditions, the well-known 18- parameter-based interpolation scheme by Pian and Tong (1986) seems to be optimal in the three-dimensional case, at least in the framework of linear elasticity. Here the individual interpolation vectors are given by

$$\begin{aligned}
\widehat{\mathbb{L}}_{11} &= (1,\, \eta,\, \zeta,\, \eta\zeta) \, , \\
\widehat{\mathbb{L}}_{22} &= (1,\, \xi,\, \zeta,\, \xi\zeta) \, , \\
\widehat{\mathbb{L}}_{33} &= (1,\, \xi,\, \eta,\, \xi\eta) \, , \\
\widehat{\mathbb{L}}_{12} &= (1,\, \zeta) \, , \\
\widehat{\mathbb{L}}_{23} &= (1,\, \xi) \, , \\
\widehat{\mathbb{L}}_{13} &= (1,\, \eta) \, .
\end{aligned} \tag{8}$$

In the view of the nonlinear extension of the assumed stress formulations an additional interpolation scheme with 24 parameter will be considered

$$\begin{aligned}
\widehat{\mathbb{L}}_{11} &= (1,\, \eta,\, \zeta,\, \eta\zeta) \, , \\
\widehat{\mathbb{L}}_{22} &= (1,\, \xi,\, \zeta,\, \xi\zeta) \, , \\
\widehat{\mathbb{L}}_{33} &= (1,\, \xi,\, \eta,\, \xi\eta) \, , \\
\widehat{\mathbb{L}}_{12} &= (1,\, \xi,\, \eta,\, \zeta,\, \xi\zeta,\, \eta\zeta) \, , \\
\widehat{\mathbb{L}}_{23} &= (1,\, \xi,\, \eta,\, \zeta,\, \xi\eta,\, \xi\zeta) \, , \\
\widehat{\mathbb{L}}_{13} &= (1,\, \xi,\, \eta,\, \zeta,\, \xi\eta,\, \eta\zeta) \, .
\end{aligned} \tag{9}$$

Due to the additional unknown parameters for the shear stresses, an additional stiffness in bending dominated situations can already be anticipated. The transformation from the isoparametric domain to the reference configuration for the stresses is described by

$$\boldsymbol{\sigma}_h = \boldsymbol{J}_0 \widehat{\boldsymbol{\sigma}}_h \boldsymbol{J}_0^T , \tag{10}$$

where the Jacobian matrix, describing the mapping between the isoparametric coordinates $\boldsymbol{\xi}$ and the reference coordinates $\boldsymbol{X}$, follows as

$$\boldsymbol{J} = \frac{\partial \boldsymbol{X}(\boldsymbol{\xi})}{\partial \boldsymbol{\xi}} \quad \text{and} \quad \boldsymbol{J}_0 = \left.\frac{\partial \boldsymbol{X}(\boldsymbol{\xi})}{\partial \boldsymbol{\xi}}\right|_{\boldsymbol{\xi}=0} . \tag{11}$$

In order to pass the patch test, it is necessary to use the values of the Jacobian at the origin $\{\xi, \eta, \zeta\} = \{0, 0, 0\}$ as it is discussed in Pian and Sumihara (1984) and Pian and Tong (1986). The discrete stresses defined in the global Cartesian frame can be given in matrix notation

$$\underline{\boldsymbol{\sigma}}_h = \mathbb{L}\,\underline{\boldsymbol{d}}_\sigma \tag{12}$$

with $\mathbb{L} = \underline{\boldsymbol{T}}\,\widehat{\mathbb{L}}$ where the transformation matrix $\underline{\boldsymbol{T}}$ is given by

$$\underline{\boldsymbol{T}} = \begin{bmatrix} J_{11_0}^2 & J_{12_0}^2 & J_{13_0}^2 & 2J_{11_0}J_{12_0} \\ J_{21_0}^2 & J_{22_0}^2 & J_{23_0}^2 & 2J_{21_0}J_{22_0} \\ J_{31_0}^2 & J_{32_0}^2 & J_{33_0}^2 & 2J_{31_0}J_{32_0} \\ J_{11_0}J_{21_0} & J_{12_0}J_{22_0} & J_{13_0}J_{23_0} & J_{12_0}J_{21_0} + J_{11_0}J_{22_0} \\ J_{21_0}J_{31_0} & J_{22_0}J_{32_0} & J_{23_0}J_{33_0} & J_{22_0}J_{31_0} + J_{21_0}J_{32_0} \\ J_{11_0}J_{31_0} & J_{12_0}J_{32_0} & J_{13_0}J_{33_0} & J_{12_0}J_{31_0} + J_{11_0}J_{32_0} \end{bmatrix} \cdots$$

$$\cdots \begin{bmatrix} 2J_{12_0}J_{13_0} & 2J_{11_0}J_{13_0} \\ 2J_{22_0}J_{23_0} & 2J_{21_0}J_{23_0} \\ 2J_{32_0}J_{33_0} & 2J_{31_0}J_{33_0} \\ J_{13_0}J_{22_0} + J_{12_0}J_{23_0} & J_{13_0}J_{21_0} + J_{11_0}J_{23_0} \\ J_{23_0}J_{32_0} + J_{22_0}J_{33_0} & J_{23_0}J_{31_0} + J_{21_0}J_{33_0} \\ J_{13_0}J_{32_0} + J_{12_0}J_{33_0} & J_{13_0}J_{31_0} + J_{11_0}J_{33_0} \end{bmatrix} . \tag{13}$$

The discretized counterparts of the weak forms in Equation (4) can thus be rewritten for a typical element by

$$\begin{aligned} \delta\underline{\boldsymbol{d}}_\sigma^T \Bigg(\underbrace{\int_T \mathbb{L}^T \mathbb{B}\,\mathrm{d}V}_{\boldsymbol{k}_{\sigma u}^e} \underline{\boldsymbol{d}}_u - \underbrace{\int_T \mathbb{L}^T \mathbb{C}^{-1} \mathbb{L}\,\mathrm{d}V}_{\boldsymbol{k}_{\sigma\sigma}^e} \underline{\boldsymbol{d}}_\sigma \Bigg) &= 0 \\ \delta\underline{\boldsymbol{d}}_u^T \Bigg(\underbrace{\int_T \mathbb{B}^T \mathbb{L}\,\mathrm{d}V}_{\underline{\boldsymbol{k}}_{u\sigma}^e} \underline{\boldsymbol{d}}_\sigma \underbrace{- \int_T \mathbb{N}^T \underline{\boldsymbol{f}}\,\mathrm{d}V - \int_{\partial T} \mathbb{N}^T \bar{\underline{\boldsymbol{t}}}\,\mathrm{d}A}_{\underline{\boldsymbol{r}}_u^e} \Bigg) &= 0 . \end{aligned} \tag{14}$$

The system of equations is as follows:

$$\sum_{T^e \in T} \begin{bmatrix} \delta\underline{\boldsymbol{d}}_\sigma \\ \delta\underline{\boldsymbol{d}}_u \end{bmatrix}^T \left(\begin{bmatrix} \underline{\boldsymbol{k}}_{\sigma\sigma}^e & \underline{\boldsymbol{k}}_{\sigma u}^e \\ \underline{\boldsymbol{k}}_{u\sigma}^e & \underline{\boldsymbol{0}} \end{bmatrix} \begin{bmatrix} \underline{\boldsymbol{d}}_\sigma \\ \underline{\boldsymbol{d}}_u \end{bmatrix} + \begin{bmatrix} \underline{\boldsymbol{0}} \\ \underline{\boldsymbol{r}}_u^e \end{bmatrix} \right) = 0 , \tag{15}$$

which has the typical saddle point structure. Since the stresses are interpolated elementwise and no continuity over element patches has to be satisfied, a static condensation may be applied. Therefore, the degrees of freedom associated to the stresses are already solved on individual element level with respect to the displacements. This leads to a global system of equations where only the degrees of freedom associated to the displacements have to be solved. The condensed form for a typical element is given by

$$\delta \underline{\boldsymbol{d}}_u^T \big(\underbrace{-\underline{\boldsymbol{k}}_{u\sigma}^e \, (\underline{\boldsymbol{k}}_{\sigma\sigma}^e)^{-1} \, \underline{\boldsymbol{k}}_{\sigma u}^e}_{\underline{\boldsymbol{k}}_{\text{red}}^e} \, \underline{\boldsymbol{d}}_u + \underline{\boldsymbol{r}}_u^e \big) = 0 \,. \tag{16}$$

The resulting system of equations yields $\underline{\boldsymbol{K}} \, \Delta \underline{\boldsymbol{D}} = \underline{\boldsymbol{R}}$ with $\mathbf{A}$ as an assembling operator, num_{ele} as the number of elements and

$$\underline{\boldsymbol{K}} = \mathop{\mathbf{A}}_{e=1}^{num_{ele}} \underline{\boldsymbol{k}}_{\text{red}}^e \quad \text{and} \quad \underline{\boldsymbol{R}} = \mathop{\mathbf{A}}_{e=1}^{num_{ele}} \underline{\boldsymbol{r}}_u^e, \,. \tag{17}$$

In particular the static condensation has two main advantages. First, the system size is reduced magnificent and in addition the system matrix becomes positive definite which is beneficial for many solving strategies.

2.1 Equivalences to EAS Formulations

The enhanced assumed strain elements, which can be attributed to the pioneering works of Simo and Rifai (1990) and Pantuso and Bathe (1995), are based on a Hu-Washizu-type variational framework. The main idea is an enhancement of the conforming strain approximation by an assumed strain field. Assumed stress and assumed strain elements possess a very close relationship (together with the family of nonconforming finite elements), which can also result in an equivalence under certain conditions. This observation has been first published by Andelfinger and Ramm (1993) and was later analyzed by Yeo and Lee (1996), Bischoff et al. (1999) and Djoko et al. (2006). Following their conclusions, it can be shown in linear elasticity that (under the assumption of a piecewise constant Jacobian) the EAS element with 15 enhanced modes (referred to Pantuso and Bathe (1995)) is equivalent to the assumed stress element with 18 stress parameters. The assumed stress element with 24 stress parameters has an equivalent EAS counterpart based on 9 enhanced modes, recently proposed by Krischok and Linder (2016).

3 Assumed Stress Elements for Hyperelasticity

In the following sections the previously described element technology is extended to the hyperelastic framework. Therefore some foundations of continuum mechanics are briefly introduced.

3.1 Continuum Mechanics

Let $\mathcal{B} \subset \mathbb{R}^3$, parametrized in $\boldsymbol{X}$, be the body of interest in the reference configuration and $\mathcal{B}_t \subset \mathbb{R}^3$, parametrized in $\boldsymbol{x}$, the body in the current configuration. The nonlinear deformation map $\boldsymbol{\varphi}_t : \mathcal{B} \rightarrow \mathcal{B}_t$ at time $t \in \mathbb{R}_+$ maps points of $\mathcal{B}$ onto points of $\mathcal{B}_t$, i.e., $\boldsymbol{\varphi}_t : \boldsymbol{X} \mapsto \boldsymbol{x}$. Let $\boldsymbol{e}_a = \boldsymbol{e}^a$ and $\boldsymbol{E}_A = \boldsymbol{E}^A$ denote the Cartesian base vectors in the actual and in the reference placement, respectively. Thus, we arrive at the simple expressions $G_{AB} = G^{AB} = \delta_{AB}$ for Lagrangian and $g_{ab} = g^{ab} = \delta_{ab}$ for the Eulerian metric tensors. The basic kinematical quantity, the deformation gradient $\boldsymbol{F} = \nabla_X \boldsymbol{\varphi}_t(\boldsymbol{X}) = \mathrm{Grad}\boldsymbol{\varphi}_t(\boldsymbol{X})$, is given by

$$\boldsymbol{F} = F^a{}_A \, \boldsymbol{e}_a \otimes \boldsymbol{E}^A \quad \text{with} \quad F^a{}_A = \frac{\partial x^a}{\partial X^A} \,. \tag{18}$$

An important deformation measure for the construction of free energy functions is the covariant right Cauchy–Green tensor:

$$\boldsymbol{C} := \boldsymbol{F}^T \boldsymbol{F} = C_{AB} \, \boldsymbol{E}^A \otimes \boldsymbol{E}^B \quad \text{with} \quad C_{AB} = F^a{}_A \, \delta_{ab} \, F^b{}_B \,. \tag{19}$$

The corresponding nonlinear strain measure is denoted as Green–Lagrange strain tensor and given by

$$\boldsymbol{E} = \frac{1}{2}(\boldsymbol{C} - \boldsymbol{I}), \tag{20}$$

where $\boldsymbol{I}$ is the second-order identity tensor. In hyperelasticity we postulate the existence of a so-called Helmholtz free energy function $\psi = \psi(\boldsymbol{F})$, here defined per unit reference volume, depending solely on the deformation gradient. We consider perfect elastic materials, which means that the internal dissipation $\mathcal{D}_{int}$ is zero for every admissible process. Therefore, the Clausius–Planck inequality reduces to

$$\mathcal{D}_{int} = \boldsymbol{P} : \dot{\boldsymbol{F}} - \dot{\psi} = (\boldsymbol{P} - \frac{\partial \psi}{\partial \boldsymbol{F}}) : \dot{\boldsymbol{F}} \geq 0 \,, \tag{21}$$

which has to be fulfilled for all possible thermodynamic processes. Here $\dot{\boldsymbol{F}}$ denotes the material time derivative of the deformation gradient. Thus, we conclude

Table 1 Kinematic and constitutive quantities

Symbol	Continuum mechanical description
$\boldsymbol{u}$	Displacement vector
$\boldsymbol{F} = \boldsymbol{1} + \text{Grad}\boldsymbol{u}$	Deformation gradient
$\boldsymbol{C} = \boldsymbol{F}^T\boldsymbol{F}$	Right Cauchy–Green tensor
$\boldsymbol{E} = \frac{1}{2}(\boldsymbol{C} - \boldsymbol{I})$	Green–St.Venant strain tensor
χ	Complementary stored energy
ψ	Helmholtz free energy
$\boldsymbol{S}$	Second Piola–Kirchhoff stress tensor
$\boldsymbol{P} = \boldsymbol{F}\boldsymbol{S}$	First Piola–Kirchhoff stress tensor
$\boldsymbol{\tau} = \boldsymbol{P}\boldsymbol{F}^T$	Kirchhoff stress tensor
$\boldsymbol{\sigma} = (\det \boldsymbol{F})^{-1}\boldsymbol{\tau}$	Cauchy stress tensor

$$\boldsymbol{P} = \frac{\partial \psi}{\partial \boldsymbol{F}} =: \partial_{\boldsymbol{F}} \psi \,, \tag{22}$$

where $\boldsymbol{P}$ denotes the first Piola–Kirchhoff stress tensor. Applying the chain rule and assuming $\psi = \psi(\boldsymbol{C})$, we obtain the expression

$$\boldsymbol{S} = 2\,\frac{\partial \psi}{\partial \boldsymbol{C}} =: 2\,\partial_{\boldsymbol{C}} \psi \,, \tag{23}$$

for the symmetric second Piola–Kirchhoff stress tensor $\boldsymbol{S}$ (Table 1).

3.2 *Assumed Stress Elements in Hyperelasticity*

In the following we focus first on the case where a complementary stored energy function $\chi(\boldsymbol{S})$, which describes the constitutive equation in the form

$$\partial_{\boldsymbol{S}} \chi(\boldsymbol{S}) := \boldsymbol{E} \,, \tag{24}$$

exists. For example, in the case of St. Venant type nonlinear elasticity $\boldsymbol{E} = \mathbb{C}^{-1} : \boldsymbol{S}$, we simply obtain

$$\chi(\boldsymbol{S}) = \frac{1}{2}\boldsymbol{S} : \mathbb{C}^{-1} : \boldsymbol{S} \,. \tag{25}$$

The balance of momentum closes, together with suitable boundary conditions, the set of equations for the boundary value problem in hyperelasticity

$$\text{Div}\,\boldsymbol{P} + \boldsymbol{f} = \boldsymbol{0} \quad \text{on} \quad \mathcal{B}, \tag{26}$$

where $\boldsymbol{f}$ denotes the body force vector and Div the divergence operator with respect to $\boldsymbol{X}$. The solution for $\boldsymbol{u}$ and $\boldsymbol{S}$ of (24) and (26) together with suitable boundary conditions is equivalent to a stationary point of the Hellinger–Reissner functional

$$\mathcal{F}(\boldsymbol{S},\boldsymbol{u}) = \int_{\mathcal{B}} (\boldsymbol{S} : \boldsymbol{E} - \chi(\boldsymbol{S}))\,\mathrm{d}V + \mathcal{F}^{ext}(\boldsymbol{x})\,, \tag{27}$$

with the external potential $\Pi^{ext}(\boldsymbol{u})$ given by

$$\mathcal{F}^{ext}(\boldsymbol{u}) = -\int_{\mathcal{B}} \boldsymbol{f}\cdot\boldsymbol{u}\,\mathrm{d}V - \int_{\partial\mathcal{B}_t} \bar{\boldsymbol{t}}\cdot\boldsymbol{u}\,\mathrm{d}A\,, \tag{28}$$

where $\bar{\boldsymbol{t}}$ denotes the prescribed traction vector on the Neumann boundary. In order to find a stationary point of the functional, we have to calculate the roots of the first variations with respect to the unknown fields $\boldsymbol{u}$ and $\boldsymbol{S}$. In detail, we obtain

$$\begin{aligned} G_u &:= \delta_u\mathcal{F} = \int_{\mathcal{B}} \delta\boldsymbol{E} : \boldsymbol{S}\,\mathrm{d}V - \int_{\mathcal{B}} \delta\boldsymbol{u}\cdot\boldsymbol{f}\,\mathrm{d}V - \int_{\partial\mathcal{B}_t} \delta\boldsymbol{u}\cdot\bar{\boldsymbol{t}}\,\mathrm{d}A = 0\,,\\ G_S &:= \delta_S\mathcal{F} = \int_{\mathcal{B}} (\delta\boldsymbol{S} : (\boldsymbol{E} - \partial_S\chi(\boldsymbol{S}))\,\mathrm{d}V = 0\,, \end{aligned} \tag{29}$$

with the virtual deformation $\delta\boldsymbol{u}$, the virtual stress field $\delta\boldsymbol{S}$. Furthermore, the virtual strains are defined by $\delta\boldsymbol{E} = \frac{1}{2}(\delta\boldsymbol{F}^T\boldsymbol{F} + \boldsymbol{F}^T\delta\boldsymbol{F})$ with $\delta\boldsymbol{F} = \nabla_X\delta\boldsymbol{u}$.

The discretized weak forms $G_u^h = \sum_e G_u^e$ and $G_S^h = \sum_e G_S^e$ for a typical element appear with $\delta\underline{\boldsymbol{E}} = \mathbb{B}\,\delta\boldsymbol{d}$, where $\mathbb{B}$ is a suitable matrix containing the derivatives of the shape functions as

$$\begin{aligned} G_u^e &= \delta\underline{\boldsymbol{d}}^T \int_{\mathcal{B}^e} \mathbb{B}^T\,\underline{\boldsymbol{S}}\,\mathrm{d}V - \delta\underline{\boldsymbol{d}}^T \int_{\mathcal{B}^e} \mathbb{N}^T\,\underline{\boldsymbol{f}}\,\mathrm{d}V - \delta\underline{\boldsymbol{d}}^T \int_{\partial\mathcal{B}_t^e} \mathbb{N}^T\underline{\bar{\boldsymbol{t}}}\,\mathrm{d}A\,,\\ G_S^e &= \delta\underline{\boldsymbol{\beta}}^T \int_{\mathcal{B}^e} \mathbb{L}^T(\underline{\boldsymbol{E}} - \partial_S\chi(\underline{\boldsymbol{S}}))\,\mathrm{d}V\,. \end{aligned} \tag{30}$$

In general cases, where no complementary stored energy function is known, the partial derivative $\partial_S\chi(\underline{\boldsymbol{S}})$ has to be computed iteratively. The Green–Lagrange strain tensor is given by the approximation of the displacement field, whereas $\partial_S\chi(\underline{\boldsymbol{S}}) =: \underline{\boldsymbol{E}}^{cons}$ is implicitly given by the constitutive equation. Therefore, we compute in each integration point $\underline{\boldsymbol{E}}^{cons}$ by the evaluation of the residual

$$\boldsymbol{r}(\underline{\boldsymbol{E}}^{cons};\underline{\boldsymbol{S}}) = \underline{\boldsymbol{S}} - \partial_{\underline{\boldsymbol{E}}}\psi(\underline{\boldsymbol{E}})|_{\underline{\boldsymbol{E}}^{cons}} \approx \boldsymbol{0} \tag{31}$$

at fixed $\underline{\boldsymbol{S}}$. Thus we have to update

$$\underline{\boldsymbol{E}}^{cons} \Leftarrow \underline{\boldsymbol{E}}^{cons} + \underbrace{[\partial^2_{\underline{\boldsymbol{E}}\underline{\boldsymbol{E}}}\psi(\underline{\boldsymbol{E}})|_{\underline{\boldsymbol{E}}^{cons}}]^{-1}}_{=:\,\mathbb{D}}\,\boldsymbol{r}(\underline{\boldsymbol{E}}^{cons};\underline{\boldsymbol{S}}) \tag{32}$$

until $\|\boldsymbol{r}(\underline{\boldsymbol{E}}^{cons}; \underline{\boldsymbol{S}})\| \approx 0$. The linearization of the weak forms, $\text{Lin}G^e = G^e(\boldsymbol{d}, \boldsymbol{\beta}) + \Delta G^e(\Delta \boldsymbol{d}, \Delta \boldsymbol{\beta})$, yields the increments

$$\Delta G_u^e = \delta \underline{\boldsymbol{d}}^T \int_{\mathcal{B}^e} \underline{\Xi}\, \underline{\boldsymbol{S}}\, \mathrm{d}V\, \Delta \underline{\boldsymbol{d}} + \delta \underline{\boldsymbol{d}}^T \int_{\mathcal{B}^e} \underline{\mathbb{B}}^T \underline{\mathbb{L}}\, \mathrm{d}V\, \Delta \underline{\boldsymbol{\beta}}\,,$$
$$\Delta G_S^e = \delta \underline{\boldsymbol{\beta}}^T \int_{\mathcal{B}^e} \underline{\mathbb{L}}^T \underline{\mathbb{B}}\, \mathrm{d}V\, \Delta \underline{\boldsymbol{d}} - \delta \underline{\boldsymbol{\beta}}^T \int_{\mathcal{B}^e} \underline{\mathbb{L}}^T \underline{\mathbb{D}}\, \underline{\mathbb{L}}\, \mathrm{d}V\, \Delta \underline{\boldsymbol{\beta}}\,, \tag{33}$$

where $\underline{\Xi}$ is defined by $\Delta \underline{\boldsymbol{B}} = \underline{\Xi}\, \Delta \underline{\boldsymbol{d}}$. We introduce for convenience the element matrices and right-hand side vectors

$$\underline{\boldsymbol{K}}_{uu}^e := \int_{\mathcal{B}^e} \underline{\Xi}\, \underline{\boldsymbol{S}}\, \mathrm{d}V,\ \underline{\boldsymbol{K}}_{uS}^e := \int_{\mathcal{B}^e} \underline{\mathbb{L}}^T \underline{\mathbb{B}}\, \mathrm{d}V,\ \underline{\boldsymbol{K}}_{SS}^e := \int_{\mathcal{B}^e} \underline{\mathbb{L}}^T \underline{\mathbb{D}}\, \underline{\mathbb{L}}\, \mathrm{d}V\,,$$
$$\underline{\boldsymbol{r}}_u^e := \int_{\mathcal{B}^e} \underline{\mathbb{B}}^T \underline{\boldsymbol{S}}\, \mathrm{d}V - \int_{\mathcal{B}^e} \underline{\mathbb{N}}^T \underline{\boldsymbol{f}}\, \mathrm{d}V - \int_{\partial \mathcal{B}_t^e} \underline{\mathbb{N}}^T \underline{\bar{\boldsymbol{t}}}\, \mathrm{d}A \text{ and} \tag{34}$$
$$\underline{\boldsymbol{r}}_S^e := \int_{\mathcal{B}^e} \underline{\mathbb{L}}^T (\underline{\boldsymbol{E}} - \underline{\boldsymbol{E}}^{cons})\, \mathrm{d}V\,.$$

This yields the system of equations as

$$\text{Lin}G^e = \begin{bmatrix} \delta \underline{\boldsymbol{d}}^{eT} \\ \delta \underline{\boldsymbol{\beta}}^{eT} \end{bmatrix} \left(\begin{bmatrix} \underline{\boldsymbol{K}}_{uu}^e & \underline{\boldsymbol{K}}_{uS}^{e\ T} \\ \underline{\boldsymbol{K}}_{uS}^e & \underline{\boldsymbol{K}}_{SS}^e \end{bmatrix} \begin{bmatrix} \Delta \underline{\boldsymbol{d}} \\ \Delta \underline{\boldsymbol{\beta}} \end{bmatrix} + \begin{bmatrix} \underline{\boldsymbol{r}}_u^e \\ \underline{\boldsymbol{r}}_S^e \end{bmatrix} \right). \tag{35}$$

Table 2 sketches the nested algorithmic treatment for a typical element for the case that a complementary stored energy is not known.

Assembling over the number of elements num_{ele} leads to the global system of equations

$$\mathop{\mathsf{A}}_{e=1}^{num_{ele}} \begin{bmatrix} \delta \underline{\boldsymbol{d}}^{eT} \\ \delta \underline{\boldsymbol{\beta}}^{eT} \end{bmatrix} \left(\begin{bmatrix} \underline{\boldsymbol{K}}_{uu}^e & \underline{\boldsymbol{K}}_{uS}^{e\ T} \\ \underline{\boldsymbol{K}}_{uS}^e & \underline{\boldsymbol{K}}_{SS}^e \end{bmatrix} \begin{bmatrix} \Delta \underline{\boldsymbol{d}} \\ \Delta \underline{\boldsymbol{\beta}} \end{bmatrix} + \begin{bmatrix} \underline{\boldsymbol{r}}_u^e \\ \underline{\boldsymbol{r}}_S^e \end{bmatrix} \right)$$
$$= \delta \underline{\boldsymbol{D}} (\underline{\boldsymbol{K}}\, \Delta \underline{\boldsymbol{D}} + \underline{\boldsymbol{R}}) = 0 \tag{36}$$

and therefore the nodal unknowns are computed via

$$\Delta \underline{\boldsymbol{D}} = -\underline{\boldsymbol{K}}^{-1} \underline{\boldsymbol{R}}\,. \tag{37}$$

Due to the elementwise discontinuous interpolation of the stresses, the unknowns $\Delta \underline{\boldsymbol{\beta}}$ in (35) can already be eliminated on element level. This leads to a global system of equations with the same number of unknowns, and almost the same computational cost, as a displacement based trilinear element.

Table 2 Nested algorithmic treatment for a single element

ELEMENT LOOP

(1) Update displacements and stresses (Newton iteration k+1)

$\underline{d} = \underline{d}_n^{(k)} + \Delta\underline{d}$, $\underline{\beta} = \underline{\beta}_n^{(k)} + \Delta\underline{\beta}$

INTEGRATION LOOP

(2) Compute stresses $\boldsymbol{S}$ and Green-Lagrange strain tensor $\boldsymbol{E}$ at each Gauss Point:

$\underline{S} = \mathbb{L}\,\underline{\beta}$, $\underline{E} = \mathbb{B}\,\underline{d}$,

Read from history: $\underline{E}^{\text{cons}}$

CONSTITUTIVE LOOP

(3) Compute residuum: $\boldsymbol{r}(\underline{E}^{\text{cons}};\, \underline{S}) = \underline{S} - \partial_{\underline{E}}\psi(\underline{E})\Big|_{\underline{E}^{cons}}$

(4) Update: $\underline{E}^{\text{cons}} = \underline{E}^{\text{cons}} + \underline{\mathbb{D}} : \boldsymbol{r}(\underline{E}^{\text{cons}},\, \underline{S})$

with $\underline{\mathbb{D}} = \left(\partial^2_{\underline{E}\,\underline{E}}\psi(\underline{E})\Big|_{\underline{E}^{cons}}\right)^{-1}$

(5) Check convergence

IF $\|\boldsymbol{r}(\underline{E}^{\text{cons}};\, \underline{S})\| \leq tol$

THEN Update History $\underline{E}^{\text{cons}}$ and exit CONSTITUTIVE LOOP

(6) Check divergence

IF $n_{\text{iter}} > n_{\text{tol}}$ THEN Stop Calculation

(7) Determine and export element stiffness and rhs-vector

4 Remarks on EAS Methods for Finite Deformations

In the following numerical examples, the proposed assumed stress elements will be compared with the well known and established EAS element formulations. Since these elements are only considered for validation and comparison, the whole EAS framework will not be explained for the sake of brevity. The interested reader is referred to Simo et al. (1993) where the implementational aspects of the elements are discussed in detail.

The crucial step is the additive enhancement of the deformation gradient

$$\boldsymbol{F} = \boldsymbol{I} + \text{Grad}\,\boldsymbol{u} + \boldsymbol{F}_0\widetilde{\boldsymbol{F}} \tag{38}$$

with

$$F_0 = I + (\text{Grad}\, u)|_{\xi=0} \quad \text{and} \quad \widetilde{F} = \frac{\det J_0}{\det J} J_0^{-T} \widetilde{F}_\xi \, J_0^{-1}, \tag{39}$$

whereas the multiplication of the enhanced deformation gradient $\widetilde{F}$ by F_0 ensures objectivity of the formulation as discussed in Glaser and Armero (1997). It remains to choose a discretization of the enhanced part of the deformation gradient in the isoparametric space. In the literature several different discretizations are discussed and all of them might have their justification. However, in order to validate the assumed stress elements in the finite deformation range, we focus on the two elements which are equivalent to assumed stress discretizations in the linear elastic framework. This is on the one hand side the EAS formulation with 15 enhanced modes (Pantuso and Bathe (1995)) where the enhanced part is given by

$$\widetilde{F}_\xi = \begin{bmatrix} \beta_1\xi + \beta_{10}\xi\eta + \beta_{11}\xi\zeta & \beta_4\eta & \beta_6\zeta \\ \beta_7\xi & \beta_2\eta + \beta_{12}\eta\xi + \beta_{13}\eta\zeta & \beta_5\zeta \\ \beta_9\xi & \beta_8\eta & \beta_3\zeta + \beta_{14}\zeta\xi + \beta_{15}\zeta\eta \end{bmatrix}. \tag{40}$$

On the other hand, the EAS formulation with nine enhanced modes (Krischok and Linder (2016)) explicitly given by

$$\widetilde{F}_\xi = \begin{bmatrix} \beta_1\xi + \beta_4\xi\eta + \beta_5\xi\zeta & 0 & 0 \\ 0 & \beta_2\eta + \beta_6\eta\xi + \beta_7\eta\zeta & 0 \\ 0 & 0 & \beta_3\zeta + \beta_8\zeta\xi + \beta_9\zeta\eta \end{bmatrix}. \tag{41}$$

5 Numerical Examples

In the following, the proposed assumed stress finite elements will be validated in a couple of numerical examples in the hyperelastic framework. In particular, the reliability of the results will be verified. In addition, the performance regarding to the computational cost as well as the robustness regarding to the load step size and the maximal deformation is compared to well known and established finite element formulations.

The underlying strain energy is of Neo-Hookean type and given by

$$\psi = \frac{\mu}{2}(\text{tr}\, C - 3) + \frac{\lambda}{4}(J^2 - 1) - (\frac{\lambda}{2} + \mu)\ln(J), \tag{42}$$

where $\lambda = \frac{E\nu}{(1+\nu)(1-2\nu)}$ and $\mu = \frac{E}{2(1+\nu)}$ are the Lamé constants with E as the Young's modulus and ν as the Poisson's ratio.

5.1 Bending Plate

The superior performance of the elements of Pian and Tong or alternatively Pantuso and Bathe are well known and one of the main reasons for their success. The first numerical example considers bending dominated problem of a clamped plate with a moment at the free end. The geometry, material parameters and boundary conditions are depicted in Fig. 1. It should be noted, that the material is chosen to be nearly incompressible, in order to study the element behavior in the crucial regime.

Small Deformations First the equivalence of the assumed stress elements and their corresponding enhanced assumed strain elements is illustrated. Figure 2 depicts a study of convergence of the u_2-displacement over the number of elements. In order to obtain results in the small deformation framework, we considered the results after the first Newton iteration, using the elements for the finite deformation framework. This procedure ensures, that the comparison of the elements in the small and finite deformation range is based on the same implementation. It can be recognized that the result of the assumed stress element with 18 stress parameters (AS-18) is similar

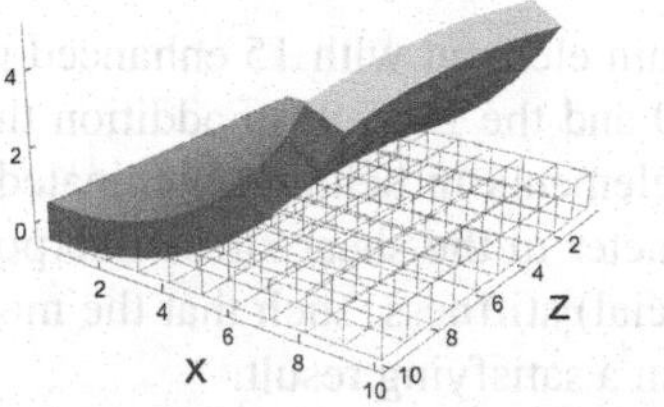

Material parameter:

$E = 200$
$\nu = 0.499$

Boundary Conditions:
$x = 0$:
$u_1 = 0$
$u_2 = 0$
$u_3 = 0$
$x = 10$:
$\bar{t}(y) = (-50y, 0, 0)^T$

Fig. 1 Bending Plate; An exemplary reference mesh depicted by the grid and the deformed body depicted by the solid figure

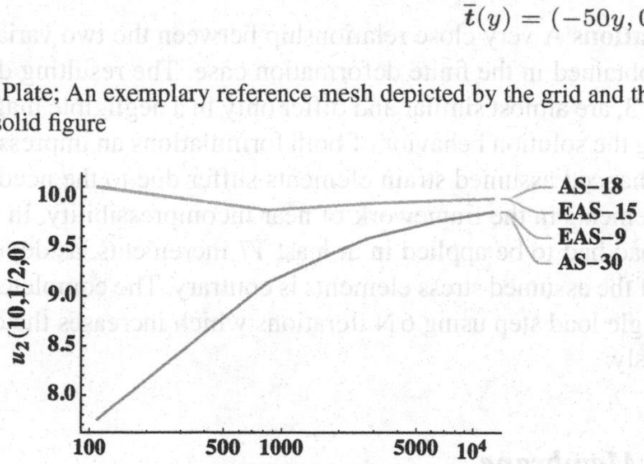

Fig. 2 Bending Plate, Small Deformations: Convergence study of the u_2-displacements over the number of elements. The equivalence of the Assumed Stress and the Enhanced Assumed Strain elements in the small deformation regime can be recognized

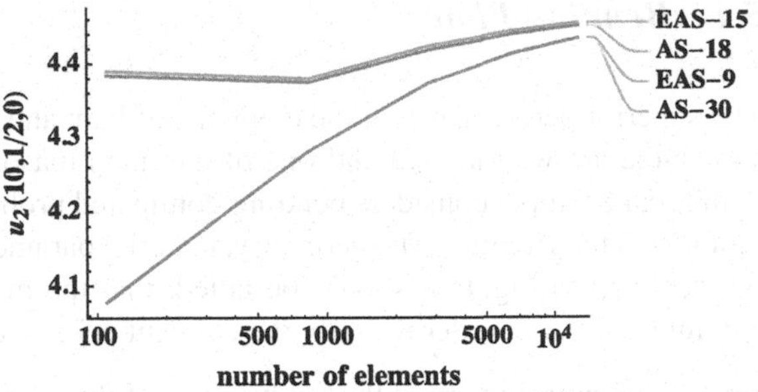

Fig. 3 Bending Plate, Finite Deformations: Convergence study of the u_2-displacements over the number of elements. The results of the assumed stress elements and their corresponding enhanced assumed strain counterparts are almost similar

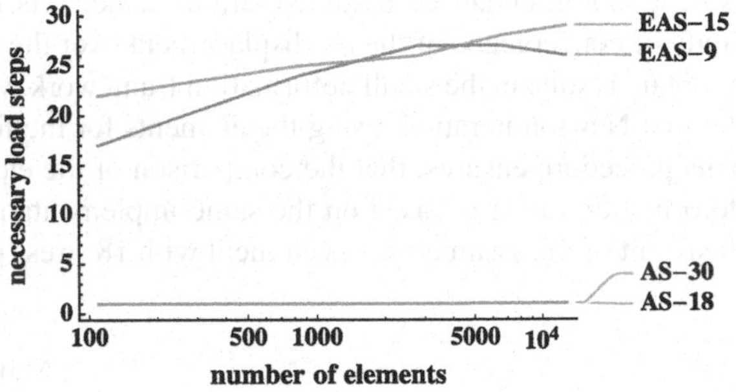

Fig. 4 Bending Plate, Finite Deformations: Convergence study of the number of necessary load steps over the number of elements. The assumed stress elements perform significantly better than their corresponding enhanced assumed strain counterparts

to the result of the enhanced assumed strain element with 15 enhanced parameters (EAS-15). The same holds for the AS-30 and the EAS-9. In addition the superior performance of the AS-18 and EAS-15 elements in bending dominated problems is illustrated. Due to the additional parameter in the shear stress interpolation, the AS-30 element contains additional (artificial) stiffness, such that the more degrees of freedom are necessary in order to obtain a satisfying result.

Finite Deformations A very close relationship between the two variational formulations is also obtained in the finite deformation case. The resulting displacements, depicted in Fig. 3, are almost similar and differ only in a negligible magnitude. However, comparing the solution behavior of both formulations an impressive difference is obtained. Enhanced assumed strain elements suffer due to the need for relatively small load increments in the framework of near incompressibility. In this particular example, the load had to be applied in at least 17 increments, as depicted in Fig. 4. The behavior of the assumed stress elements is contrary. The complete load could be applied in a single load step using 6 N iterations which increases the computational effort enormously.

5.2 *Cook's Membrane*

The classical benchmark problem of the Cook's membrane is considered as a numerical example. Figure 5 depicts the geometry, material parameter and boundary conditions of the specific problem.

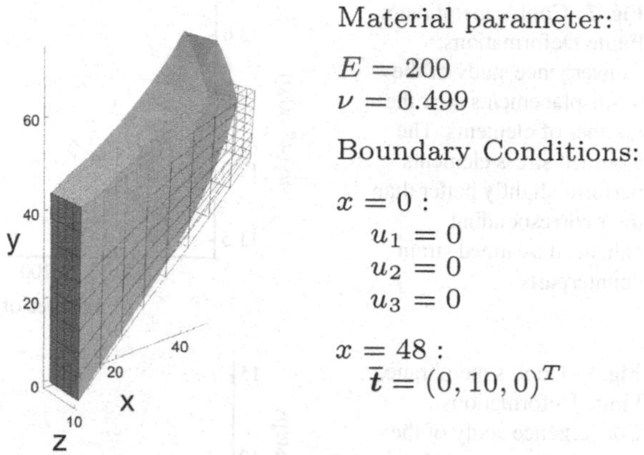

Fig. 5 Cook's membrane: An exemplary reference mesh depicted by the grid and the deformed body depicted by the solid figure

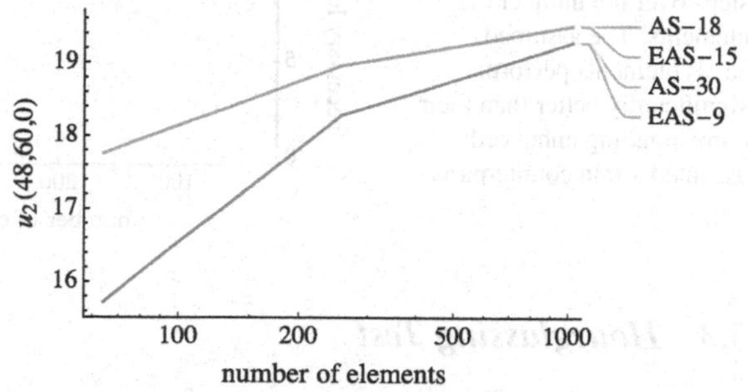

Fig. 6 Cook's membrane, Small Deformations: Convergence study of the u_2-displacements over the number of elements. The equivalence of the assumed stress and the enhanced assumed strain elements in the small deformation regime can be recognized

Small Deformations First, the equivalence of the assumed stress elements and their corresponding enhanced assumed strain element counterparts is illustrated again. Figure 6 depicts a study of convergence of the u_2-displacement over the number of elements. It should be noted, that for this numerical example, the elements are not equal from the analytical point of view, since the elements are not parallelogram shaped, which is a necessary condition for the analytical equivalence. Nonetheless, the difference between the connected elements is negligible.

Finite Deformations Considering the same setup in the framework of finite deformations yield the displacement convergence as depicted in Fig. 7. It can be noticed that, qualitatively the assumed stress elements perform slightly better than their enhanced assumed strain counterparts, regarding to the displacement convergence. In addition to that slight benefit of the proposed assumed stress elements, they show again a significant better behavior regarding to the load step size. The necessary load steps for the considered numerical example are plotted over the number of elements in Fig. 8.

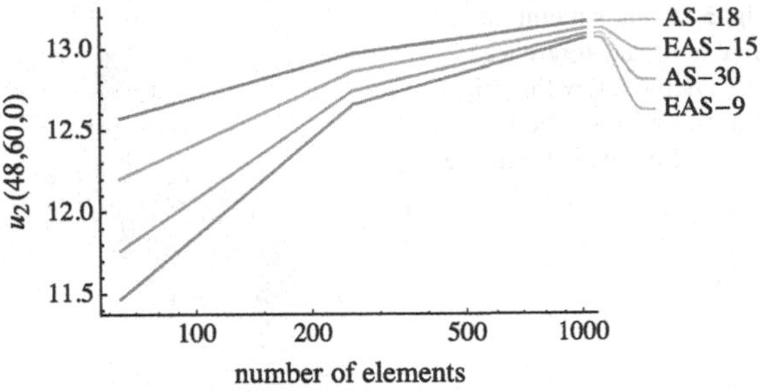

Fig. 7 Cook's membrane, Finite Deformations: Convergence study of the u_2-displacements over the number of elements. The assumed stress elements perform slightly better than their corresponding enhanced assumed strain counterparts

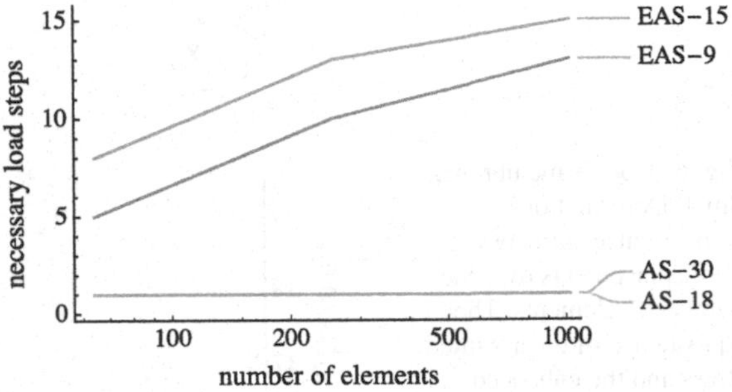

Fig. 8 Cook's membrane, Finite Deformations: Convergence study of the number of necessary load steps over the number of elements. The assumed stress elements perform significantly better than their corresponding enhanced assumed strain counterparts

5.3 Hourglassing Test

Nonphysical instabilities have been detected in compression tests for a variety of enhanced assumed strain elements in the nonlinear regime, see, e.g., Wriggers and Reese (1996). The investigation of these unphysical free energy modes have been subject to many following publications, see, e.g., de Souza Neto et al. (1995) and Wall et al. (2000). These artificial modes are known in the literature as hourglassing modes and constitute a nonunique displacement field for the corresponding state of equilibrium. A classical numerical example in order to depict potential hourglass instabilities is the displacement driven compression of a simple supported cube. The geometry and boundary conditions are depicted in Fig. 9. Investigated is the evolution of the smallest eigenvalue of the stiffness matrix over the applied displacement until a critical load level is achieved. Figure 10 depicts the this study for the case of 8^3 elements. It can be recognized, that the behavior for AS-30 and EAS-9 is similar, detecting a critical loading for an applied displacement at the top of $\overline{u} = -22.3$. On the other side, the AS-18 and EAS-15 depict qualitatively the same evolution of the smallest eigenvalue. The critical loading is distinctly away from $\overline{u} = -22.3$ and the value of the smallest eigenvalue is oscillating prior to the critical loading. An investigation of the eigenmode corresponding to the smallest eigenvalue of the last load step prior to the critical load step, is depicted in Fig. 11. Whereas the AS-30 and EAS-9 depict the physically reasonable buckling mode a nonphysical hourglass mode is captured in case of the AS-18 and the EAS-15.

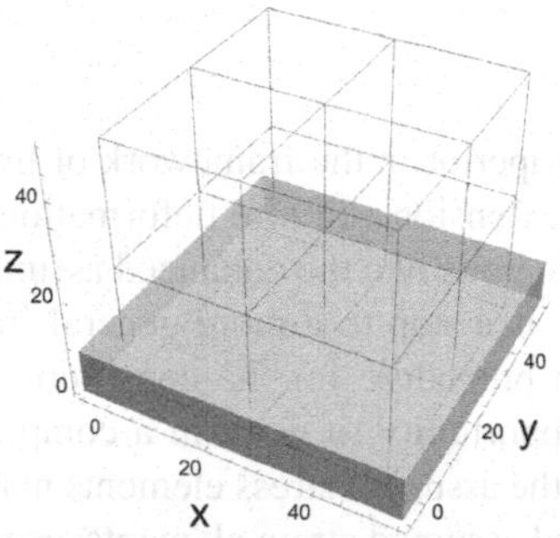

Boundary Conditions:

$x = 25 \wedge z = 0:$
$u_1 = 0$

$x = 25 \wedge y = 25 \wedge z = 25:$
$u_2 = 0$

$z = 0:$
$u_3 = 0 \; z = 50:$
$u_3 = \overline{u}$

Fig. 9 Hourglassing test: Reference geometry depicted by the grid, deformed body depicted by the solid figure. A prescribed displacement $\overline{u}$ is applied at the top

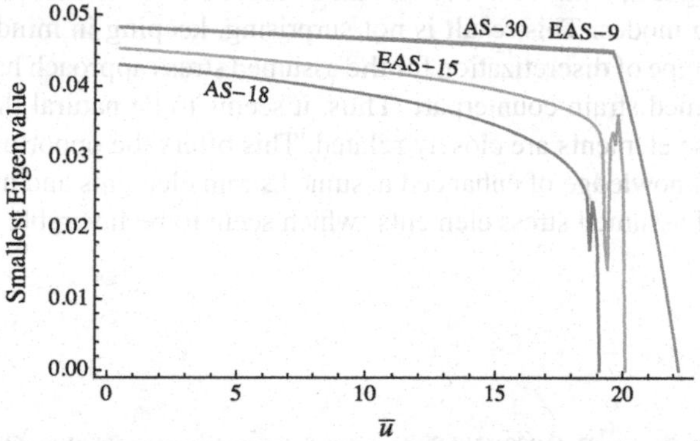

Fig. 10 Hourglassing test: Convergence study of the smallest eigenvalue over the magnitude of applied displacement at the top edge

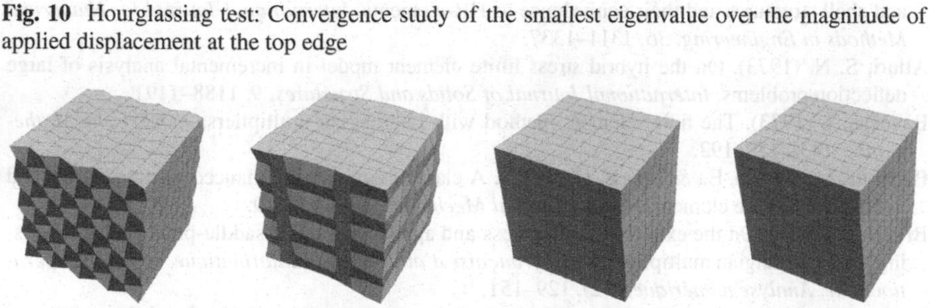

Fig. 11 Hourglassing test: Corresponding Eigenmode at critical loading (corresponding smallest eigenvalue < 0.002). The AS-18 and EAS-15 depict unphysical hourglassing mode, whereas AS-30 and EAS-9 do not suffer due to hourglassing in this numerical example

6 Conclusion

Assumed stress elements are known to be superior in the framework of linear elasticity. However, their non-straightforward extension to finite deformations led the ensuing research preferring different formulations like the enhanced assumed strain approach. The proposed work can be seen as a step to a more general framework of assumed stresses. The iterative solution procedure for the constitutive relation between stresses and strains, admits the opportunity to work in a complementary setting. The numerical examples show that the assumed stress elements may outperform the common and widely used enhanced assumed strain elements, especially in the framework of near incompressibility. The main advantage is constituted by the large deformations which can be applied per load step. In the numerical example of the hourglassing test, it has been shown that assumed stress elements may suffer due to hourglassing modes. This result is not surprising, keeping in mind that in linear elasticity each type of discretization for the assumed stress approach has a equivalent enhanced assumed strain counterpart. Thus, it seems to be natural that the characteristics of these elements are closely related. This offers the opportunity to use the widely gained knowledge of enhanced assumed strain elements and utilize it for the construction of assumed stress elements, which seem to be favorable.

References

Andelfinger, U., & Ramm, E. (1993). EAS-elements for two-dimensional, three-dimensional, plate and shell structures and their equivalence to HR-elements. *International Journal for Numerical Methods in Engineering*, *36*, 1311–1337.

Atluri, S. N. (1973). On the hybrid stress finite element model in incremental analysis of large deflection problems. *International Journal of Solids and Structures*, *9*, 1188–1191.

Babuška, I. (1973). The finite element method with Lagrangian multipliers. *Numerische Mathematik*, *20*(3), 179–192.

Bischoff, M., Ramm, E., & Braess, D. (1999). A class of equivalent enhanced assumed strain and hybrid stress finite elements. *Computational Mechanics*, *22*, 443–449.

Brezzi, F. (1974). On the existence, uniqueness and approximation of saddle-point problems arising from Lagrangian multipliers. *Revue française d'automatique, informatique, recherche opérationnelle. Analyse numérique*, *8*(2), 129–151.

de Souza Neto, E. A., Peric, D., Huang, G. C., & Owen, D. R. J. (1995). Remarks on the stability of enhanced strain elements in finite elasticity and elastoplasticity. *Communications in Numerical Methods in Engineering*, *11*(11), 951–961.

Djoko, J. K., Lamichhane, B. P., Reddy, B. P., & Wohlmuth, B. I. (2006). Conditions for equivalence between the Hu-Washizu and related formulations, and computational behavior in the incompressible limit. *Computer Methods in Applied Mechanics and Engineering*, *195*, 4161–4178.

Fraejis de Veubeke, B. (1965). *Stress analysis*, chapter Displacement ant equilibrium models in finite element methods (pp. 145–197). John Wiley & Sons.

Glaser, S., & Armero, F. (1997). On the formulation of enhanced strain finite elements in finite deformations. *Engineering Computations*, *14*, 759–791.

Hellinger, E. (1913). Encyklopädie der mathematischen wissenschaften mit einschluss ihrer anwendungen. In *Encyklopdie der mathematischen Wissenschaften mit Einschluss ihrer Anwendungen, 4*.
Hu, H. C. (1955). On some variational principles in the theory of elasticity and the theory of plasticity. *Science Sinica, 4*, 33–54.
Krischok, A., & Linder, C. (2016). On the enhancement of low-order mixed finite element methods for the large deformation analysis of diffusion in solids. *International Journal for Numerical Methods in Engineering, 106*, 278–297.
Ladyzhenskaya, O. (1969). *The mathematical theory of viscous incompressible flow* (Vol. 76). Gordon and Breach New York.
Ogden, R. W. (1984). *Non-linear elastic deformations*. Dover Publications.
Pantuso, D., & Bathe, K.-J. (1995). A four-node quadrilateral mixed-interpolated element for solids and fluids. *Mathematical Models and Methods in Applied Sciences (M3AS), 5*(8), 1113–1128.
Pian, T. H. H. (1964). Derivation of element stiffness matrices by assumed stress distribution. *AIAA Journal, 20*, 1333–1336.
Pian, T. H. H., & Chen, D.-P. (1982). Alternative ways for formulation of hybrid stress elements. *International Journal for Numerical Methods in Engineering, 18*, 1679–1684.
Pian, T. H. H., & Sumihara, K. (1984). A rational approach for assumed stress finite elements. *International Journal for Numerical Methods in Engineering, 20*, 1685–1695.
Pian, T. H. H., & Tong, P. (1986). Relations between incompatible displacement model and hybrid stress model. *International Journal for Numerical Methods in Engineering, 22*, 173–181.
Prange, G. (1916). *Das Extremum der Formänderungsarbeit*. Technische Hochschule Hannover: Habilitationsschrift.
Reissner, E. (1950). On a variational theorem in elasticity. *Journal of Mathematical Physics, 29*, 90–95.
Schröder, J., Klaas, O., Stein, E., & Miehe, C. (1997). A physically nonlinear dual mixed finite element formulation. *Computer Methods in Applied Mechanics and Engineering, 144*, 77–92.
Schröder, J., Igelbüscher, M., Schwarz, A., & Starke, G. (2017). A Prange-Hellinger-Reissner type finite element formulation for small strain elasto-plasticity. *Computer Methods in Applied Mechanics and Engineering, 317*, 400–418.
Simo, J. C., & Rifai, M. S. (1990). A class of mixed assumed strain methods and the method of incompatible modes. *International Journal for Numerical Methods in Engineering, 29*, 1595–1638.
Simo, J. C., Armero, F., & Taylor, R. L. (1993). Assumed enhanced strain tri-linear elements for 3D finite deformation problems. *Computer Methods in Applied Mechanics and Engineering, 110*, 359–386.
Simo, J. C., Kennedy, J. G., & Taylor, R. L. (1989). Complementary mixed finite element formulations for elastoplasticity. *Computer Methods in Applied Mechanics and Engineering, 74*, 177–206.
Viebahn, N., Schröder, J., & Wriggers, P. (2019). An extension of assumed stress finite elements to a general hyperelastic framework. *Advanced Modeling and Simulation in Engineering Sciences, 6*, 9. https://doi.org/10.1186/s40323-org-0133-7.
Wall, W. A., Bischoff, M., & Ramm, E. (2000). A deformation dependent stabilization technique, exemplified by eas elements at large strains. *Computer Methods in Applied Mechanics and Engineering, 188*, 859–871.
Washizu, K. (1955, March). On the variational principles of elasticity and plasticity. Technical Report 25–18, Aeroelastic and Structures Research Laboratory, Massachusetts Institute of Technology, Cambridge.
Wilson, E. L., Taylor, R. L., Doherty, W. P., & Ghabaussi, J. (1973). Incompatible displacement models. In S. J. Fenves et al. (Eds.), *Numerical and computer methods in structural mechanics* (pp. 43–57). New York: Academic Press

Wriggers, P. (2009). Mixed finite element methods - theory and discretization. In C. Carstensen, & P. Wriggers (Eds.), *Mixed finite element technologies*, volume 509 of *CISM International Centre for Mechanical Sciences*. Springer.
Wriggers, P., & Reese, S. (1996). A note on enhanced strain methods for large deformations. *Computer Methods in Applied Mechanics and Engineering*, *135*, 201–209.
Yeo, S. T., & Lee, B. C. (1996). Equivalence between enhanced assumed strain method and assumed stress hybrid method based on the Hellinger-Reissner principle. *International Journal for Numerical Methods in Engineering*, *39*, 3083–3099.
Zienkiewicz, O. C., Qu, S., Taylor, R. L., & Nakazawa, S. (1986). The patch test for mixed formulations. *International Journal for Numerical Methods in Engineering*, *23*, 1873–1883.

A Fully Nonlinear Beam Model of Bernoulli–Euler Type

Paulo de Mattos Pimenta, Sascha Maassen, Cátia da Costa e Silva and Jörg Schröder

Abstract This work presents a geometrically exact Bernoulli–Euler rod model. In contrast to Pimenta (1993b), Pimenta and Yojo (1993), Pimenta (1996), Pimenta and Campello (2001), where the hypothesis considered was Timoshenko's, this approach is based on the Bernoulli–Euler theory for rods, so that transversal shear deformation is not accounted for. Energetically conjugated cross-sectional stresses and strains are defined. The fact that both the first Piola–Kirchhoff stress tensor and the deformation gradient appear again as primary variables is also appealing. A straight reference configuration is assumed for the rod, but, in the same way, as in Pimenta (1996), Pimenta and Campello (2009), initially curved rods can be accomplished, if one regards the initial configuration as a stress-free deformed state from the straight position. Consequently, the use of convective non-Cartesian coordinate systems is not necessary, and only components on orthogonal frames are employed. A cross section is considered to undergo a rigid body motion and parameterization of the rotation field is done by the rotation tensor with the Rodrigues formula that makes the updating of the rotational variables very simple. This parametrization can be seen in Pimenta et al. (2008), Campello et al. (2011). A simple formula for the incremental Rodrigues parameters in function of the displacements derivative and the torsion angle is also settled down. A 2-node finite element with Cubic Hermitian interpolation for the displacements, together with a linear approximation for the torsion angle, is displayed within the usual Finite Element Method, leading to adequate C_1 continuity.

P. de Mattos Pimenta (✉) · C. da Costa e Silva
Polytechnic School, University of São Paulo, São Paulo, Brazil
e-mail: ppimenta@usp.br

S. Maassen · J. Schröder
University of Duisburg-Essen, Essen, Germany
e-mail: j.schroeder@uni-due.de

J. Schröder and P. de Mattos Pimenta (eds.), *Novel Finite Element Technologies for Solids and Structures*, CISM International Centre for Mechanical Sciences 597,
https://doi.org/10.1007/978-3-030-33520-5_5

1 Introduction

The first objective of this work is to present a geometrically exact Bernoulli–Euler rod formulation and its finite element implementation. The class of admissible motions, that follows from this assumption, is obtained by imposing that the cross sections of the rod, that are initially orthogonal to the chosen axis, remain rigid and orthogonal to it after deformation. Thus, the transversal shear deformation is not accounted for. This theory is called geometrically exact because no approximation is employed after the basic kinematical assumption made. Displacements and rotations can be unlimited large. The Bernoulli–Euler formulation for rods is analogous to the Kirchhoff–Love's for shells presented in Viebahn et al. (2016), Pimenta et al. (2010).

As framework one uses the theory presented in Pimenta (1993b), Pimenta and Yojo (1993), which is now constrained to obey the Bernoulli–Euler assumption. This approach defines energetically conjugated generalized cross-sectional stress and strains. Besides their practical importance, cross-sectional quantities make the derivation of equilibrium equations easier, as well as the achievement of the corresponding tangent bilinear form, which is always symmetric for hyper-elastic materials and conservative loadings, even far from an equilibrium state.

The models are implemented using the finite element method with cubic Hermitian polynomial interpolation on the displacements and linear Lagrangian interpolation for the considered torsion degree of freedom. Usually in shear deformable rod theories, one needs to worry about shear-locking. With Bernoulli–Euler assumption, the shear deformation is not accounted for in the initial kinematics, therefore, there is no shear-locking. So, there is no need for reduced numerical integration or any other techniques to bypass this problem. Since only initially straight elements are considered here, membrane locking is not an issue too. Linear elastic constitutive equations for small strains are considered in the numerical examples of this paper. A forthcoming paper will address the issue of finite strain elastic and elastic–plastic constitutive equations.

As mentioned before, a straight reference configuration is assumed for the rod. Initially, curved rods can then be regarded as a stress-free deformation from this configuration. This approach was already employed for rods and shells in Pimenta (1996), Pimenta and Campello (2009). It precludes the use of convective non-Cartesian coordinate systems and other complicate entities like Christoffel symbols and fundamental forms. It simplifies, as well, the comprehension of tensor quantities, since only components on orthogonal systems are employed.

Throughout the text, italic Greek or Latin lowercase letters $(a, b, \ldots, \alpha, \beta, \ldots)$ denote scalars, bold italic Greek or Latin lowercase letters $(\boldsymbol{a}, \boldsymbol{b}, \ldots, \boldsymbol{\alpha}, \boldsymbol{\beta}, \ldots)$ denote vectors and bold italic Greek or Latin capital letters $(\boldsymbol{A}, \boldsymbol{B}, \ldots)$ denote second-order tensors in a three-dimensional Euclidean space. Summation convention over repeated indices is adopted in the entire text, whereby Greek indices range from 1 to 2, while Latin indices range from 1 to 3. $\|\boldsymbol{v}\| = \sqrt{\boldsymbol{v} \cdot \boldsymbol{v}}$ is the is the norm of vector $\boldsymbol{v}$, where $\cdot$ denotes de scalar product of two vectors. The operator $\otimes$ denotes the dyadic or tensor product of two vectors. For instance, $\boldsymbol{a} \otimes \boldsymbol{b}$ is a second-order tensor such that $(\boldsymbol{a} \otimes \boldsymbol{b})\boldsymbol{c} = (\boldsymbol{b} \cdot \boldsymbol{c})\boldsymbol{a}$. Note that $(\boldsymbol{a} \otimes \boldsymbol{b})^T = (\boldsymbol{b} \otimes \boldsymbol{a})$, where $(\bullet)^T$ denotes the transpose. The operator axial $(\bullet)$ is such that, if $\boldsymbol{v} = \text{axial}\ (\boldsymbol{V})$, with $\boldsymbol{V}$ skew-symmetric, then $\boldsymbol{V}\boldsymbol{x} = \boldsymbol{v} \times \boldsymbol{x}, \forall \boldsymbol{x}$, where $\times$ denotes the cross product of two vectors. If $\boldsymbol{v} = \text{axial}\ (\boldsymbol{V})$, then $\boldsymbol{V} = \text{Skew}\ (\boldsymbol{v})$, with $\boldsymbol{V}$ skew-symmetric.

Rod models are of great interest in structural mechanics and flexible multibody systems. The first works on bending problems date back to Bernoulli investigating deflections of beams and Euler published the first systematic treatment of elastic curves. A full history can be seen in Timoshenko (1953). After these first discoveries, many authors wrote about rods. Until the 60s, most of these works were restricted to linear kinematics.

With the advent of computers, nonlinear problems started to be addressed. First as plane problems Reissner (1972, 1973) and then as three-dimensional ones Antman (1974), Whirman and De Silva (1974), Argyris (1982). The first geometrically exact problems in three-dimensional space were addressed by Simo (1985), resulting in a nonsymmetric tangent matrix far from the equilibrium state. Many authors solved geometrically exact rod problems based on this work, as to name just a few Simo and Vu-Quoc (1986, 1991), Simo (1992), Simo et al. (1992).

As Campello (2000) pointed out, it is evident that these early theories did not have rigor and precision in their conceptualizations, mainly because they are derived from simplifications imposed in the theories of three-dimensional solids.

Pimenta (1993b), Pimenta and Yojo (1993) presented a geometrically exact rod theory in three-dimensional space with the Fréchet derivative of the weak form of the equilibrium being exact and the rotations in three-dimensional space treated in a consistent and convenient way through the Euler–Rodrigues formula. Many authors extended these geometrically exact rod models to incorporate general cross-sectional in-plane changes and out-of-plane warping Pimenta and Campello (2003), distortion of the cross section Sokolov et al. (2015).

All those geometrically exact rod models are not constrained to obey the Bernoulli–Euler assumption as it is done herein. This class of rod models has drawn some attention in the last few years. Boyer and Primault (2004) present a geometrically exact nonlinear Euler–Bernoulli model for the special case of beams with circular cross sections and a straight initial configuration, in Boyer et al. (2011) the same theory is applied to cable dynamics. In the present approach, one can have arbitrary cross sections, the initial configuration is also straight, initially curved rods could be accomplished in the same it way as in Pimenta (1996, Pimenta and Campello 2009), if one regards the initial configuration as a stress-free deformed state from the plane position, this will be subject of future work. Bauer et al. (2016) extend Boyer and Primault (2004) into a nonlinear isogeometric spatial Bernoulli–Euler rod theory that is treated spatially curved and a rotation around the centerline of the rod is adopted as a degree of freedom, that also differs from this work, as it can be seen later. Greco and Cuomo (2013, 2016) have made some advances in nonlinear Bernoulli–Euler rod theory. They use an isogeometric approach. Meier et al. (2014, 2017) has a similar approach for the geometrically exact Bernoulli–Euler rod theory, in terms of initial kinematics configuration, but presents different parameterizations for the rotation. He also indicates two portions of motion on the beam axes, and 4 degrees of freedom, but they connect the elements through as usual in the finite element method, which imposes a continuous rotational degree of freedom. This cannot be true in many examples and is not consistent with the theory. Meier et al. (2016) extend Meier et al. (2014), a geometrically exact beam theory was developed considering discrete Bernoulli hypothesis of rigid cross sections that remain orthogonal to the chosen axis during deformation. Meier et al. (2016) focus on the development of finite element formulations that are capable of accurately modeling the dynamics of slender components and their contact interaction with circular cross sections. All the papers referred to above describes the rotation in a different way we do in this work.

Bernoulli–Euler theory can be widely applied to engineering problems. It can be used in the aerospace industry, oil drilling rods, robot arms and for rib-reinforced shells that are common in aerospace, naval and automobile industry. The hypothesis can be used whenever the rods are slender.

It is proposed a novel interpolation scheme for the rotation field representing the cross-sectional orientation, which is based on Rodrigues parameters and obeys the Bernoulli–Euler constraint. This formulation has continuous displacement degrees of freedom and can have discontinuous degrees of freedom for the derivatives of the displacements and the rotation. The connection between elements is enforced by the de Rodrigues parameter for the rotation being equal on both connecting ends. This is an advantage because one can address sudden changes of cross section or material along the rod, an example that is shown later in Sect. 6. And, there is the opportunity, in general, for the rod element to be used together with a Kirchhoff–Love shell element.

2 Geometrically Exact Bernoulli–Euler Rod Theory

2.1 *Kinematics*

It is assumed at the outset that the rod is straight at the initial configuration, which is used as a reference. This formulation can be directly used for straight finite elements. The case of initially curved rods, which can be used for initially curved finite elements, can be treated as in Pimenta (1996) and is subject to future work. Let $\{\boldsymbol{e}_1^r, \boldsymbol{e}_2^r, \boldsymbol{e}_3^r\}$ be an orthogonal system placed at the reference (or initial) configuration of the rod. The vectors $\boldsymbol{e}_\alpha^r$, $\alpha = 1, 2$, are placed on the cross sections of the rod, which are orthogonal to the axis at that configuration. Thus, $\boldsymbol{e}_3^r$ is orthogonal to this plane and tangent to the rod axis.

The position of the rod material points in the reference configuration can be described by

$$\boldsymbol{\xi} = \boldsymbol{\zeta} + \boldsymbol{r}^r, \tag{1}$$

where the vector

$$\boldsymbol{\zeta} = \zeta \boldsymbol{e}_3^r, \tag{2}$$

describes the rod axis at reference configuration and $\boldsymbol{r}^r$ is the director given by

$$\boldsymbol{r}^r = \xi_\alpha \boldsymbol{e}_\alpha^r. \tag{3}$$

One introduces the axial coordinate $\zeta = (\boldsymbol{\zeta} - \boldsymbol{\zeta}_0) \cdot \boldsymbol{e}_3^r$, $\zeta \in \Omega = (0, \ell)$, where ℓ is the rod length at reference configuration and $\boldsymbol{\zeta}_0$ is the position of the axis for $\zeta = 0$. The boundary of the domain Ω is denoted by Γ. Herein, Γ contains the two ends of the rod, i.e., $\Gamma = \{0, \ell\}$. $A \subset \mathbb{R}^2$ is the cross-sectional domain at the reference configuration. The contour of A is denoted by C. Coordinates $\xi_\alpha = \boldsymbol{r}^r \cdot \boldsymbol{e}_\alpha^r$ are such that $\{\xi_1,\ \xi_2\} \in A$. Thus, $\xi_1,\ \xi_2$ and ζ build a Cartesian coordinate system.

At the current configuration, according to Fig. 1, the position of the material points is given by

$$\boldsymbol{x} = \boldsymbol{z} + \boldsymbol{r}, \tag{4}$$

where $\boldsymbol{z} = \hat{\boldsymbol{z}}(\zeta)$ describes the position of the rod axis at the current configuration and $\boldsymbol{r}$ is the current director given by

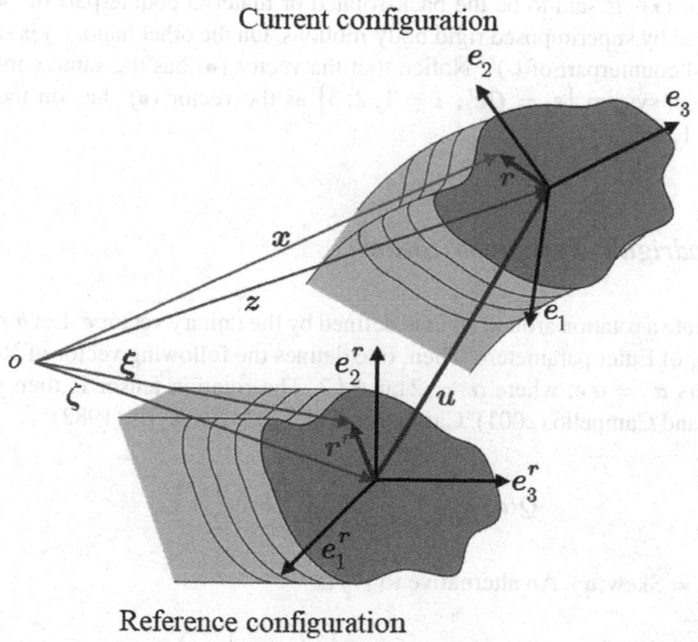

Fig. 1 Rod description and basic kinematical quantities

$$r = Qr^r, \tag{5}$$

where $Q = \hat{Q}(\zeta)$ is the cross-sectional rotation tensor.

The Bernoulli–Euler assumption states that the plane cross sections are subjected to a rigid body motion and remain orthogonal to the axis. After deformation the triad $\{e_1^r, e_2^r, e_3^r\}$ is transformed to $\{e_1, e_2, e_3\}$ at current configuration. e_3 is orthogonal to the cross sections and tangent to rod axis, while $e_\alpha, \alpha = 1, 2$ remain attached to the cross sections. The axis of the rod at current configuration is defined by the axis placement. The vector e_3 is defined by the axis as well, but the unitary vectors $e_\alpha,\ \alpha = 1, 2$ the cross sections are not. They can be rotated around the rod axis and need an additional parameter, which is called herein torsion parameter. It can also be regarded as *a rotation around a moving axis*. We denote this scalar by φ. It follows that the rotation tensor can be expressed by

$$Q = \hat{Q}(e_3, \varphi). \tag{6}$$

Note that no cross-sectional change is assumed. A general Bernoulli–Euler theory that incorporates cross section in-plane and out-of-plane changes will be presented in a coming work under preparation.

Remark 1: Back-Rotated Or Material Vectors
The following notation for vectors in $\mathbb{R}^3$ is used, $(\bullet) = Q(\bullet)^r \Leftrightarrow (\bullet)^r = Q^T(\bullet)$. The vector $(\bullet)^r$ is said to be the back-rotated or material counterpart of $(\bullet)$ and is not affected by superimposed rigid body motions. On the other hand, $(\cdot)$ is said to be the spatial counterpart of $(\cdot)^r$. Notice that the vector $(\bullet)$ has the same components on the local system $\{e_i = Qe_i^r,\ i = 1, 2, 3\}$ as the vector $(\bullet)^r$ has on the system $\{e_i^r,\ i = 1, 2, 3\}$.

2.2 *Rodrigues Parameterization*

Let θ denote a rotation around an axis defined by the unitary vector e. Let $\boldsymbol{\theta}$ represent the vector of Euler parameters. Then, one defines the following vector of Rodrigues parameters $\boldsymbol{\alpha} = \alpha\, e$, where $\alpha = 2 \tan \theta / 2$. The rotation tensor is then given by Pimenta and Campello (2001), Campello et al. (2003), Argyris (1982)

$$\hat{Q}(\boldsymbol{\alpha}) = \left(I - \frac{1}{2}A\right)^{-1}\left(I + \frac{1}{2}A\right), \tag{7}$$

where $A = \mathrm{Skew}(\boldsymbol{\alpha})$. An alternative to (7) is

$$\hat{Q}(\boldsymbol{\alpha}) = I + \frac{4}{4 + \alpha^2}\left(A + \frac{1}{2}A^2\right), \tag{8}$$

where $\alpha^2 = \boldsymbol{\alpha} \cdot \boldsymbol{\alpha}$.

For the spin vector

$$\boldsymbol{\omega} = \text{axial}(\boldsymbol{\Omega}), \quad \text{where} \quad \boldsymbol{\Omega} = \dot{\boldsymbol{Q}}\boldsymbol{Q}^T, \tag{9}$$

the following relation holds

$$\boldsymbol{\omega} = \boldsymbol{\Xi}\dot{\boldsymbol{\alpha}}, \quad \text{where} \quad \boldsymbol{\Xi} = \frac{4}{4+\alpha^2}\left(\boldsymbol{I} + \frac{1}{2}\boldsymbol{A}\right), \tag{10}$$

which has been derived for the first time in Pimenta and Campello (2001).

2.3 *Incremental Description of the Rotation*

The use of Rodrigues parameters is restricted to $-\pi < \theta < \pi$. To overcome this drawback, we describe the rotation by the incremental approach, as in Pimenta et al. (2008). This limitation is then restricted to a load increment in Statics or to a time increment in Dynamics.

Let $(\cdot)_i$ and $(\cdot)_{i+1}$ denote a quantity $(\cdot)$ at instants t_i and t_{i+1}, respectively. And let $(\cdot)_\Delta$ be an incremental quantity. Thus, one gets for the rotation tensor the following relations

$$\begin{aligned} &\boldsymbol{Q}_{i+1} = \boldsymbol{Q}_\Delta \boldsymbol{Q}_i, \text{ where} \\ &\boldsymbol{Q}_{i+1} = \hat{\boldsymbol{Q}}(\boldsymbol{\alpha}_{i+1}), \; \boldsymbol{Q}_\Delta = \hat{\boldsymbol{Q}}(\boldsymbol{\alpha}_\Delta) \quad \text{and} \quad \boldsymbol{Q}_i = \hat{\boldsymbol{Q}}(\boldsymbol{\alpha}_i). \end{aligned} \tag{11}$$

We recall the following result by Rodrigues, which is probably the most relevant result by him,

$$\boldsymbol{\alpha}_{i+1} = \frac{4}{4 - \boldsymbol{\alpha}_i \cdot \boldsymbol{\alpha}_\Delta}\left(\boldsymbol{\alpha}_i + \boldsymbol{\alpha}_\Delta - \frac{1}{2}\boldsymbol{\alpha}_i \times \boldsymbol{\alpha}_\Delta\right). \tag{12}$$

In the incremental description one has for the spin vector

$$\boldsymbol{\omega} = \text{axial}(\boldsymbol{\Omega}), \quad \text{where} \quad \boldsymbol{\Omega} = \dot{\boldsymbol{Q}}_\Delta \boldsymbol{Q}_\Delta^T, \tag{13}$$

the following relation

$$\boldsymbol{\omega} = \boldsymbol{\Xi}_\Delta \dot{\boldsymbol{\alpha}}_\Delta, \quad \text{where} \quad \boldsymbol{\Xi}_\Delta = \frac{4}{4+\alpha_\Delta^2}\left(\boldsymbol{I} + \frac{1}{2}\boldsymbol{A}_\Delta\right), \tag{14}$$

where $\boldsymbol{A}_\Delta = \text{Skew}(\boldsymbol{\alpha}_\Delta)$ and $\alpha_\Delta^2 = \boldsymbol{\alpha}_\Delta \cdot \boldsymbol{\alpha}_\Delta$.

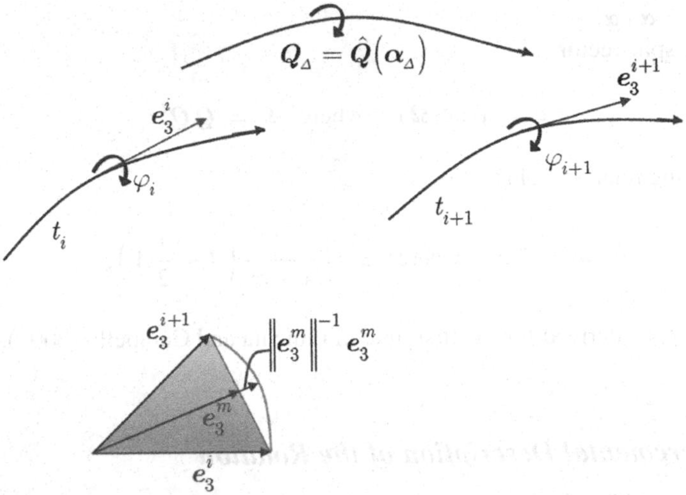

Fig. 2 Description of the incremental motion of the vector that describes de axis of the rod

At instants t_i and t_{i+1} triad $\{\boldsymbol{e}_1, \boldsymbol{e}_2, \boldsymbol{e}_3\}$ is denoted by $\{\boldsymbol{e}_1^i, \boldsymbol{e}_2^i, \boldsymbol{e}_3^i\}$ and $\{\boldsymbol{e}_1^{i+1}, \boldsymbol{e}_2^{i+1}, \boldsymbol{e}_3^{i+1}\}$, respectively. We denote the incremental torsion parameter by φ_Δ. This is schematically shown in Fig. 2. From (7) and $\boldsymbol{e}_3^{i+1} = \boldsymbol{Q}_\Delta \boldsymbol{e}_3^i$, one arrives at the important result below

$$\boldsymbol{e}_3^{i+1} - \boldsymbol{e}_3^i = \boldsymbol{\alpha}_\Delta \times \boldsymbol{e}_3^m, \quad \text{where} \quad \boldsymbol{e}_3^m = \frac{1}{2}(\boldsymbol{e}_3^{i+1} + \boldsymbol{e}_3^i). \tag{15}$$

We remark that $\boldsymbol{e}_3^m$ is not a unitary vector, but $\|\boldsymbol{e}_3^m\|^{-1} \boldsymbol{e}_3^m$ is. Now we state

$$\boldsymbol{\alpha}_\Delta = \|\boldsymbol{e}_3^m\|^{-2} (\boldsymbol{e}_3^i \times \boldsymbol{e}_3^{i+1}) + \varphi_\Delta \|\boldsymbol{e}_3^m\|^{-1} \boldsymbol{e}_3^m. \tag{16}$$

We can show that (16), after some algebraic manipulation, preserves $(15)_1$. Note that

$$\varphi_\Delta = \|\boldsymbol{e}_3^m\|^{-1} \boldsymbol{\alpha}_\Delta \cdot \boldsymbol{e}_3^m. \tag{17}$$

Thus, assuming that the configuration at t_i is known, the incremental rotation tensor in (11) can be expressed as

$$\boldsymbol{Q}_\Delta = \hat{\boldsymbol{Q}}_\Delta (\boldsymbol{e}_3^{i+1}, \varphi_\Delta), \tag{18}$$

which is the incremental counterpart of (6).

Meier et al. (2014) uses a similar rotation parameter, but their conclusion on this matter is that an Hermite interpolation of the relative angle φ_Δ'(s) would lead to a non-objective element formulation. φ_Δ in Meier et al. (2014) is used within the context of the "smallest rotation" triad, but not within the Rodrigues parameterization.

Remark 2: Alternative Definition of Incremental Torsion Parameter
In place of (16), we could define

$$\boldsymbol{\alpha}_\Delta = \left\|\boldsymbol{e}_3^m\right\|^{-2}\left(\boldsymbol{e}_3^i \times \boldsymbol{e}_3^{i+1}\right) + 2\tan\frac{\varphi_\Delta}{2}\left\|\boldsymbol{e}_3^m\right\|^{-1}\boldsymbol{e}_3^m,$$

which seems to be more adequate when the rotation is a torsion around a fixed axis. (16) simplifies the resulting equations. We recall that both are equal to second order with respect to the torsion parameter φ_Δ.

Remark 3: Objectivity
Objectivity is a major issue for any formulation dealing with large deformations. We remark that the rod theory presented herein entirely fulfills objectivity requirements in the sense of continuum mechanics. Objectivity of the strain and stress measures is assured since one uses only material (back-rotated) quantities in the constitutive equations and thus are invariant under superposed rigid body motions. An analytical proof of this property is straightforward and will be omitted here. It is easy to verify that (16) is also objective. The torsion parameter φ_Δ is a scalar, therefore it is objective as well, as (17) show.

Remark 4: Path-Dependency
It is also important to mention that by adopting an incremental description for the rotations, this description for the rotation turns out to be path-dependent. But one should keep in mind that path-dependency is a natural consequence when the time variable is discretized in the framework of a time-stepping scheme, which is mandatory for the numerical simulation of dynamical processes. Moreover, upon time increment refinement the dependence on the history of incrementation decreases and in an asymptotic manner path-independence is reached, as also discussed in Pimenta et al. (2008), Campello et al. (2011), Crisfield and Jelenic (1999). This is an obvious consequence, since (16) is numerically consistent.

2.4 Strains

According to the Bernoulli–Euler assumption the unitary vector $\boldsymbol{e}_3$ is given by

$$\boldsymbol{e}_3 = \left\|\boldsymbol{z}'\right\|^{-1}\boldsymbol{z}', \tag{19}$$

whereby the following notation for derivative along the axis has been defined

$$(\bullet)' = \frac{d(\bullet)}{d\zeta}. \tag{20}$$

e_3 is tangent to the rod axis in the current configuration and orthogonal to the cross sections (19) makes this formulation different from Pimenta (1993b), Pimenta and Yojo (1993) and the geometrically exact theory is constrained ab initio to obey the Bernoulli–Euler assumption.

Displacements of the points on the rod axis are defined by

$$\boldsymbol{u} = \boldsymbol{z} - \boldsymbol{\zeta}. \tag{21}$$

Note also that

$$\boldsymbol{z}' = \boldsymbol{e}_3^r + \boldsymbol{u}' \quad \text{and} \quad \boldsymbol{z}'' = \boldsymbol{u}''. \tag{22}$$

Analogously to (9), the curvature vector of the axis at the current configuration is given by

$$\boldsymbol{\kappa} = \text{axial}\left(\boldsymbol{Q}'\boldsymbol{Q}^T\right). \tag{23}$$

Since $\left(\dot{\boldsymbol{Q}}\right)' = \left(\boldsymbol{Q}'\right)^{\cdot}$, one has

$$\boldsymbol{\omega}' = \dot{\boldsymbol{\kappa}} - \boldsymbol{\omega} \times \boldsymbol{\kappa}. \tag{24}$$

Time differentiation of $\boldsymbol{\kappa}^r = \boldsymbol{Q}^T\boldsymbol{\kappa}$ leads to $\dot{\boldsymbol{\kappa}}^r = \boldsymbol{Q}^T(\dot{\boldsymbol{\kappa}} - \boldsymbol{\omega} \times \boldsymbol{\kappa})$. Hence, from (24), one arrives at the important relation displayed below

$$\dot{\boldsymbol{\kappa}}^r = \boldsymbol{Q}^T\boldsymbol{\omega}'. \tag{25}$$

The deformation gradient can then be expressed by

$$\boldsymbol{F} = \boldsymbol{Q}\boldsymbol{F}^r, \tag{26}$$

where

$$\boldsymbol{F}^r = \boldsymbol{I} + \boldsymbol{\gamma}^r \otimes \boldsymbol{e}_3^r \tag{27}$$

is the back-rotated deformation gradient, $\boldsymbol{I}$ is the identity tensor and

$$\boldsymbol{\gamma}^r = \boldsymbol{\eta}^r + \boldsymbol{\kappa}^r \times \boldsymbol{r}^r \tag{28}$$

are back-rotated cross section strains. In (28) the following generalized back-rotated strain has been introduced

$$\eta^r = Q^T z' - e_3^r. \tag{29}$$

We remark that

$$\eta^r \cdot e_\alpha^r = \eta^r \times e_3^r = 0. \tag{30}$$

due to the Bernoulli–Euler assumption. Note that

$$\eta^r = \varepsilon e_3^r, \quad \text{where} \quad \varepsilon = \|z'\| - 1 \tag{31}$$

(28) and (29) are the back-rotated counterparts of the following cross-sectional generalized strains

$$\gamma = \eta + \kappa \times r \quad \text{and} \quad \eta = z' - e_3. \tag{32}$$

From (19) and $(22)_1$, it follows that $e_3 = \hat{e}_3(u')$, which together with (18) leads to

$$\alpha_\Delta = \hat{\alpha}_\Delta(u'_{i+1}, \varphi_\Delta). \tag{33}$$

Hence, one may write

$$\dot{\alpha}_\Delta = W\dot{u}' + w\dot{\varphi}_\Delta, \tag{34}$$

where

$$W = \frac{\partial \alpha_\Delta}{\partial u'} \quad \text{and} \quad w = \frac{\partial \alpha_\Delta}{\partial \varphi_\Delta}. \tag{35}$$

With the aid of (14), (35) and (34), the spin vector can be written as

$$\omega = \Xi_\Delta W\dot{u}' + \Xi_\Delta w\dot{\varphi}_\Delta. \tag{36}$$

On the other hand, the curvature vector needs to be updated at instant t_{i+1} from the curvature vector at instant κ_i. From (23) one gets

$$\kappa_{i+1} = \text{axial}\left(Q'_{i+1} Q_{i+1}^T\right) = \text{axial}\left(\left(Q_\Delta Q_i\right)' Q_i^T Q_\Delta^T\right) \tag{37}$$

This delivers

$$\kappa_{i+1} = Q_\Delta \kappa_i + \Xi_\Delta \alpha'_\Delta. \tag{38}$$

Introducing (33) in (38), and using definitions (35), one gets

$$\boldsymbol{\alpha}'_\Delta = \boldsymbol{W}\boldsymbol{u}'' + \boldsymbol{w}\varphi'_\Delta, \tag{39}$$

Similarly, the back-rotated curvature vector at instant t_{i+1} is given by

$$\boldsymbol{\kappa}^r_{i+1} = \boldsymbol{\kappa}^r_i + \boldsymbol{Q}^T_{i+1}\boldsymbol{\Xi}_\Delta\boldsymbol{\alpha}'_\Delta. \tag{40}$$

The derivatives in (35) are now displayed below,

$$\begin{aligned}\boldsymbol{W} = &\left[\left\|\boldsymbol{e}^m_3\right\|^{-2}\boldsymbol{E}^i_3 - \left\|\boldsymbol{e}^m_3\right\|^{-4}\left(\boldsymbol{e}^i_3 \times \boldsymbol{e}^{i+1}_3\right) \otimes \boldsymbol{e}^m_3\right]\left\|\boldsymbol{z}'\right\|^{-1}\boldsymbol{M}^b \\ &+ \frac{1}{2}\varphi_\Delta\left\|\boldsymbol{e}^m_3\right\|^{-1}\left(\boldsymbol{I} - \left\|\boldsymbol{e}^m_3\right\|^{-2}\boldsymbol{e}^m_3 \otimes \boldsymbol{e}^m_3\right)\left\|\boldsymbol{z}'\right\|^{-1}\boldsymbol{M}^b\end{aligned} \tag{41}$$

and

$$\boldsymbol{w} = \left\|\boldsymbol{e}^m_3\right\|^{-1}\boldsymbol{e}^m_3. \tag{42}$$

In (41), one has introduced $\boldsymbol{E}^i_3 = \mathrm{Skew}\left(\boldsymbol{e}^i_3\right)$, and $\boldsymbol{M}^b = \boldsymbol{I} - \boldsymbol{e}^{i+1}_3 \otimes \boldsymbol{e}^{i+1}_3$. Note that $\left(\boldsymbol{M}^b\right)^k = \boldsymbol{M}^b$, so that $\boldsymbol{W}$ in (35) has following property

$$\boldsymbol{W}\boldsymbol{M}^b = \boldsymbol{W}. \tag{43}$$

Note that, with assistance from (15), one gets

$$\boldsymbol{\Xi}_\Delta\boldsymbol{w} = \frac{4}{4+\alpha^2_\Delta}\left\|\boldsymbol{e}^m_3\right\|^{-1}\boldsymbol{e}^{i+1}_3. \tag{44}$$

Hence, in place of (36) and (40), one has

$$\boldsymbol{\omega} = \boldsymbol{\omega}^m + \boldsymbol{\omega}^b \quad \text{and} \quad \boldsymbol{\kappa}^r_{i+1} = \boldsymbol{\kappa}^r_i + \boldsymbol{Q}^T_{i+1}\left(\boldsymbol{\kappa}^m_\Delta + \boldsymbol{\kappa}^b_\Delta\right), \tag{45}$$

respectively, where, with aid of (43), one has

$$\begin{aligned}\boldsymbol{\omega}^m &= \left(\frac{4}{4+\alpha^2_\Delta}\left\|\boldsymbol{e}^m_3\right\|^{-1}\dot{\varphi}_\Delta\right)\boldsymbol{e}^{i+1}_3, \quad \boldsymbol{\omega}^b = \boldsymbol{\Xi}_\Delta\boldsymbol{W}\boldsymbol{M}^b\dot{\boldsymbol{u}}', \\ \boldsymbol{\kappa}^m_\Delta &= \left(\frac{4}{4+\alpha^2_\Delta}\left\|\boldsymbol{e}^m_3\right\|^{-1}\varphi'_\Delta\right)\boldsymbol{e}^{i+1}_3 \quad \text{and} \quad \boldsymbol{\kappa}^b_\Delta = \boldsymbol{\Xi}_\Delta\boldsymbol{W}\boldsymbol{M}^b\boldsymbol{u}''.\end{aligned} \tag{46}$$

Remark 5: Variance of the Axis Position
It is remarked that $\boldsymbol{\gamma}^r \cdot \boldsymbol{e}^r_\alpha = 0$ only at the chosen rod axis ($\xi_\alpha = 0$). Therefore, the Bernoulli–Euler theory is not invariant with respect to the axis position. One can show that the axis should be placed on the cross-section shear centers.

Remark 6: Axial and Transversal Parts of a Vector
According to (45), one defines the axial (membrane) and transversal (bending) parts of a vector $\boldsymbol{v}$ by $\boldsymbol{v}^m = (\boldsymbol{e}_3 \otimes \boldsymbol{e}_3)\boldsymbol{v}$ and $\boldsymbol{v}^b = \boldsymbol{M}^b \boldsymbol{v}$, respectively. (46) shows that only the bending parts of $\boldsymbol{u}'$ and $\boldsymbol{u}''$ affect the spin and incremental curvature vectors, respectively.

Remark 7: Number of Turns Around a Moving Axis
The number of turns around the moving axis $\boldsymbol{e}_3$ can be computed through

$$N = \sum_i \frac{\varphi_\Delta}{2\pi}. \tag{47}$$

(47) allows us to count the number of turns that a cross section did from the initial to the current configuration.

2.5 Strain Rates

The velocity gradient is given by time differentiation of (26)

$$\dot{\boldsymbol{F}} = \boldsymbol{\Omega}\boldsymbol{F} + \boldsymbol{Q}\big(\dot{\boldsymbol{\gamma}}^r \otimes \boldsymbol{e}_3^r\big), \tag{48}$$

where

$$\dot{\boldsymbol{\gamma}}^r = \dot{\boldsymbol{\eta}}^r + \dot{\boldsymbol{\kappa}}^r \times \boldsymbol{r}^r. \tag{49}$$

Finally, from (25) and (40), one gets

$$\dot{\boldsymbol{\kappa}}^r_{i+1} = \boldsymbol{Q}^T_{i+1}\boldsymbol{\omega}'_{i+1} = \boldsymbol{Q}^T_i \boldsymbol{Q}^T_\Delta (\boldsymbol{\Xi}_\Delta \dot{\boldsymbol{\alpha}}_\Delta)'. \tag{50}$$

Hence, one may write

$$\dot{\boldsymbol{\kappa}}^r_{i+1} = \boldsymbol{Q}^T_i \boldsymbol{Q}^T_\Delta \boldsymbol{\Xi}'_\Delta \dot{\boldsymbol{\alpha}}_\Delta + \boldsymbol{Q}^T_i \boldsymbol{Q}^T_\Delta \boldsymbol{\Xi}_\Delta \dot{\boldsymbol{\alpha}}'_\Delta. \tag{51}$$

On the other hand, time differentiation of (29) yields

$$\dot{\boldsymbol{\eta}}^r = \boldsymbol{Q}^T \dot{\boldsymbol{u}}' + \dot{\boldsymbol{Q}}^T \boldsymbol{z}' = \boldsymbol{Q}^T \big(\dot{\boldsymbol{u}}' + \boldsymbol{z}' \times \boldsymbol{\omega}\big). \tag{52}$$

Thus, with $\boldsymbol{Z}' = \text{Skew}\big(\boldsymbol{z}'\big)$, one may write

$$\dot{\boldsymbol{\eta}}^r_{i+1} = \boldsymbol{Q}^T_{i+1}\big(\dot{\boldsymbol{u}}'_{i+1} + \boldsymbol{Z}'_{i+1}\boldsymbol{\Xi}_\Delta \dot{\boldsymbol{\alpha}}_\Delta\big). \tag{53}$$

2.6 Stresses

Let the 1st Piola–Kirchhoff stress tensor be expressed by its columns as follows

$$\boldsymbol{P} = \boldsymbol{\tau}_i \otimes \boldsymbol{e}_i^r = \boldsymbol{Q}\left(\boldsymbol{\tau}_i^r \otimes \boldsymbol{e}_i^r\right). \tag{54}$$

One can now introduce the back-rotated 1st Piola-Kirchhoff stress tensor by

$$\boldsymbol{P}^r = \boldsymbol{Q}^T \boldsymbol{P} = \boldsymbol{\tau}_i^r \otimes \boldsymbol{e}_i^r, \tag{55}$$

where

$$\boldsymbol{\tau}_i^r = \boldsymbol{Q}^T \boldsymbol{\tau}_i, \quad i = 1, 2, 3, \tag{56}$$

are the back-rotated nominal stress vectors.

The following cross-sectional resultants are obtained by integration of the stresses $\boldsymbol{\tau} = \boldsymbol{\tau}_3$ on the cross section

$$\boldsymbol{n} = \int_A \boldsymbol{\tau} dA \quad \text{and} \quad \boldsymbol{m} = \int_A (\boldsymbol{r} \times \boldsymbol{\tau}) dA. \tag{57}$$

$\boldsymbol{n}$ are the true forces and $\boldsymbol{m}$ are the true moments that are acting on a cross section. The axial (membrane) and transversal (bending) parts of the force $\boldsymbol{n}$ are expressed by

$$\boldsymbol{n}^m = (\boldsymbol{e}_3 \otimes \boldsymbol{e}_3)\boldsymbol{n} = N\boldsymbol{e}_3 \quad \text{and} \quad \boldsymbol{n}^b = \boldsymbol{M}^b \boldsymbol{n} = V_\alpha \boldsymbol{e}_\alpha, \tag{58}$$

respectively, where $N = \boldsymbol{n} \cdot \boldsymbol{e}_3$ and $V_\alpha = \boldsymbol{n} \cdot \boldsymbol{e}_\alpha$ are the normal and shear forces that are acting on the cross section, respectively.

Their back-rotated counterparts are

$$\boldsymbol{n}^r = \boldsymbol{Q}^T \boldsymbol{n} \quad \text{and} \quad \boldsymbol{m}^r = \boldsymbol{Q}^T \boldsymbol{m}. \tag{59}$$

Hence, one may also write

$$\boldsymbol{n}^r = \int_A \boldsymbol{\tau}^r dA \quad \text{and} \quad \boldsymbol{m}^r = \int_A \left(\boldsymbol{r}^r \times \boldsymbol{\tau}^r\right) dA. \tag{60}$$

$\boldsymbol{n}^r$ and $\boldsymbol{m}^r$ are the back-rotated cross section forces and moments, respectively. The back-rotated counterparts of (58) are

$$\boldsymbol{n}^{mr} = N\boldsymbol{e}_3^r \quad \text{and} \quad \boldsymbol{n}^{br} = V_\alpha \boldsymbol{e}_\alpha^r, \tag{61}$$

For the bending moments and the torsion moment, $M_\alpha = \boldsymbol{m} \cdot \boldsymbol{e}_\alpha = \boldsymbol{m} \cdot \boldsymbol{e}_\alpha^r$ and $T = \boldsymbol{m} \cdot \boldsymbol{e}_3 = \boldsymbol{m}^r \cdot \boldsymbol{e}_3^r$ are written, respectively. Hence, one has

$$\boldsymbol{n} = V_\alpha \boldsymbol{e}_\alpha + N\boldsymbol{e}_3 \quad \text{and} \quad \boldsymbol{m} = M_\alpha \boldsymbol{e}_\alpha + T\boldsymbol{e}_3, \quad \text{or}$$
$$\boldsymbol{n}^r = V_\alpha \boldsymbol{e}_\alpha^r + N\boldsymbol{e}_3^r \quad \text{and} \quad \boldsymbol{m}^r = M_\alpha \boldsymbol{e}_\alpha^r + T\boldsymbol{e}_3^r. \tag{62}$$

2.7 Kinetics

From (54) and (48) and the angular momentum balance $\boldsymbol{P}\boldsymbol{F}^T : \boldsymbol{\Omega} = 0$, one gets the following result:

$$\boldsymbol{P} : \dot{\boldsymbol{F}} = \boldsymbol{\tau}^r \cdot \dot{\boldsymbol{\gamma}}^r. \tag{63}$$

(63) is the stress power per unit of reference volume. Introducing (49) in (63) and after some manipulation with the cross product, one gets

$$\boldsymbol{P} : \dot{\boldsymbol{F}} = \boldsymbol{\tau}^r \cdot \dot{\boldsymbol{\eta}}^r + \left(\boldsymbol{r}^r \times \boldsymbol{\tau}^r\right) \cdot \dot{\boldsymbol{\kappa}}^r. \tag{64}$$

Note that $\boldsymbol{\tau}_\alpha^r$ are powerless in this model. With the aid of the definitions (60), the integration of (64) over the cross section furnishes

$$\int_A (\boldsymbol{P} : \dot{\boldsymbol{F}}) dA = \boldsymbol{n}^r \cdot \dot{\boldsymbol{\eta}}^r + \boldsymbol{m}^r \cdot \dot{\boldsymbol{\kappa}}^r. \tag{65}$$

(65) is the stress power per unit length of the reference axis. It is important to remark that $\boldsymbol{n}^r$, $\boldsymbol{m}^r$, $\boldsymbol{\eta}^r$ and $\boldsymbol{\kappa}^r$ are not affected by superimposed rigid body motions. Regarding $(61)_1$, one has

$$\boldsymbol{n}^r \cdot \dot{\boldsymbol{\eta}}^r + \boldsymbol{m}^r \cdot \dot{\boldsymbol{\kappa}}^r = \boldsymbol{n}^{mr} \cdot \dot{\boldsymbol{\eta}}^r + \boldsymbol{m}^r \cdot \dot{\boldsymbol{\kappa}}^r.. \tag{66}$$

The internal power on the domain Ω is then given by

$$P_{\text{int}}^{\Omega} = \int_{\Omega} \left(\boldsymbol{n}^r \cdot \dot{\boldsymbol{\eta}}^r + \boldsymbol{m}^r \cdot \dot{\boldsymbol{\kappa}}^r\right) d\Omega. \tag{67}$$

On the other hand, the external power on the same domain can be expressed by

$$P_{\text{ext}}^{\Omega} = \int_{\Omega} \left[\int_C (\bar{\boldsymbol{t}} \cdot \dot{\boldsymbol{x}}) dC + \int_A (\bar{\boldsymbol{b}} \cdot \dot{\boldsymbol{x}}) dA \right] d\Omega, \tag{68}$$

where $\bar{\boldsymbol{t}}$ is the surface traction per unit reference area that is prescribed on the lateral surface of the rod and $\bar{\boldsymbol{b}}$ is the body force per unit reference volume. The time differentiation of (4) yields

$$\dot{\boldsymbol{x}} = \dot{\boldsymbol{u}} + \boldsymbol{\omega} \times \boldsymbol{r}. \tag{69}$$

With following definitions

$$\bar{\boldsymbol{n}}^{\Omega} = \int_C \bar{\boldsymbol{t}} dC + \int_A \bar{\boldsymbol{b}} dA \text{ and } \bar{\boldsymbol{m}}^{\Omega} = \int_C (\boldsymbol{r} \times \bar{\boldsymbol{t}}) dC + \int_A (\boldsymbol{r} \times \bar{\boldsymbol{b}}) dA, \tag{70}$$

Together with (69), one may write

$$P_{\text{ext}}^{\Omega} = \int_{\Omega} (\bar{\boldsymbol{n}}^{\Omega} \cdot \dot{\boldsymbol{u}} + \bar{\boldsymbol{m}}^{\Omega} \cdot \boldsymbol{\omega}) d\Omega = \int_{\Omega} (\bar{\boldsymbol{n}}^{\Omega} \cdot \dot{\boldsymbol{u}} + \boldsymbol{\Xi}_{\Delta}^{T} \bar{\boldsymbol{m}}^{\Omega} \cdot \dot{\boldsymbol{\alpha}}_{\Delta}) d\Omega. \tag{71}$$

$\bar{\boldsymbol{n}}^{\Omega}$ is the applied external force per unit length at reference configuration and $\bar{\boldsymbol{m}}^{\Omega}$ is the applied external moment per unit length at reference configuration. Introducing (34) in (71), it furnishes

$$P_{\text{ext}}^{\Omega} = \int_{\Omega} \left(\bar{\boldsymbol{n}}^{\Omega} \cdot \dot{\boldsymbol{u}} + \bar{\boldsymbol{\mu}}^{\Omega} \cdot \dot{\boldsymbol{u}}' + \bar{\mu}^{\Omega} \dot{\varphi}_{\Delta} \right) d\Omega, \tag{72}$$

where

$$\bar{\boldsymbol{\mu}}^{\Omega} = \boldsymbol{W}^{T} \boldsymbol{\Xi}_{\Delta}^{T} \bar{\boldsymbol{m}}^{\Gamma} \quad \text{and} \quad \bar{\mu}^{\Omega} = \boldsymbol{w} \cdot \boldsymbol{\Xi}_{\Delta}^{T} \bar{\boldsymbol{m}}^{\Gamma} \tag{73}$$

are the pseudo-bending-moments and the pseudo-torsion-moments per unit reference length applied along the rod, respectively. Note that $\bar{\boldsymbol{\mu}}^{\Omega b} = \bar{\boldsymbol{\mu}}^{\Omega}$ and $\bar{\boldsymbol{\mu}}^{\Omega} \cdot \dot{\boldsymbol{u}}' = \bar{\boldsymbol{\mu}}^{\Omega} \cdot \left(\dot{\boldsymbol{u}}' \right)^{b}$.

Similarly to (70), one defines

$$\bar{\boldsymbol{n}}^{\Gamma} = \int_C \bar{\boldsymbol{t}} dC + \int_A \bar{\boldsymbol{b}} dA \text{ and } \bar{\boldsymbol{m}}^{\Gamma} = \int_C (\boldsymbol{r} \times \bar{\boldsymbol{t}}) dC + \int_A (\boldsymbol{r} \times \bar{\boldsymbol{b}}) dA. \tag{74}$$

Thus, with the aid of (69), one may write for the rod ends

$$P_{\text{ext}}^{\Gamma} = \left(\bar{\boldsymbol{n}}^{\Gamma} \cdot \dot{\boldsymbol{u}} + \bar{\boldsymbol{m}}^{\Gamma} \cdot \boldsymbol{\omega} \right)_{\Gamma} = \left(\bar{\boldsymbol{n}}^{\Gamma} \cdot \dot{\boldsymbol{u}} + \boldsymbol{\Xi}_{\Delta}^{T} \bar{\boldsymbol{m}}^{\Gamma} \cdot \dot{\boldsymbol{\alpha}}_{\Delta} \right)_{\Gamma}. \tag{75}$$

$\bar{\boldsymbol{n}}^{\Gamma}$ and $\bar{\boldsymbol{m}}^{\Gamma}$ are the applied external forces and moments at rod ends, respectively. In (75), the notation $(\cdot)_{\Gamma} = (\cdot)_{\zeta=\ell} - (\cdot)_{\zeta=0}$ has been introduced. With (34), (71) furnishes

$$P_{\text{ext}}^{\Gamma} = \left(\bar{\boldsymbol{n}}^{\Gamma} \cdot \dot{\boldsymbol{u}} + \bar{\boldsymbol{\mu}}^{\Gamma} \cdot \dot{\boldsymbol{u}}' + \bar{\mu}^{\Gamma} \dot{\varphi}_{\Delta} \right)_{\Gamma}, \tag{76}$$

where

$$\bar{\boldsymbol{\mu}}^{\Gamma} = \boldsymbol{W}^{T} \boldsymbol{\Xi}_{\Delta}^{T} \bar{\boldsymbol{m}}^{\Gamma} \quad \text{and} \quad \bar{\mu}^{\Gamma} = \boldsymbol{w} \cdot \boldsymbol{\Xi}_{\Delta}^{T} \bar{\boldsymbol{m}}^{\Gamma} \tag{77}$$

are the pseudo-bending-moments and the pseudo-torsion-moments applied on the rod ends, respectively. Note that $\bar{\boldsymbol{\mu}}^{\Gamma b} = \bar{\boldsymbol{\mu}}^{\Gamma}$.

2.8 Weak Form of the Local Equilibrium Equation

The internal virtual work on a domain $\Omega \subset \mathbb{R}$ is given by

$$\delta W_{\text{int}}^{\Omega} = \int_{\Omega} (\boldsymbol{n}^{mr} \cdot \delta\boldsymbol{\eta}_{i+1}^{r} + \boldsymbol{m}^{r} \cdot \delta\boldsymbol{\kappa}_{i+1}^{r}) d\Omega, \tag{78}$$

while the external virtual work on the domain $\Omega \subset \mathbb{R}$ is, in a similar manner, given by

$$\delta W_{\text{ext}}^{\Omega} = \int_{\Omega} (\bar{\boldsymbol{n}}^{\Omega} \cdot \delta\boldsymbol{u}_{i+1} + \boldsymbol{\Xi}_{\Delta}^{T} \bar{\boldsymbol{m}}^{\Omega} \cdot \delta\boldsymbol{\alpha}_{\Delta}) d\Omega, \tag{79}$$

where

$$\begin{aligned} \delta\boldsymbol{\eta}_{i+1}^{r} &= \boldsymbol{Q}_{i+1}^{T}(\delta\boldsymbol{u}'_{i+1} + \boldsymbol{Z}'_{i+1}\boldsymbol{\Xi}_{\Delta}\delta\boldsymbol{\alpha}_{\Delta}), \\ \delta\boldsymbol{\kappa}_{i+1}^{r} &= \boldsymbol{Q}_{i+1}^{T}(\boldsymbol{\Xi}_{\Delta}\delta\boldsymbol{\alpha}_{\Delta})' \quad \text{and} \\ \delta\boldsymbol{\alpha}_{\Delta} &= \boldsymbol{W}\delta\boldsymbol{u}'_{i+1} + \boldsymbol{w}\delta\varphi_{\Delta}. \end{aligned} \tag{80}$$

Introducing (80) in (78), one gets

$$\begin{aligned} \delta W_{\text{int}}^{\Omega} = &\int_{\Omega} \boldsymbol{n}^{m} \cdot (\delta\boldsymbol{u}'_{i+1} + \boldsymbol{Z}'_{i+1}\boldsymbol{\Xi}_{\Delta}\delta\boldsymbol{\alpha}_{\Delta}) d\Omega \\ &+ \int_{\Omega} \boldsymbol{m} \cdot (\boldsymbol{\Xi}_{\Delta}(\boldsymbol{W}\delta\boldsymbol{u}'_{i+1} + \boldsymbol{w}\delta\varphi_{\Delta}))' d\Omega. \end{aligned} \tag{81}$$

Similarly, from (79) one arrives at

$$\delta W_{\text{ext}}^{\Omega} = \int_{\Omega} (\bar{\boldsymbol{n}}^{\Omega} \cdot \delta\boldsymbol{u} + \bar{\boldsymbol{\mu}}^{\Omega} \cdot \delta\boldsymbol{u}' + \bar{\mu}^{\Omega}\delta\varphi_{\Delta}) d\Omega. \tag{82}$$

The rod local equilibrium equations are obtained by applying the Virtual Work Theorem as follows:

$$\delta W_{\text{int}}^{\Omega} - \delta W_{\text{ext}}^{\Omega} = \delta W_{\text{ext}}^{\Gamma}, \quad \forall \delta\boldsymbol{u}, \delta\varphi_{\Delta} \text{ in } \Omega. \tag{83}$$

where $\delta W_{\text{ext}}^{\Gamma}$ is the external virtual work on the boundary Γ, which is given by

$$\delta W_{\text{ext}}^{\Gamma} = (\bar{\boldsymbol{n}}^{\Gamma} \cdot \delta\boldsymbol{u} + \bar{\boldsymbol{\mu}}^{\Gamma} \cdot \delta\boldsymbol{u}' + \bar{\mu}^{\Gamma}\delta\varphi_{\Delta})_{\Gamma}. \tag{84}$$

Introducing (81) and (82) in (83), and taking into account that $\boldsymbol{z'}_{i+1} \times \boldsymbol{n}^m = \boldsymbol{o}$, one gets

$$\int_\Omega \left(\boldsymbol{n}^m \cdot \delta\boldsymbol{u'}_{i+1} + \boldsymbol{m} \cdot \left(\boldsymbol{\Xi}_\Delta\left(\boldsymbol{W}\delta\boldsymbol{u'}_{i+1} + \boldsymbol{w}\delta\varphi_\Delta\right)\right)'\right)d\Omega + \\ - \int_\Omega \left(\bar{\boldsymbol{n}}^\Omega \cdot \delta\boldsymbol{u}_{i+1} + \bar{\boldsymbol{\mu}}^\Omega \cdot \delta\boldsymbol{u}' + \bar{\mu}^\Omega \delta\varphi_\Delta\right)d\Omega = \delta W^\Gamma_{\text{ext}}. \tag{85}$$

By integration by parts of (85), one obtains

$$-\int_\Omega \left(\left((\boldsymbol{n}^m)' + \bar{\boldsymbol{n}}^\Omega\right) \cdot \delta\boldsymbol{u}_{i+1}\right)d\Omega + \\ - \int_\Omega \left(\left(\boldsymbol{W}^T\boldsymbol{\Xi}^T_\Delta\boldsymbol{m}' + \bar{\boldsymbol{\mu}}^\Omega\right) \cdot \delta\boldsymbol{u'}_{i+1} + \left(\boldsymbol{w} \cdot \boldsymbol{\Xi}^T_\Delta\boldsymbol{m}' + \bar{\mu}^\Omega\right)\delta\varphi_\Delta\right)d\Omega \\ + \left(\boldsymbol{n}^m \cdot \delta\boldsymbol{u}_{i+1} + \boldsymbol{m} \cdot \boldsymbol{\Xi}_\Delta\left(\boldsymbol{W}\delta\boldsymbol{u'}_{i+1} + \boldsymbol{w}\delta\varphi_\Delta\right)\right)\Big|_\Gamma = \delta W^\Gamma_{\text{ext}} \tag{86}$$

and again on (86), one arrives at

$$-\int_\Omega \left[\left(\tilde{\boldsymbol{n}}' + \bar{\boldsymbol{n}}^\Omega\right) \cdot \delta\boldsymbol{u}_{i+1}\right]d\Omega - \int_\Omega \left(\left(\boldsymbol{w} \cdot \boldsymbol{\Xi}^T_\Delta\boldsymbol{m}' + \bar{\boldsymbol{\mu}}^\Omega\right)\delta\varphi_\Delta\right)d\Omega \\ + \left(\tilde{\boldsymbol{n}} \cdot \delta\boldsymbol{u}_{i+1} + \boldsymbol{m} \cdot \boldsymbol{\Xi}_\Delta\left(\boldsymbol{W}\delta\boldsymbol{u'}_{i+1} + \boldsymbol{w}\delta\varphi_\Delta\right)\right)\Big|_\Gamma = \delta W^\Gamma_{\text{ext}}, \tag{87}$$

where

$$\tilde{\boldsymbol{n}} = \boldsymbol{n}^m - \boldsymbol{W}^T\boldsymbol{\Xi}^T_\Delta\boldsymbol{m}' - \bar{\boldsymbol{\mu}}^\Omega. \tag{88}$$

By standard arguments of Calculus of Variation, (87) delivers the following local equilibrium equations in Ω

$$\tilde{\boldsymbol{n}}' + \bar{\boldsymbol{n}}^\Omega = \boldsymbol{o} \quad \text{and} \quad \boldsymbol{w} \cdot \boldsymbol{\Xi}^T_\Delta\boldsymbol{m}' + \bar{\boldsymbol{\mu}}^\Omega = 0. \tag{89}$$

It remains the following boundary term on Γ

$$\left(\tilde{\boldsymbol{n}} \cdot \delta\boldsymbol{u}_{i+1} + \boldsymbol{m} \cdot \boldsymbol{\Xi}_\Delta\left(\boldsymbol{W}\delta\boldsymbol{u'}_{i+1} + \boldsymbol{w}\delta\varphi_\Delta\right)\right)_\Gamma \\ = \left(\bar{\boldsymbol{n}}^\Gamma \cdot \delta\boldsymbol{u} + \bar{\boldsymbol{\mu}}^\Gamma \cdot \delta\boldsymbol{u}' + \bar{\mu}^\Gamma\delta\varphi_\Delta\right)_\Gamma. \tag{90}$$

Thus, one can conclude that the natural (Neumann) boundary conditions are

$$\bar{\boldsymbol{n}}^\Gamma = \tilde{\boldsymbol{n}}, \quad \bar{\boldsymbol{\mu}}^\Gamma = \boldsymbol{W}^T\boldsymbol{\Xi}^T_\Delta\boldsymbol{m} \quad \text{and} \quad \bar{\mu}^\Gamma = \boldsymbol{w} \cdot \boldsymbol{\Xi}^T_\Delta\boldsymbol{m}, \tag{91}$$

while the essential (Dirichlet) boundary conditions are

$$\boldsymbol{u} = \bar{\boldsymbol{u}}, \quad (\boldsymbol{u}')^b = (\bar{\boldsymbol{u}}')^b \quad \text{and} \quad \varphi_\Delta = \bar{\varphi}_\Delta. \tag{92}$$

2.9 Statics

The rod local equilibrium equations can be directly derived by Statics (see, for example, Pimenta (1993a), Pimenta and Yojo (1993)). They are displayed below

$$\boldsymbol{n}' + \bar{\boldsymbol{n}}^{\Omega} = \boldsymbol{o} \quad \text{and} \quad \boldsymbol{m}' + \boldsymbol{z}' \times \boldsymbol{n} + \bar{\boldsymbol{m}}^{\Omega} = \boldsymbol{o}. \tag{93}$$

From $(93)_2$, one gets $\boldsymbol{z}' \times \boldsymbol{n} = -\left(\boldsymbol{m}' + \bar{\boldsymbol{m}}^{\Omega}\right)$, which, with the aid of (58), i.e., $\boldsymbol{n} = \boldsymbol{n}^m + \boldsymbol{n}^b$, and $\boldsymbol{z}' \times \boldsymbol{n}^m = \boldsymbol{o}$, leads to the result below

$$\boldsymbol{z}' \times \boldsymbol{n}^b = -\left(\boldsymbol{m}' + \bar{\boldsymbol{m}}^{\Omega}\right). \tag{94}$$

From (94), with $\boldsymbol{n}^b = V_\alpha \boldsymbol{e}_\alpha$, one can derive

$$\begin{aligned} \boldsymbol{e}_\beta \cdot \left(\boldsymbol{z}' \times \boldsymbol{n}^b\right) &= \left\|\boldsymbol{z}'\right\| V_\alpha \left(\boldsymbol{e}_\beta \cdot \boldsymbol{e}_3 \times \boldsymbol{e}_\alpha\right) \\ &= \varepsilon_{\alpha\beta} \left\|\boldsymbol{z}'\right\| V_\alpha = -\boldsymbol{e}_\beta \cdot \left(\boldsymbol{m}' + \bar{\boldsymbol{m}}^{\Omega}\right), \end{aligned} \tag{95}$$

where $\varepsilon_{\alpha\beta} = \boldsymbol{e}_\alpha \cdot \boldsymbol{e}_\beta \times \boldsymbol{e}_3$ is a permutation symbol. From (95), one arrives at

$$V_\alpha = -\left\|\boldsymbol{z}'\right\|^{-1} \varepsilon_{\alpha\beta} \boldsymbol{e}_\beta \cdot \left(\boldsymbol{m}' + \bar{\boldsymbol{m}}^{\Omega}\right). \tag{96}$$

An alternative to (96) is

$$\begin{aligned} \boldsymbol{n}^b = V_\alpha \boldsymbol{e}_\alpha &= -\left\|\boldsymbol{z}'\right\|^{-1} \left[\varepsilon_{\alpha\beta} \boldsymbol{e}_\beta \cdot \left(\boldsymbol{m}' + \bar{\boldsymbol{m}}^{\Omega}\right)\right] \boldsymbol{e}_\alpha = \\ &= -\left\|\boldsymbol{z}'\right\|^{-1} \left(\boldsymbol{e}_1 \otimes \boldsymbol{e}_2 - \boldsymbol{e}_2 \otimes \boldsymbol{e}_1\right)\left(\boldsymbol{m}' + \bar{\boldsymbol{m}}^{\Omega}\right). \end{aligned} \tag{97}$$

From (97), with the assistance of $\boldsymbol{e}_2 \otimes \boldsymbol{e}_1 - \boldsymbol{e}_1 \otimes \boldsymbol{e}_2 = \mathrm{Skew}(\boldsymbol{e}_3)$, one arrives at

$$\boldsymbol{n}^b = \left\|\boldsymbol{z}'\right\|^{-1} \boldsymbol{e}_3 \times \left(\boldsymbol{m}' + \bar{\boldsymbol{m}}^{\Omega}\right) = \left\|\boldsymbol{z}'\right\|^{-2} \boldsymbol{z}' \times \left(\boldsymbol{m}' + \bar{\boldsymbol{m}}^{\Omega}\right). \tag{98}$$

3 Elastic Constitutive Equations

Only elastic small strains have been considered in this work. In a later work under preparation, other constitutive equations will be considered. If the rod axis is placed along with the cross-sectional shear centers, the following linear elastic constitutive equation for small strain isotropic elasticity can be adopted :

$$\boldsymbol{\sigma}^r = \boldsymbol{D}\boldsymbol{\varepsilon}^r, \tag{99}$$

where

$$\boldsymbol{\sigma}^r = \begin{bmatrix} \boldsymbol{n}^{mr} \\ \boldsymbol{m}^r \end{bmatrix}, \quad \boldsymbol{\varepsilon}^r = \begin{bmatrix} \boldsymbol{\eta}^r \\ \boldsymbol{\kappa}^r \end{bmatrix} \quad \text{and} \quad \boldsymbol{D} = \begin{bmatrix} \boldsymbol{D}_{\eta\eta} & \boldsymbol{D}_{\eta\kappa} \\ \boldsymbol{D}_{\kappa\eta} & \boldsymbol{D}_{\kappa\kappa} \end{bmatrix}. \tag{100}$$

The strain energy per unit reference length, in this case, is given by

$$\psi = \frac{1}{2}\boldsymbol{\varepsilon}^r \cdot \boldsymbol{D}\boldsymbol{\varepsilon}^r. \tag{101}$$

In $(100)_3$, one has

$$\begin{aligned} \boldsymbol{D}_{\eta\eta} &= EA\boldsymbol{e}_3^r \otimes \boldsymbol{e}_3^r \\ \boldsymbol{D}_{\eta\kappa} &= ES_\alpha \boldsymbol{e}_3^r \otimes \boldsymbol{e}_\alpha^r = \boldsymbol{D}_{\kappa\eta}^T \text{ and} \\ \boldsymbol{D}_{\kappa\kappa} &= EJ_{\alpha\beta}\boldsymbol{e}_\alpha^r \otimes \boldsymbol{e}_\beta^r + GJ_T \boldsymbol{e}_3^r \otimes \boldsymbol{e}_3^r, \end{aligned} \tag{102}$$

where E is the elasticity modulus, G is the shear modulus, A is the cross-sectional area, J_T is the cross-sectional torsion constant, $S_\alpha = \varepsilon_{\alpha\beta}\int_A \xi_\beta dA$ are the cross-sectional static moments and $J_{\alpha\beta} = \varepsilon_{\alpha\gamma}\varepsilon_{\beta\delta}\int_A \xi_\gamma \xi_\delta dA$ are the cross-sectional inertia moments. It is recalled that J_T is given by

$$J_T = J_O - \int_A \varepsilon_{\alpha\beta}\xi_\beta \phi_{,\alpha} dA, \tag{103}$$

where $\phi = \hat{\phi}(\xi_\alpha)$ is the St.Venant warping function, $\phi_{,\alpha} = \partial\phi/\partial\xi_\alpha$ and

$$J_O = \int_A \xi_\alpha \xi_\alpha dA = J_{11} + J_{22} \tag{104}$$

is the cross-sectional polar inertia moment. For circular or annular sections, with the origin at the barycenter $S_\alpha = 0$, $J_{12} = J_{21} = 0$, $\phi = 0$ and $J_T = J_O$. For bisymmetrical cross sections with the origin at the barycenter and $\boldsymbol{e}_\alpha^r$ along the principal axes of the cross section, one has $S_\alpha = 0$, $J_{12} = J_{21} = 0$, and J_T given by (103).

Remark 8: Strain Energy Density
According to (40) and (100), we may write

$$\psi = \hat{\psi}(\boldsymbol{u}, \varphi_\Delta). \tag{105}$$

4 Finite Element Implementation

The simulations can be performed within the AceFEM finite element software. Both AceGen and AceFEM programs are developed and maintained by Joze Korelc (University of Ljubljana). The interested reader is referred to Korelc and Wriggers (2016).

Within the class of conservative problems, only the formulation of the total potential energy is required, which can be given by

$$\Pi = \sum_e \left(\Pi^e_{\text{int}} + \Pi^e_{\text{ext}}\right), \tag{106}$$

where $(\cdot)^e$ is the contribution of each element $e = 1, 2, \ldots N_{elements}$. The strain energy of an element is

$$\Pi^e_{\text{int}} = \int_{\Omega^e} \psi \, d\Omega, \tag{107}$$

with ψ given by (101). Regarding (105), we may write

$$\Pi^e_{\text{int}} = \hat{\Pi}^e_{\text{int}}(\boldsymbol{u}, \varphi_\Delta) \tag{108}$$

The potential energy of an element, for the case of constant forces and constant pseudo-moments along the rods, is given by

$$\Pi^e_{\text{ext}} = -\int_{\Omega^e} \left(\bar{\boldsymbol{n}}^\Omega \cdot \boldsymbol{u} + \bar{\boldsymbol{\mu}}^\Omega \cdot \boldsymbol{u}' + \bar{\mu}^\Omega \varphi_\Delta\right) d\Omega. \tag{109}$$

In view of (109), we may write

$$\Pi^e_{\text{ext}} = \hat{\Pi}^e_{\text{ext}}(\boldsymbol{u}, \varphi_\Delta) \tag{110}$$

According to (92), for a smooth axis at reference configuration, the finite element approximation must be continuous for $\boldsymbol{u}$, $\left(\boldsymbol{u}'\right)^b$ and φ_Δ. If there is no sudden cross-sectional change, sudden material change nor concentrated loads at a connection node, a C_1 interpolation for the displacements $\boldsymbol{u}$ and a C_0 interpolation for the incremental rotation φ_Δ and a standard connection between elements is adequate. This has been done in [46]. In these cases, the element can be directly employed with the usual finite element tying procedure. However, a continuity of $\left(\boldsymbol{u}'\right)^m$, i.e. the axial part of $\boldsymbol{u}'$, is also achieved, what is not required by the theory. Moreover, the imposition of general Dirichlet boundary conditions can be complicated.

For the general case, i.e., for nonsmooth axis, multiple connections or for the case of cross section or material change from an element to the other, the connection of elements must be carefully performed. The appropriate connection can be generally formulated by imposing the equality of $\boldsymbol{u}$ and $\boldsymbol{\alpha}_\Delta$ at connecting ends.

Herein, the connection is achieved in a more standard way, as follows. The proposed element has only 2 nodes. Displacements $\boldsymbol{u}$ are approximated by cubic Hermitian polynomials, as usual

$$\boldsymbol{u}(\zeta) = N_1^u \boldsymbol{u}_1 + N_1^{u'} \boldsymbol{u}'_1 + N_2^u \boldsymbol{u}_2 + N_2^{u'} \boldsymbol{u}'_2, \tag{111}$$

where

$$N_1^u = 1 - \frac{3\zeta^2}{\ell^2} + \frac{2\zeta^3}{\ell^3},\ N_1^{u'} = \zeta - \frac{2\zeta^2}{\ell} + \frac{\zeta^3}{\ell^2},$$
$$N_2^u = \frac{3\zeta^2}{\ell^2} - \frac{2\zeta^3}{\ell^3} \text{ and } N_2^{u'} = \frac{\zeta^3}{\ell^2} - \frac{\zeta^2}{\ell}. \tag{112}$$

At nodes $I = 1, 2$, from the nodal values $\boldsymbol{\alpha}_{\Delta I}$ and ε_I^{i+1}, we get

$$\boldsymbol{e}_{iI}^{i+1} = \boldsymbol{Q}_{\Delta I} \boldsymbol{e}_{iI}^i = \hat{\boldsymbol{Q}}(\boldsymbol{\alpha}_{\Delta I}) \boldsymbol{e}_{iI}^i, \tag{113}$$

$$\varphi_{\Delta I} = \left\| \boldsymbol{e}_{3I}^m \right\|^{-1} \left(\boldsymbol{\alpha}_{\Delta I} \cdot \boldsymbol{e}_{3I}^m \right) \tag{114}$$

and

$$\boldsymbol{u}'^{i+1}_I = \left(1 + \varepsilon_I^{i+1}\right) \boldsymbol{e}_{3I}^{i+1} - \boldsymbol{e}_3^r. \tag{115}$$

Along the rod, we compute $\boldsymbol{u}(\zeta)$ and $\boldsymbol{u}'(\zeta)$ with the aid of (111). When necessary, we obtain $\boldsymbol{e}_3^{i+1}(\zeta)$ with the help from (19) and $\boldsymbol{\alpha}_\Delta(\zeta)$ with the assistance from (16) together with the following linear approximation:

$$\begin{array}{c} \varphi_\Delta(\zeta) = \varphi_{\Delta 1} N_1^\varphi(\zeta) + \varphi_{\Delta 2} N_2^\varphi(\zeta), \quad \text{where} \\ N_1^\varphi(\zeta) = 1 - \frac{\zeta}{\ell} \ \text{ and } \ N_2^\varphi(\zeta) = \frac{\zeta}{\ell}. \end{array} \tag{116}$$

This 2-node finite element has 7 DOFs, namely, $\boldsymbol{u}$, $\boldsymbol{\alpha}_\Delta$ and ε, at each extremity, but only $\boldsymbol{u}$ and $\boldsymbol{\alpha}_\Delta$ can be shared with neighboring elements.

Remark 9: Quadratic Approximation for φ_Δ
A quadratic approximation for the incremental torsion angle φ_Δ could also be used, but an extra mid-length node with a DOF for φ_Δ will be needed. This can be interesting for coupling with Kirchhoff–Love shell elements and to achieve a better convergence.

5 Conclusions

The geometrically exact rod formulation presented in Pimenta (1993b), Pimenta and Yojo (1993) was extended to a Bernoulli–Euler-type rod. Thereby, the present work is based on rotational parametrization via the Rodrigues rotation vector, which is used to propose an incremental update that a priory fulfills the shear rigidity constraint. It also introduces the ability to formulate inter-element connections in a more flexible manner. As in Pimenta (1993b), Pimenta and Yojo (1993), the approach has defined energetically conjugated generalized cross-sectional stress and strains based on the concept of a cross section. Besides their practical importance, cross-sectional quantities make the derivation of equilibrium equations easy, as well as the achievement of the corresponding tangent bilinear form, which is always symmetric for hyper-elastic materials and conservative loadings, even far from an equilibrium state.

A straight reference configuration was assumed for the rod on this work. Initially, curved rods are then regarded as a stress-free deformation from the straight configuration. This approach was already employed for rods in Pimenta (1996) and for shells in Pimenta and Campello (2009). It precludes the use of convective non-Cartesian coordinate systems and simplifies the comprehension of tensor quantities, since only components on orthogonal systems are employed.

Some examples were computed to show the capabilities of the formulation presented. As exposed throughout the paper, some examples with this Bernoulli–Euler rod theory were compared to benchmark problems and presented satisfying results. This formulation shows great promises and can be used to accurately describe the stresses, strains, displacements of flexible structures with great efficiency.

The derived beam formulation will be implemented in a finite element framework and investigated in various aspects. The authors aim to consider non-straight reference configurations on the element level in future studies. Further work is planned on extending the formulation to a pointwise approach, to incorporate general three-dimensional material laws. Consideration of out of plane warping is on the schedule as well.

Acknowledgements The author P. M. Pimenta acknowledges the support by CNPq under the grant 308142/2018-7 as well as expresses his acknowledgement to the Alexander von Humboldt Foundation for the Georg Forster Award that made possible his stays at the Universities of Duisburg-Essen and Hannover in Germany in the quadrennium 2015–2018 as well as to the French and Brazilian Governments for the Chair CAPES-Sorbonne that made possible his stay at "Sorbonne Universités" during the year of 2016 on a leave from the University of São Paulo. The third author gratefully acknowledges the Federal Institute of Science and Technology Education of São Paulo for financial support. The second and fourth authors gratefully acknowledge support by the Mercator Research Centre Ruhr in the Project "Mikromechanische Modellierung der Materialumformung zur Vorhersage der anisotropen Verfestigung" (Pr-2015-0049).

References

Antman, S. S. (1974). Kirchhoff's problem for nonlinearly elastic rods. *Quarterly Of Applied Mathematics*, 221–240.

Argyris, J. H., & Symeonidis, Sp. (1981). Nonlinear finite element analysis of elastic systems under non-conservative loading-natural formulation. part I. quasistatic problems. *Computational Methods in Applied Mechanics and Engineering, 58*, 75–123.

Argyris, J. H. (1982). An excursion into large rotations. *Computer Methods in Applied Mechanics and Engineering, 32*(1–3), 85–155.

Armero, F., & Valverde, J. (2012). Invariant Hermitian finite elements for thin Kirchhoff rods. I: The linear plane case. *Computer Methods in Applied Mechanics and Engineering, 213*, 427–457.
Bauer, A. M., Breitenberger, M., Philipp, B., Wüchner, R., & Bletzinger, K.-U. (2016). Nonlinear isogeometric spatial Bernoulli beam. *Computer Methods in Applied Mechanics and Engineering, 303,* 101–127.
Boyer, F., & Primault, D. (2004). Finite element of slender beams in finite transformations: A geometrically exact approach. *International Journal for Numerical Methods in Engineering, 59*(5), 669–702.
Boyer, F. G., De Nayer, Leroyer, A., & Visonneau, M. (2011). Geometrically exact Kirchhoff beam theory: Application to cable dynamics. *Journal of Computational and Nonlinear Dynamics, 6,* 1–14.
Campello, E. (2000). Análise não-linear de perfis metálicos conformados a frio. 106 f. Dissertação (Mestrado) - Curso de Engenharia Civil, Estruturas, Escola Politécnica da Universidade de São Paulo, São Paulo.
Campello, E. M. B., Pimenta, P. M., & Wriggers, P. (2003). A triangular finite shell element based on a fully nonlinear shell formulation. *Computational Mechanics, 31*(6), 505–518.
Campello, E. M. B., Pimenta, P. M., & Wriggers, P. (2011). An exact conserving algorithm for nonlinear dynamics with rotational DOFs and general hyperelasticity. Part 2: shells. *Computational Mechanics, 48*(2r), 195–211.
Crisfield, M. A., & Jelenic, G. (1999). Objectivity of strain measures in the geometrically-exact three-dimensional beam theory and its finite element implementation. *Proceedings of the Royal Society of London, 455,* 1125–1147.
Greco, L., & Cuomo, M. (2013). B-Spline interpolation of Kirchhoff-Love space rods. *Computer Methods in Applied Mechanics and Engineering, 256,* 251–269.
Greco, L., & Cuomo, M. (2016). An isogeometric implicit G1 mixed finite element for Kirchhoff space rods. *Computer Methods in Applied Mechanics and Engineering, 298*, 325–349 (Elsevier).
Korelc, J., & Wriggers, P. (2016). *Automation of finite element methods*. Springer.
Meier, C., Popp, A., & Wall, W. A. (2014). An objective 3D large deformation finite element formulation for geometrically exact curved Kirchhoff rods. *Computer Methods in Applied Mechanics and Engineering, 278,* 445–478.
Meier C., Grill M. J., Wall W. A., & Popp A. (2016). Geometrically exact beam elements and smooth contact schemes for the modeling of fiber-based materials and structures. *International Journal of Solids and Structures.*
Meier, C., Popp, A., & Wall, W. A. (2017). Geometrically exact finite element formulations for slender beams: Kirchhoff-Love theory vs. Simo-Reissner theory. *Archives of Computational Methods in Engineering*, 1–81.
Pimenta P. M. (1993a). On a geometrically-exact finite-strain shell model. In *Proceedings of the 3rd Pan-American Congress on Applied Mechanics*, III PACAM, São Paulo.
Pimenta P. M. (1993b). On a geometrically-exact finite-strain rod model. In *Proceedings of the 3rd Pan-American Congress on Applied Mechanics*, III PACAM, São Paulo.
Pimenta P. M. (1996). Geometrically-exact analysis of initially curved rods. In *Advances in Computational Techniques for Structural Engineering* (Edinburgh, U.K., v. 1, 99–108). Edinburgh: Civil-Comp Press.
Pimenta P. M., & Campello E. M. B. (2001). Geometrically nonlinear analysis of thin-walled space frames. In: *Proceedings of the Second European Conference on Computational Mechanics*, II ECCM, Cracow, Poland.
Pimenta, P. M., & Campello, E. M. B. (2003). A fully nonlinear multi-parameter rod model incorporating general cross-section in-plane changes and out-of-plane warping. *Latin-American Journal of Solids and Structures, 1*(1), 119–140.
Pimenta, P. M., & Campello, E. M. B. (2009). Shell curvature as an initial deformation: geometrically exact finite element approach. *International Journal for Numerical Methods in Engineering, 78,* 1094–1112.

Pimenta, P. M., & Yojo, T. (1993). Geometrically-exact analysis of spatial frames. *Applied Mechanics Reviews, ASME, New York, 46*(11), 118–128.

Pimenta, P. M., Campello, E. M. B., & Wriggers, P. (2008). An exact conserving algorithm for nonlinear dynamics with rotational DOFs and general hyperelasticity. Part 1: Rods. *Computational Mechanics, 42*(5), 715–732.

Pimenta P. M., Almeida Neto E. S., & Campello E. M. B. (2010). A fully nonlinear thin shell model of kirchhoff-love type. In: P. De Mattos Pimenta, & P. Wriggers (Eds.), *New trends in thin structures: Formulation, optimization and coupled problems. CISM international centre for mechanical sciences* (vol 519). Vienna: Springer.

Reissner, E. (1972). On one-dimensional finite-strain beam theory: The plane problem. *Journal of Applied Mathematics and Physics, 23*(5), 795–804.

Reissner, E. (1973). On one-dimensional large-displacement finite-strain beam theory. *Studies in Applied Mathematics, 52*(2), 87–95.

Simo, J. C. (1985). A finite strain beam formulation. The three-dimensional dynamics. Part I. *Computer Methods in Applied Mechanics and Engineering, 49*(1), 55–70.

Simo, J. C. (1992). The (symmetric) Hessian for geometrically nonlinear models in solid mechanics: Intrinsic definition and geometric interpretation. *Computer Methods in Applied Mechanics and Engineering*, 189–200.

Simo, J. C., & Vu-Quoc, L. (1986). A three dimensional finite-strain rod model. Part II: Computational aspects. *Computer Methods in Applied Mechanics and Engineering, 58*(1), 79–116.

Simo, J. C., & Vu-Quoc, L. (1991). A geometrically-exact rod model incorporating shear and torsion-warping deformation. *International Journal of Solids and Structures, 27*(3), 371–393.

Simo, J. C., Fox, D. D., & Hughes, T. J. R. (1992). Formulations of finite elasticity with independent rotations. *Computer Methods in Applied Mechanics and Engineering*. Holland, 277–288.

Sokolov I., Krylov S., & Harari I. (2015) Extension of non-linear beam models with deformable cross sections. *Computational Mechanics*, 999–1021.

Timoshenko, S. P. (1953). *History of Strength of Materials: With a brief account of the history of theory of elasticity and theory of structures* (p. 452). New York: McGraw-Hill.

Viebahn, N., Pimenta, P. M., & Schroeder, J. (2016) A simple triangular finite element for nonlinear thin shells - Statics, Dynamics and anisotropy. *Computational Mechanics*, online.

Whirman, A. B., & De Silva, C. N. (1974, December). An exact solution in a nonlinear theory of rods. *Journal of Elasticity*. Netherlands, 265–280.

Isogeometric Analysis of Solids in Boundary Representation

Sven Klinkel and Margarita Chasapi

Abstract In this chapter, we present boundary-oriented numerical methods to analyze three-dimensional solid structures. For the analysis, the original geometry of the solid is employed according to the isogeometric paradigm. For the parametrization of the domain, the idea of the scaled boundary finite element method is adopted. Hence, the boundary of the solid is sufficient to describe the entire domain. The presented approaches employ analytical and numerical solution methods such as the Galerkin and collocation methods. To illustrate the applicability in the analysis procedure, three formulations are elaborated and demonstrated by means of numerical examples. The advantages compared to standard numerical methods are discussed thoroughly.

1 Introduction

Typically solids are designed by the boundary representation modeling technique in computer-aided design (CAD) software (Stroud 2006). From the analysis point of view, the finite element method (FEM) is the most popular numerical technique. The geometry and the displacement response of the structure are approximated by Lagrange basis functions. This leads in general to an approximation of the geometry, which accordingly affects the accuracy of deformation results (Cottrell et al. 2009). To circumvent the geometrical approximation error, an exact description from the CAD model could be employed. This is the idea of the isogeometric analysis, which was introduced by Hughes et al. (2005). The main concept is to employ the same NURBS basis functions in order to describe the geometry and to approximate the displacements. However, for three-dimensional solids a three-dimensional tensor–product structure of NURBS objects must be adopted in isogeometric analysis in

The financial support of the German Research Foundation (DFG) under Grant No. KL1345/10-1 is gratefully acknowledged.

S. Klinkel (✉) · M. Chasapi
RWTH Aachen University, Aachen, Germany
e-mail: klinkel@lbb.rwth-aachen.de

J. Schröder and P. de Mattos Pimenta (eds.), *Novel Finite Element Technologies for Solids and Structures*, CISM International Centre for Mechanical Sciences 597,
https://doi.org/10.1007/978-3-030-33520-5_6

order to parameterize the physical domain (Cottrell et al. 2009; Düster et al. 2008; Temizer et al. 2012; Rank et al. 2012). Such a trivariate tensor–product structure, however, is not defined in the CAD model. In CAD, only the boundary surfaces of the solid are defined. A classical volumetric discretization of the inner domain becomes, therefore, a complicated task. This observation motivated the development of numerical formulations in which the solid is defined by its boundary, and only this boundary is used for isogeometric analysis. These so-called *boundary-oriented solid formulations* combine the advantages of boundary-oriented methods and isogeometric analysis.

Currently, the most well-known boundary-oriented methods are the boundary element method (BEM) and the scaled boundary finite element method (SB-FEM). The latter one is a special kind of fundamental solution-less boundary element method, which was introduced by Song and Wolf (1997, 1998). The basic idea lies on a boundary scaling technique. In the analysis, the solid is defined by its boundary and a scaling center. The scaling center is chosen in a zone from which the total boundary of the solid is visible (Song and Wolf 1997). The scaling center C will, in general, be located inside the domain. A radial scaling parameter ξ is introduced to conduct the scaling process. Hence, $\xi = 1$ represents the boundary of the solid and $\xi = 0$ denotes the scaling center, while $0 < \xi < 1$ describes a certain point inside the domain. Scaling the boundary of the solid with respect to the specified scaling center yields the solid, see Fig. 1. In the analysis, only the tensor–product structure of the boundary is employed, which is different from the "polar mesh" suggested by Bazilevs et al. (2014). In the SB-FEM approach, it is distinguished between parameters in the circumferential direction and in the radial scaling direction. The weak form of equilibrium is only enforced in the circumferential direction. In the scaling direction, the equilibrium is strongly applied. In the framework of linear elasticity, a second-order ordinary differential equation (ODE) is obtained in terms of the scaling parameter. In the circumferential direction a finite element approximation is employed, which utilizes the Lagrange basis functions (Song and Wolf 1997, 1998) or the NURBS basis functions as investigated by Lin et al. (2014), Natarajan et al. (2015),

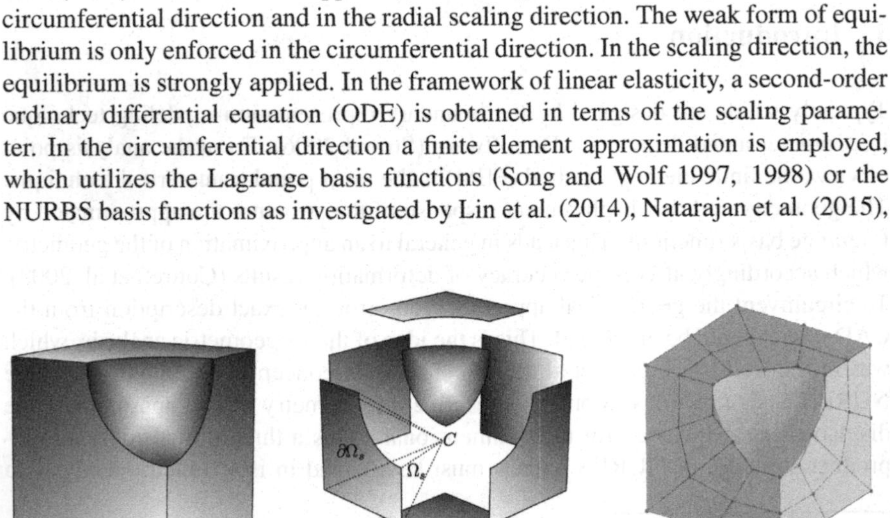

(a) Geometry (b) Boundary surfaces (c) Control polygon

Fig. 1 Geometry and control net of cube with spherical intersection. The geometry is created in CAD with the boundary representation modeling technique

Klinkel et al. (2015), and Chen et al. (2015, 2016) for the description of the geometry and the displacement. The Lagrange basis functions will lead to an approximation of the geometry. For linear elastic problems, the second-order ODE can be solved analytically or numerically. Analytical approaches include the eigenvalue method and the matrix function solution (Song and Wolf 1998). For the eigenvalue method, by introducing a dual vector form of the differential equation, the second-order ODE is reduced to first-order ODE according to Song and Wolf (1998) and Song (2004). Then, the eigenvalue problem of the first-order ODE is solved, which leads to the displacement response of the domain. An extension to nonlinear problems was proposed by Lin and Liao (2011), Ooi et al. (2014) and Behnke et al. (2014). The former one suggested an approach for nonlinear SB-FEM based on the homotopy analysis method. The latter studies are based on nonlinear shape functions derived from the solution of linear problems, which are employed for the nonlinear analysis. Besides the analytical approaches, a NURBS-based collocation approach has been proposed to solve the ODE numerically by Klinkel et al. (2015) for 2D and by Chen et al. (2015) for 3D problems. For this numerical approach, certain approximation is made for the choice of the first collocation point due to the numerical instability arising at the scaling center. Furthermore, a NURBS-based Galerkin approach has been proposed by Chen et al. (2016) to solve elasticity problems of boundary-represented solids. Moreover, Chasapi and Klinkel (2018) proposed the treatment of nonlinear problems by employing the approximation with NURBS and the Galerkin method for the solution in scaling direction.

In this chapter, boundary-oriented numerical methods are presented to solve the elasticity problem of solids in boundary representation. The chapter summaizes the main results of the publications Klinkel et al. (2015), Chen et al. (2015, 2016) and Chasapi and Klinkel (2018). The boundary scaling technique is employed to describe the solid. Thus, the boundary is exactly described in isogeometric analysis. Three numerical approaches will be demonstrated: the semi- analytical method, the NURBS-based hybrid collocation-Galerkin method and the NURBS- based Galerkin method. In the first two approaches, the weak form of equilibrium is enforced in the circumferential direction. The response in radial scaling direction is derived from the eigenvalue method and the collocation method accordingly. In the last approach, the weak form of equilibrium is employed in the radial scaling and circumferential direction. In all cases, NURBS basis functions are employed for the description of the boundary geometry as well as for the approximation of the displacements at the boundary. The displacement response in the radial scaling direction is approximated by one-dimensional NURBS basis functions for the numerical solution. Each approach results in a global system of equations, the solution of which yields the displacement response at the boundary surfaces and in the interior domain.

The outline of the chapter is as follows. In Sect. 2, the parametrization is presented. Section 3 provides the governing equations for linear elasticity of 3D problems. In Sect. 4, methods for the numerical approximation are presented. First, the basics of B-splines and NURBS as interpolation functions are illustrated. Moreover, a semi-analytical approach based on the eigenvalue method in radial scaling direction is given. Here, the derivation of the scaled boundary finite element equation is addressed. Furthermore, a NURBS-based collocation approach is presented. Here, NURBS basis functions are employed for the approximation, whereas the collocation method yields the solution in radial scaling direction. Finally, a NURBS-based Galerkin approach is presented. Here, the weak form of equilibrium discretized with NURBS is applied in all parametric directions. In Sect. 5, numerical examples are presented to evaluate the accuracy of the numerical methods. Suggestions for the optimum choice of the polynomial degree of collocation NURBS and the number of collocation points are provided. Furthermore, comparisons to the standard FEM and isogeometric analysis are given.

2 Parametrization

In this Section, the basic concept of the transformation of the geometry is provided. The main idea is based on the scaled boundary finite element method as proposed by Song and Wolf (1997, 1998). For the transformation, a radial scaling parameter is introduced to define the geometry of the solid. The boundary of the solid is thus scaled with respect to a scaling center C, see Fig. 2. The coordinates of C are denoted as $\hat{\boldsymbol{x}}_0$. The scaling center is defined such that the total boundary of the solid is visible (Song and Wolf 1997). The radial scaling parameter ξ runs from the scaling center toward the boundary, where $\xi = 0$ corresponds to the scaling center C and $\xi = 1$ describes the boundary of the solid. The total domain is partitioned into sectional domains $\Omega = \cup_{s=1}^{nsec} \Omega_s$. Each sectional domain is parametrized in the circumferential direction to describe the boundary $\partial\Omega_s$.

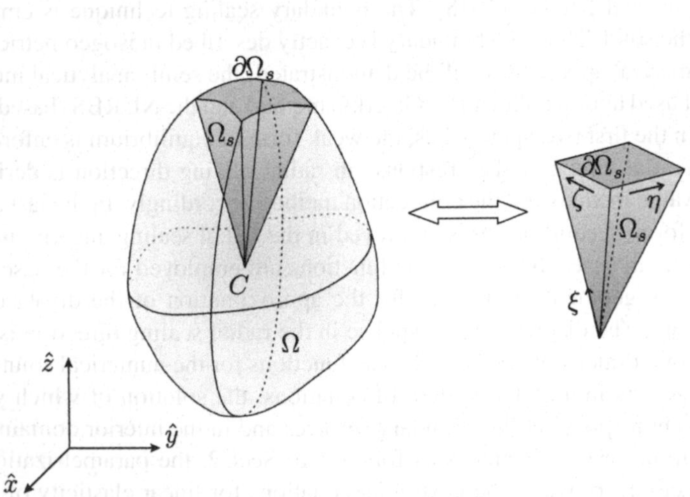

Fig. 2 The three-dimensional domain Ω and the sectional domain Ω_s in the physical space and the parameter space

For 3D problems, the boundary of each sectional domain Ω_s is a surface, see Fig. 2, and parametrized in the circumferential direction with η and ζ. It holds $0 \leq \eta \leq 1$ and $0 \leq \zeta \leq 1$. The scaling center C is defined as $\hat{\boldsymbol{x}}_0 = (\hat{x}_0, \hat{y}_0, \hat{z}_0)^T$. The position of a point on the boundary surfaces is denoted by $\boldsymbol{x}_s = (x_s, y_s, z_s)^T$ and a point in the interior of the solid is described by $\hat{\boldsymbol{x}}_s = (\hat{x}_s, \hat{y}_s, \hat{z}_s)^T$. Let $\boldsymbol{N}_s(\eta, \zeta)$ be a matrix of shape functions employed to describe the boundary surfaces. An arbitrary point on the boundary surfaces or in the domain is given as

$$\boldsymbol{x}_s = \boldsymbol{N}_s(\eta, \zeta)\,\boldsymbol{X} \quad \text{on } \partial\Omega_s, \qquad \hat{\boldsymbol{x}}_s = \hat{\boldsymbol{x}}_0 + \xi(\boldsymbol{N}_s(\eta, \zeta)\,\boldsymbol{X} - \hat{\boldsymbol{x}}_0) \quad \text{in } \Omega_s. \tag{1}$$

Here, we employ the NURBS basis functions to define the geometry of the boundary surfaces. This conforms ideally to the boundary representation modeling technique used in CAD. The vector $\boldsymbol{X}$ represents the coordinates of the control points on the boundary. Its dimension is $nst = 3 \cdot n_{bs}$, where n_{bs} is the number of control points on the boundary.

Considering Eq. (1) yields the Jacobian matrix

$$\boldsymbol{J} = \begin{bmatrix} \frac{\partial \hat{x}_s}{\partial \xi} & \frac{\partial \hat{y}_s}{\partial \xi} & \frac{\partial \hat{z}_s}{\partial \xi} \\ \frac{\partial \hat{x}_s}{\partial \eta} & \frac{\partial \hat{y}_s}{\partial \eta} & \frac{\partial \hat{z}_s}{\partial \eta} \\ \frac{\partial \hat{x}_s}{\partial \zeta} & \frac{\partial \hat{y}_s}{\partial \zeta} & \frac{\partial \hat{z}_s}{\partial \zeta} \end{bmatrix} = \begin{bmatrix} 1 & 0 & 0 \\ 0 & \xi & 0 \\ 0 & 0 & \xi \end{bmatrix} \underbrace{\begin{bmatrix} x_s - \hat{x}_0 & y_s - \hat{y}_0 & z_s - \hat{z}_0 \\ \frac{\partial x_s}{\partial \eta} & \frac{\partial y_s}{\partial \eta} & \frac{\partial z_s}{\partial \eta} \\ \frac{\partial x_s}{\partial \zeta} & \frac{\partial y_s}{\partial \zeta} & \frac{\partial z_s}{\partial \zeta} \end{bmatrix}}_{\bar{\boldsymbol{J}}} \tag{2}$$

It results in a multiplicative decomposition of the determinant $\det \boldsymbol{J} = \xi^2 \det \bar{\boldsymbol{J}} = \xi^2 \bar{J}$. The transformation of a volume element $\mathrm{d}V$ from the physical space to the parameter space reads

$$\mathrm{d}V = \mathrm{d}\hat{x}\,\mathrm{d}\hat{y}\,\mathrm{d}\hat{z} = \hat{\boldsymbol{x}}_{s,\xi} \cdot (\hat{\boldsymbol{x}}_{s,\eta} \times \hat{\boldsymbol{x}}_{s,\zeta})\,\mathrm{d}\xi\,\mathrm{d}\eta\,\mathrm{d}\zeta = \xi^2 \bar{J}\,\mathrm{d}\xi\,\mathrm{d}\eta\,\mathrm{d}\zeta. \tag{3}$$

3 Governing Equations

The governing equations for the three-dimensional (3D) problem is formulated in the Cartesian coordinates $(\hat{x}, \hat{y}, \hat{z})$, see Fig. 2. The displacement vector is defined as $\boldsymbol{u} = \boldsymbol{u}(\hat{x}, \hat{y}, \hat{z}) = [u_{\hat{x}}, u_{\hat{y}}, u_{\hat{z}}]^T$. It is assumed that the 3D domain Ω is bounded by $\partial\Omega = \partial_u\Omega \cup \partial_t\Omega$, where $\partial_u\Omega$ is the boundary with a prescribed displacement $\bar{\boldsymbol{u}}$ and $\partial_t\Omega$ is the boundary with a prescribed traction $\bar{\boldsymbol{t}}$. Here, the Neumann boundary condition does not overlap with the Dirichlet boundary condition, that is $\partial_u\Omega \cap \partial_t\Omega = \varnothing$.

The differential equation of motion reads

$$\mathbf{D}\boldsymbol{\sigma} + \rho \boldsymbol{b} = \mathbf{0} \tag{4}$$

where ρ is the mass density, $\boldsymbol{b}$ is the body force, and $\mathbf{D}$ is the linear differential operator.

The relation between the strains $\boldsymbol{\varepsilon}$ and the displacements $\boldsymbol{u}$ is given as

$$\boldsymbol{\varepsilon} = \mathbf{D}^T \boldsymbol{u}. \tag{5}$$

The stresses and strains are related by the elasticity matrix $\mathbb{C}$

$$\boldsymbol{\sigma} = \mathbb{C}\boldsymbol{\varepsilon}. \tag{6}$$

The Dirichlet and Neumann boundary conditions read

$$\boldsymbol{u} = \bar{\boldsymbol{u}} \quad \text{on } \partial_u \Omega, \qquad \boldsymbol{n}\,\boldsymbol{\sigma} = \bar{\boldsymbol{t}} \quad \text{on } \partial_t \Omega. \tag{7}$$

The matrix $\boldsymbol{n}$ contains the components of the outward unit normal vector. Equations (4)–(7) are the general formulas for elastostatic problems. For the 3D case, the strains are denoted by $\boldsymbol{\varepsilon} = [\varepsilon_x,\ \varepsilon_y,\ \varepsilon_z,\ \gamma_{yz},\ \gamma_{xz},\ \gamma_{xy}]^T$ and the stresses as $\boldsymbol{\sigma} = [\sigma_x,\ \sigma_y,\ \sigma_z,\ \tau_{yz},\ \tau_{xz},\ \tau_{xy}]^T$. Let $\mathbf{D}$ be the differential operator

$$\mathbf{D} = \begin{bmatrix} \frac{\partial}{\partial x} & 0 & 0 & 0 & \frac{\partial}{\partial z} & \frac{\partial}{\partial y} \\ 0 & \frac{\partial}{\partial y} & 0 & \frac{\partial}{\partial z} & 0 & \frac{\partial}{\partial x} \\ 0 & 0 & \frac{\partial}{\partial z} & \frac{\partial}{\partial y} & \frac{\partial}{\partial x} & 0 \end{bmatrix}. \tag{8}$$

The elasticity matrix $\mathbb{C}$ is written as

$$\mathbb{C} = \frac{E}{(1+\nu)(1-2\nu)} \begin{bmatrix} 1-\nu & \nu & \nu & 0 & 0 & 0 \\ \nu & 1-\nu & \nu & 0 & 0 & 0 \\ \nu & \nu & 1-\nu & 0 & 0 & 0 \\ 0 & 0 & 0 & \frac{1-\nu}{2} & 0 & 0 \\ 0 & 0 & 0 & 0 & \frac{1-\nu}{2} & 0 \\ 0 & 0 & 0 & 0 & 0 & \frac{1-\nu}{2} \end{bmatrix}. \tag{9}$$

The outward unit normal vector $\boldsymbol{n}$ is given as

$$\boldsymbol{n} = \begin{bmatrix} n_{\hat{x}} & 0 & 0 & 0 & n_{\hat{z}} & n_{\hat{y}} \\ 0 & n_{\hat{y}} & 0 & n_{\hat{z}} & 0 & n_{\hat{x}} \\ 0 & 0 & n_{\hat{z}} & n_{\hat{y}} & n_{\hat{x}} & 0 \end{bmatrix} \tag{10}$$

where $n_{\hat{x}}$, $n_{\hat{y}}$, and $n_{\hat{z}}$ are the components of the outward unit normal vector on $\partial\Omega$. Employing the parametrization of Sect. 2, each section is bounded by five surfaces, see Fig. 2. The normal vectors $\boldsymbol{n}^{\xi}$, $\boldsymbol{n}^{\eta}$, and $\boldsymbol{n}^{\zeta}$ are perpendicular to the surfaces described by the parameters $(\eta,\ \zeta)$, $(\zeta,\ \xi)$, and $(\xi,\ \eta)$, respectively, see Fig. 3. The formulas for the determination of the outward normal vectors and the description of the infinitesimal surface elements $\mathrm{d}S^{\xi}$, $\mathrm{d}S^{\eta}$, $\mathrm{d}S^{\zeta}$ are given in the Appendix.

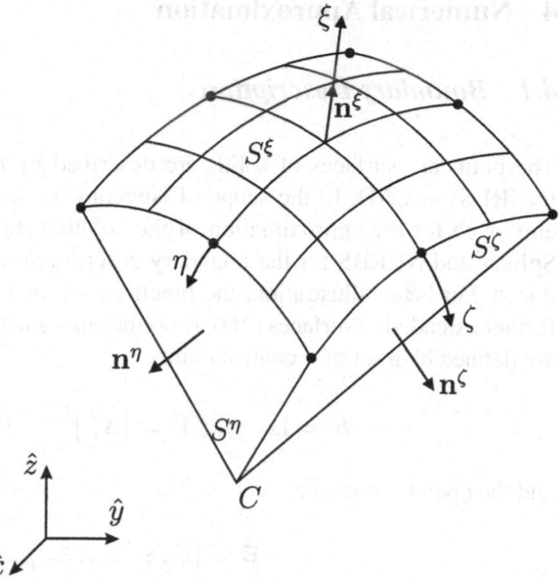

Fig. 3 The boundary of one 3D sectional domain is partitioned in $\partial\Omega_s = S^\xi \cup S^\eta \cup S^\zeta$

With the help of Eqs. (2), (8), and (A.1)–(A.3), the differential operator $\mathbf{D}$ is rewritten as

$$\mathbf{D} = \frac{1}{J}\left[\boldsymbol{b}_\xi \frac{\partial}{\partial \xi} + \frac{1}{\xi}\left(\boldsymbol{b}_\eta \frac{\partial}{\partial \eta} + \boldsymbol{b}_\zeta \frac{\partial}{\partial \zeta}\right)\right] \tag{11}$$

with the coefficient matrices

$$\boldsymbol{b}_i^T = g^i \begin{bmatrix} n_{\hat{x}}^i & 0 & 0 & 0 & n_{\hat{z}}^i & n_{\hat{y}}^i \\ 0 & n_{\hat{y}}^i & 0 & n_{\hat{z}}^i & 0 & n_{\hat{x}}^i \\ 0 & 0 & n_{\hat{z}}^i & n_{\hat{y}}^i & n_{\hat{x}}^i & 0 \end{bmatrix} \qquad (i = \xi,\ \eta,\ \zeta) \tag{12}$$

Using Eq. (12), the traction $\overline{\boldsymbol{t}} = \boldsymbol{n}\boldsymbol{\sigma}$ on any of the boundary surfaces $(\eta,\ \zeta)$, $(\zeta,\ \xi)$ and $(\xi,\ \eta)$ can be rewritten as

$$\overline{\boldsymbol{t}}^i = \frac{1}{g^i}\boldsymbol{b}_i^T \boldsymbol{\sigma} \qquad (i = \xi,\ \eta,\ \zeta) \tag{13}$$

Substituting Eq. (11) into Eqs. (5) and (6), we obtain the strains and stresses in the parameter space. However, it should be noted that there is a denominator in Eq. (11). The strains and stresses will exhibit singularity at the scaling center C, as at this point $\xi = 0$ holds. Here, the singularity does not arise from the method itself, but from the employed parametrization. The singularity will arise in the context of a solution to the strong form of the equation. To obviate the singularity in this case, we choose a tolerance in calculating the strains and stresses at the scaling center, see also Sect. 4.3.

4 Numerical Approximation

4.1 Boundary Description

The boundary surfaces of solids are described by nonuniform rational B-Splines (NURBS) in CAD. In the scope of isogeometric analysis, the same functions are employed for the approximation of the solution. In this Section, the basics of B-Splines and NURBS for the boundary description of the 3D domain will be introduced. For better illustration, the functions are first presented for curves (1D) and further extended to surfaces (2D). B-spline curves in the three-dimensional space $\mathbb{R}^3$ are defined by a set of n control points

$$\boldsymbol{B}_i = [x_i, y_i, z_i]^T = \left[\boldsymbol{X}_i^T\right]^T \qquad i = 1, \ldots, n \tag{14}$$

and the open knot vector

$$\Xi = \left\{\xi_1, \xi_2, \ldots, \xi_{n+p+1}\right\} , \tag{15}$$

where p is the polynomial degree of the B-spline basis functions. The entries ξ_i in the knot vector are nondecreasing. Intervals $\left[\xi_i, \xi_{i+1}\right]$ with $i = 1, \ldots, n + p$ are referred to as knot spans. The control points $\boldsymbol{B}_i$ are the nodal values in $\mathbb{R}^3$, which define the location in space of the B-spline curve $\boldsymbol{X}(\xi)$. The piecewise straight connection lines from $\boldsymbol{B}_i$ to $\boldsymbol{B}_{i+1}$ for $i = 1$ until $i = n - 1$ form the so-called control polygon, which is a piecewise linear approximation of the curve $\boldsymbol{X}(\xi)$, see Fig. 4. The B-spline basis functions $N_i^p(\xi)$ are defined recursively by the Cox-de Boor formula

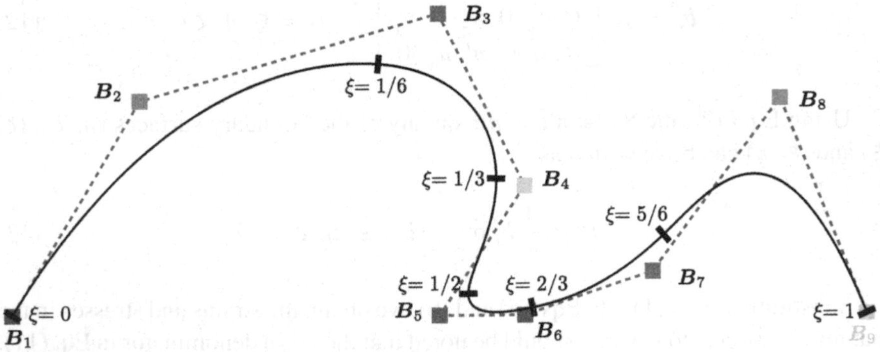

Fig. 4 Physical curve (solid black line) and control polygon (dotted red line) of a B-spline curve of order $p = 3$ with the knot vector $\Xi = \left\{0, 0, 0, 0, \frac{1}{6}, \frac{1}{3}, \frac{1}{2}, \frac{2}{3}, \frac{5}{6}, 1, 1, 1, 1\right\}$

$$
\begin{aligned}
p = 0: \qquad & N_i^0(\xi) = \begin{cases} 1 & \text{if } \xi_i \le \xi \le \xi_{i+1} \\ 0 & \text{otherwise} \end{cases} \\
p > 0: \qquad & N_i^p(\xi) = \frac{\xi - \xi_i}{\xi_{i+p} - \xi_i} N_i^{p-1}(\xi) + \frac{\xi_{i+p+1} - \xi}{\xi_{i+p+1} - \xi_{i+1}} N_{i+1}^{p-1}(\xi)\,.
\end{aligned}
\tag{16}
$$

The basis functions establish a map from the parameter space defined by the knot vector Ξ to the physical B-spline curve

$$
\boldsymbol{X}(\xi) = \sum_{i=1}^{n} N_i^p(\xi)\, \boldsymbol{B}_i \qquad \xi_1 \le \xi \le \xi_{n+p+1}. \tag{17}
$$

The support of basis functions is local and the influence of the control point $\boldsymbol{B}_i$ is limited to that interval. The number of basis functions which have influence on one knot span is given by $n_{en} = p + 1$. In the interval $[\xi_i, \xi_{i+1}]$, the basis functions N_{i-p}^p to N_i^p are nonzero.

Figure 4, a B-spline curve together with its control polygon is given. The associated basis functions are given in Fig. 5. Hereby each basis function is plotted in the same color as its associated control point. The knot values are denoted by a black stroke. The locally confined influence of the basis functions in each knot interval is clearly visible in Fig. 5. One important property for the usage of B-Splines as interpolation functions is the partition of unity

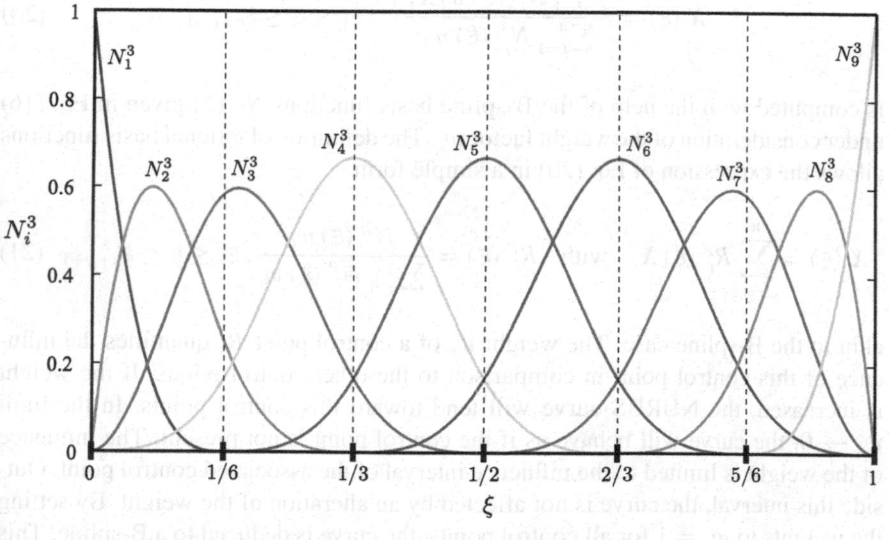

Fig. 5 Basis functions for the B-spline curve displayed in Fig. 4

$$\sum_{i=1}^{n} N_i^p(\xi) = 1 \qquad \forall \xi \in \Xi. \tag{18}$$

Further properties are the affine invariance, non-negativity, and variation diminishing property. A significant advantage of B-splines is that higher continuity allows the computation of $p - m$ continuous derivatives at knots and of an infinite number of derivatives within knot spans. Also, with the rising order of B-splines the smoothness of the curve increases in contrast to higher order Lagrange basis functions, which can entail oscillations. Univariate B-splines can be directly incorporated for the approximation of the solution in the radial scaling direction of the solid, see also Sects. 4.3 and 4.4.

NURBS curves are nonuniform rational B-splines. Their rational character allows an exact description of conic sections, such as circles. They can be understood as a projection of four-dimensional curves projected onto $\mathbb{R}^3$ (Cottrell et al. 2009). The notion of four dimensions is kept in the definition of the four-dimensional control points

$$\boldsymbol{B}_i = [x_i, y_i, z_i, w_i]^T = \left[\boldsymbol{X}_i^T, w_i\right]^T \qquad i = 1, \ldots, n. \tag{19}$$

Together with a knot vector, as given in Eq. (15), they define a NURBS curve of order p. The fourth coordinate w_i is the weight factor of the respective control point. All definitions and properties of B-splines hold accordingly, except the definition of the physical curve and the derivatives thereof. A physical point $\boldsymbol{X}(\xi)$ on the NURBS curve

$$\boldsymbol{X}(\xi) = \frac{\sum_{i=1}^{n} N_i^p(\xi)\, w_i \boldsymbol{X}_i}{\sum_{\hat{i}=1}^{n} N_{\hat{i}}^p(\xi)\, w_{\hat{i}}} \qquad \xi_1 \leq \xi \leq \xi_{n+p+1} \tag{20}$$

is computed with the help of the B-spline basis functions $N_i^p(\xi)$ given in Eq. (16) under consideration of the weight factor w_i. The definition of rational basis functions allows the expression of Eq. (20) in a simple form

$$\boldsymbol{X}(\xi) = \sum_{i=1}^{n} R_i^p(\xi)\, \boldsymbol{X}_i \quad \text{with} \quad R_i^p(\xi) = \frac{N_i^p(\xi)\, w_i}{\sum_{\hat{i}=1}^{n} N_{\hat{i}}^p(\xi)\, w_{\hat{i}}},\ \xi_1 \leq \xi \leq \xi_{n+p+1} \tag{21}$$

akin to the B-spline case. The weight w_i of a control point $\boldsymbol{B}_i$ quantifies the influence of this control point in comparison to the other control points. If the weight is increased, the NURBS curve will tend toward this control points. In the limit $w_i \to 0$, the curve will behave as if the control point is not present. The influence of the weight is limited to the influence interval of the associated control point. Outside this interval, the curve is not affected by an alteration of the weight. By setting the weights to $w_i = 1$ for all control points, the curve is deduced to a B-spline. This approach can be employed for the approximation in the scaling direction, where only straight radial lines are defined (see Sect. 4.3 and 4.4).

Now that we have gathered all necessary expressions to define one-dimensional NURBS, we can easily extend these to the two-dimensional case by employing the parametrization of the solid in Sect. 2. The geometry of the boundary surface $\partial\Omega_s$ is described by a NURBS surface, which is created by a tensor–product combination of the two knot vectors $\boldsymbol{H} = \{\eta_1, \eta_2, \ldots, \eta_{n_\eta+p+1}\}$ and $\boldsymbol{Z} = \{\zeta_1, \zeta_2, \ldots, \zeta_{n_\zeta+q+1}\}$. The orders of the basis functions along each parametric direction η and ζ are denoted by p and q, respectively. The control points $\boldsymbol{B}_{ij}$ are in general arranged in a rectangular grid called control point net. They are identified by a double index (ij) in parentheses, where the first number $i \in \{1, 2, \ldots, n_\eta\}$ denotes the position of the control point in η-direction. Analogously, $j \in \{1, 2, \ldots, n_\zeta\}$ identifies the position in ζ-direction.

The four components of the control points

$$\boldsymbol{B}_{(ij)} = \left[x_{(ij)}, y_{(ij)}, z_{(ij)}, w_{(ij)}\right]^T = \left[\boldsymbol{X}_{(ij)}^T, w_{(ij)}\right]^T \tag{22}$$

correspond to the spatial coordinates $\boldsymbol{X}_{(ij)}$ and the weight factor $w_{(ij)}$. The total number of control points is denoted by $n_{bs} = n_\eta \cdot n_\zeta$. The projection of the control point net from a four-dimensional space $\mathbb{R}^4$ to a surface embedded in the three-dimensional space $\mathbb{R}^3$ is carried out with the help of the rational NURBS basis functions $R_{(ij)}^{pq}(\eta, \zeta)$. The univariate B-spline basis functions given in Eq. (16) are used for both parametric directions and multiplied with the weight $w_{(ij)}$ to arrive at the rational NURBS surface basis functions

$$R_{(ij)}^{pq}(\eta, \zeta) = \frac{N_i^p(\eta)\, N_j^q(\zeta)\, w_{(ij)}}{\sum_{i=1}^{n_\eta} \sum_{j=1}^{n_\zeta} N_i^p(\eta)\, N_j^q(\zeta)\, w_{(ij)}}\,. \tag{23}$$

In analogy to the univariate B-spline, there are only $n_{en} = (p+1)(q+1)$ nonzero basis functions in each knot span that have an impact on the arbitrary rectangle $[\eta_i, \eta_{i+1}] \times [\zeta_j, \zeta_{j+1}]$. The number of potentially nonzero rectangles within a NURBS surface is given by $n_{el} = (n_1 - p_1)(n_2 - p_2)$. For a pair of parameters $(\eta, \zeta) \in [\eta_{i_0}, \eta_{i_0+1}] \times [\zeta_{j_0}, \zeta_{j_0+1}]$ a physical point $\boldsymbol{x}_s$ on the NURBS surface can be determined by

$$\boldsymbol{x}_s(\eta, \zeta) = \sum_{i=i_0-p}^{i_0} \sum_{j=j_0-q}^{j_0} R_{(ij)}^{pq}(\eta, \zeta)\, \boldsymbol{X}_{(ij)}\,. \tag{24}$$

Recall that this is the definition of the boundary geometry (see also Eq. 1) and keep in mind that the same definition will be employed for the approximation of the solution at the boundary (see also Eq. 27). All properties mentioned above for B-spline and NURBS curves can be carried forward to NURBS surfaces. The interested reader is referred to the studies of Piegl and Tiller (1997) as well as Cottrell et al. (2009) for more details on B-splines and NURBS.

For 3D problems, following the isogeometric concept, the displacements $\boldsymbol{u}_s(\xi = 1, \eta, \zeta)$ at the boundary surfaces are approximated with the same basis shape functions as the original geometry of the CAD model. Therefore, it holds

$$\boldsymbol{x}_s = \sum_{ij=1}^{n_{bs}} R_{(ij)}^{pq}(\eta,\,\zeta)\boldsymbol{X}_{(ij)} \qquad \boldsymbol{u}_s = \sum_{ij=1}^{n_{bs}} R_{(ij)}^{pq}(\eta,\,\zeta)\boldsymbol{U}_{(ij)}, \tag{25}$$

where $\boldsymbol{X}_{(ij)}$ defines the coordinate of the control point (ij) and n_{bs} denotes the total number of control points at the boundary surface $\partial\Omega_s$. The nodal displacement degrees of freedom are arranged akin in the vector $\boldsymbol{U}_{(ij)}$. $R_{(ij)}^{pq}(\eta,\,\zeta)$ is the NURBS basis function employed to describe the boundary surfaces, which is termed as boundary NURBS. The corresponding control points are denoted as boundary control points.

Considering Eq. (25) and rearranging all control point vectors $\boldsymbol{U}_{(ij)}$ in the vector $\boldsymbol{U}_s$, the approximation of the displacement at the boundary surface reads

$$\boldsymbol{u}_s = \underbrace{\begin{bmatrix} R_1 & 0 & 0 & R_2 & 0 & 0 & \ldots & R_{n_{bs}} & 0 & 0 \\ 0 & R_1 & 0 & 0 & R_2 & 0 & \ldots & 0 & R_{n_{bs}} & 0 \\ 0 & 0 & R_1 & 0 & 0 & R_2 & \ldots & 0 & 0 & R_{n_{bs}} \end{bmatrix}}_{\boldsymbol{N}_s(\eta,\,\zeta)} \boldsymbol{U}_s \tag{26}$$

Note that a bijective mapping holds between the subscript $I = 1, 2, \ldots, n_{bs}$ and the control point (ij). Considering Eqs. (25) and (26), the approximation of the displacements and the virtual displacements on the sectional domain Ω_s are defined as

$$\boldsymbol{u}(\xi,\,\eta,\,\zeta) = \boldsymbol{N}_s(\eta,\,\zeta)\boldsymbol{U}_s(\xi), \qquad \delta\boldsymbol{u}(\xi,\,\eta,\,\zeta) = \boldsymbol{N}_s(\eta,\,\zeta)\delta\boldsymbol{U}_s(\xi), \tag{27}$$

where $\boldsymbol{U}_s$ contains all nodal degrees of freedom in the circumferential direction of Ω_s. Accordingly, $\delta\boldsymbol{U}_s$ contains all virtual nodal displacements. An example of the interpolation in the circumferential direction of the boundary is illustrated in Fig. 6 for a 3D problem.

4.2 Scaled Boundary Finite Element Equation

The weak form of equilibrium can be derived by multiplying Eq. (4) with a test function $\delta\boldsymbol{u}$. Integration over the whole domain, application of integration by parts and consideration of the Neumann boundary condition in Eq. (7) yields the weak form

$$\sum_{s=1}^{nsec}\left(\int_{\Omega_s} \delta\boldsymbol{\varepsilon}^T\boldsymbol{\sigma}\,\mathrm{d}V - \int_{\partial\Omega_s} \delta\boldsymbol{u}^T\bar{\boldsymbol{t}}\,\mathrm{d}S - \int_{\Omega_s} \delta\boldsymbol{u}^T\rho\boldsymbol{b}\,\mathrm{d}V \right) = \boldsymbol{0}, \tag{28}$$

where $nsec$ is the total number of sectional domains Ω_s. The first term of Eq. (28) represents the internal virtual work, the second term is the external virtual work done by the boundary tractions, and the third term is the external virtual work done by

Fig. 6 Illustration of NURBS basis functions in the parameter space for 3D problems. The boundary NURBS basis functions $R_i^p(\eta)$ and $R_j^q(\zeta)$ with $p = 2$ and $q = 2$ are shown. $n_{bs} = 3 \times 3$ control points are employed in η and ζ directions

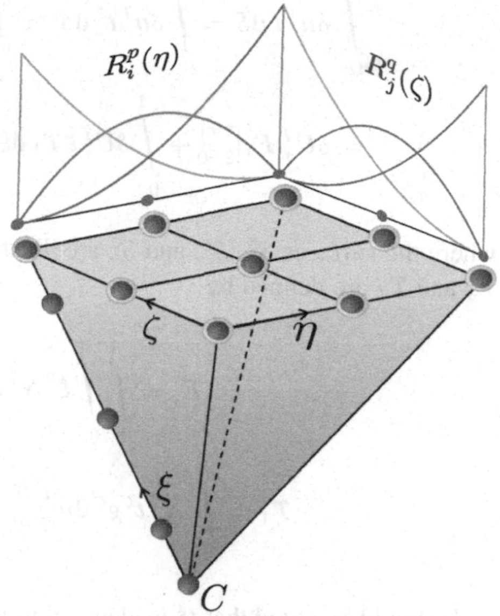

the body forces. The virtual strains are given as $\delta\boldsymbol{\varepsilon} = \mathbf{D}^T \delta\boldsymbol{u}(\xi, \eta, \zeta)$. Note that only the boundary surfaces $\partial\Omega_s$ are approximated with NURBS as described in Sect. 4.1, whereas the solution in the radial scaling direction is carried out analytically. The stress vector is computed by $\boldsymbol{\sigma} = \mathbb{C}\mathbf{D}^T \boldsymbol{u}$ using Eqs. (5) and (6). The first term in Eq. (28) is rewritten by employing integration by parts to

$$\int_{\Omega_s} \delta\boldsymbol{\varepsilon}^T \boldsymbol{\sigma} \, \mathrm{d}V = \delta\boldsymbol{U}_s^T \left(\xi^2 \boldsymbol{k}_{11} \boldsymbol{U}_{s,\xi} + \xi \boldsymbol{k}_{21} \boldsymbol{U}_s\right)\Big|_{\xi=0}^{\xi=1}$$
$$- \int_0^1 \delta\boldsymbol{U}_s^T \left[\xi^2 \boldsymbol{k}_{11} \boldsymbol{U}_{s,\xi\xi} + \xi \left(2\boldsymbol{k}_{11} + \boldsymbol{k}_{12} - \boldsymbol{k}_{21}\right) \boldsymbol{U}_{s,\xi} + \left(\boldsymbol{k}_{12} - \boldsymbol{k}_{22}\right) \boldsymbol{U}_s\right] \mathrm{d}\xi \tag{29}$$

with $(\ldots)_{,\xi} = \frac{\partial(\ldots)}{\partial\xi}$. Let $\boldsymbol{B}_1 = \frac{1}{\bar{J}} \boldsymbol{b}_\xi \boldsymbol{N}_s$ and $\boldsymbol{B}_2 = \frac{1}{\bar{J}} (\boldsymbol{b}_\eta \boldsymbol{N}_{s,\eta} + \boldsymbol{b}_\zeta \boldsymbol{N}_{s,\zeta})$ and considering Eqs. (3) and (11), the coefficient matrices are given as

$$\boldsymbol{k}_{11} = \int_0^1 \int_0^1 \boldsymbol{B}_1^T \mathbb{C} \boldsymbol{B}_1 \, \bar{J} \, \mathrm{d}\eta \, \mathrm{d}\zeta \qquad \boldsymbol{k}_{22} = \int_0^1 \int_0^1 \boldsymbol{B}_2^T \mathbb{C} \boldsymbol{B}_2 \, \bar{J} \, \mathrm{d}\eta \, \mathrm{d}\zeta$$
$$\boldsymbol{k}_{12} = \int_0^1 \int_0^1 \boldsymbol{B}_1^T \mathbb{C} \boldsymbol{B}_2 \, \bar{J} \, \mathrm{d}\eta \, \mathrm{d}\zeta \qquad \boldsymbol{k}_{21} = \int_0^1 \int_0^1 \boldsymbol{B}_2^T \mathbb{C} \boldsymbol{B}_1 \, \bar{J} \, \mathrm{d}\eta \, \mathrm{d}\zeta \, . \tag{30}$$

The second term in Eq. (28) is rewritten considering Eqs. (A.4)–(A.6) to

$$\int_{\partial\Omega_s} \delta\boldsymbol{u}^T \boldsymbol{t}\,\mathrm{d}S = \int_{S^\xi} \delta\boldsymbol{u}^T \boldsymbol{t}^\xi\,\mathrm{d}S^\xi + \int_{S^\eta} \delta\boldsymbol{u}^T \boldsymbol{t}^\eta\,\mathrm{d}S^\eta + \int_{S^\zeta} \delta\boldsymbol{u}^T \boldsymbol{t}^\zeta\,\mathrm{d}S^\zeta$$
$$= \delta\boldsymbol{U}_s^T \boldsymbol{F}_s\big|_{\xi=0}^{\xi=1} + \int_0^1 \delta\boldsymbol{U}_s^T \xi \boldsymbol{T}_1\,\mathrm{d}\xi\,, \tag{31}$$

where the surfaces S^ξ, S^η, and S^ζ are illustrated in Fig. 3. The coefficient matrices $\boldsymbol{F}_s$ and $\boldsymbol{T}_1$ are defined by

$$\boldsymbol{F}_s = \int_0^1\int_0^1 \xi^2 \boldsymbol{N}_s^T \boldsymbol{t}^\xi g^\xi\,\mathrm{d}\eta\,\mathrm{d}\zeta$$
$$\boldsymbol{T}_1 = \int_0^1 \boldsymbol{N}_s^T \boldsymbol{t}^\zeta g^\zeta\,\mathrm{d}\eta\big|_{\zeta=0}^{\zeta=1} + \int_0^1 \boldsymbol{N}_s^T \boldsymbol{t}^\eta g^\eta\,\mathrm{d}\zeta\big|_{\eta=0}^{\eta=1}\,. \tag{32}$$

It should be noted that $\boldsymbol{t}^\xi$ is identical to the prescribed traction $\bar{\boldsymbol{t}}$ on $\partial_t\Omega$ and that the force vector $\boldsymbol{F}_s$ represents the nodal forces at the control points. After assembly over all sections $\boldsymbol{T}_1$ vanishes. With the help of Eq. (3), the third term in Eq. (28) is reformulated to

$$\int_{\Omega_s} \delta\boldsymbol{u}^T \rho\boldsymbol{b}\,\mathrm{d}V = \int_0^1 \delta\boldsymbol{U}_s^T \xi^2 \boldsymbol{T}_2\,\mathrm{d}\xi \quad \text{with} \quad \boldsymbol{T}_2 = \int_0^1\int_0^1 \boldsymbol{N}_s^T \rho\boldsymbol{b}\bar{J}\,\mathrm{d}\eta\,\mathrm{d}\zeta\,. \tag{33}$$

Substituting Eqs. (29), (31), and (33) into the weak form of Eq. (28) yields

$$\Sigma_{s=1}^{nsec}\left(\delta\boldsymbol{U}_s^T\left(\xi^2\boldsymbol{k}_{11}\boldsymbol{U}_{s,\xi} + \xi\boldsymbol{k}_{21}\boldsymbol{U}_s\right)\big|_{\xi=0}^{\xi=1}\right) -$$
$$\Sigma_{s=1}^{nsec}\left(\int_0^1 \delta\boldsymbol{U}_s^T\left[\xi^2\boldsymbol{k}_{11}\boldsymbol{U}_{s,\xi\xi} + \xi\left(2\boldsymbol{k}_{11} + \boldsymbol{k}_{12} - \boldsymbol{k}_{21}\right)\boldsymbol{U}_{s,\xi} + \left(\boldsymbol{k}_{12} - \boldsymbol{k}_{22}\right)\boldsymbol{U}_s\right]\mathrm{d}\xi\right)$$
$$-\ \Sigma_{s=1}^{nsec}\left(\delta\boldsymbol{U}_s^T \boldsymbol{F}_s\big|_{\xi=0}^{\xi=1} + \int_0^1 \delta\boldsymbol{U}_s^T \xi \boldsymbol{T}_1\,\mathrm{d}\xi\right) - \Sigma_{s=1}^{nsec}\left(\int_0^1 \delta\boldsymbol{U}_s^T \xi^2 \boldsymbol{T}_2\,\mathrm{d}\xi\right) = \boldsymbol{0}. \tag{34}$$

Collecting the boundary terms and the field equations leads to the following set of equations:

$$\mathop{\mathbf{A}}_{s=1}^{nsec}\left(\boldsymbol{k}_{11}\boldsymbol{U}_{s,\xi} + \boldsymbol{k}_{21}\boldsymbol{U}_s - \boldsymbol{F}_s\right) = \boldsymbol{0} \quad \text{on } \partial_t\Omega \tag{35}$$

$$\begin{aligned}&\mathop{\mathbf{A}}_{s=1}^{nsec}\left(\xi^2\boldsymbol{k}_{11}\boldsymbol{U}_{s,\xi\xi}+\xi\left(2\boldsymbol{k}_{11}+\boldsymbol{k}_{12}-\boldsymbol{k}_{21}\right)\boldsymbol{U}_{s,\xi}+\left(\boldsymbol{k}_{12}-\boldsymbol{k}_{22}\right)\boldsymbol{U}_s\right)\\&+\mathop{\mathbf{A}}_{s=1}^{nsec}\left(\xi\boldsymbol{T}_1+\xi^2\boldsymbol{T}_2\right)=\mathbf{0}\quad\text{in }\Omega,\end{aligned}\tag{36}$$

where $\mathop{\mathbf{A}}_{s=1}^{nsec}$ is introduced as the assembly operator. Equation (36) is the so-called scaled boundary finite element equation, which is a second-order Euler-type ordinary differential equation (ODE). The displacement $\boldsymbol{U}_s$ is a function of the radial scaling parameter ξ only. Here, it is worthwhile to note that the governing equation of elasticity has been weakly enforced in the circumferential direction, see Eqs. (28) and (34), but it remains strong in the radial scaling direction as shown in Eq. (36). For linear elasticity, a unique analytical solution exists and can be computed with the eigenvalue method. The interested reader is referred to the studies of Song and Wolf (1997, 1998) for further details on the solution procedure.

4.3 NURBS-Based Hybrid Collocation-Galerkin Method

In this Section, the NURBS-based hybrid collocation-Galerkin method (NURBS-HCGM) will be presented. In the scope of this approach, the weak form of equilibrium is applied only in the circumferential direction of the boundary. In the radial scaling direction, the equation is solved numerically by employing the collocation method. NURBS basis functions approximate the response in all parametric directions. The scaled boundary finite element equation can be derived analogously to Sect. 4.2. Hereafter, the B-splines approximation and the collocation in radial scaling direction will be discussed.

B-spline approximation in scaling direction In this approach, NURBS basis functions are employed to describe the geometry of the boundary. For brevity, we will only refer to the NURBS approximation for 3D problems here. For 2D problems, the formulas for the NURBS approximation could be derived similarly.

The NURBS basis functions $R^{pq}_{(ij)}(\eta,\ \zeta)$ in the circumferential direction are adopted from the geometry model following the boundary representation modeling technique in CAD, see also Sect. 4.1. They can be enriched via order elevation or knot insertion. The geometry is, therefore, described exactly. The interpolation function $\boldsymbol{N}_s(\eta,\ \zeta)$ is employed for the approximation of the solution on the boundary as given in Eq. (27). Note that in contrast to Sect. 4.2, the sectional domain Ω_s is here approximated with NURBS basis functions on the boundary and also in the radial scaling direction. In the radial scaling direction, all weight factors are set to $w_r = 1$ since only straight lines are defined. Hence, B-splines are employed for the interpolation in the radial scaling direction. The displacement $\boldsymbol{U}_s(\xi)$ is only a function of the radial scaling parameter ξ. Hence, the univariate NURBS basis functions $R^t_r(\xi)$ are employed. The displacement in the radial scaling direction reads

$$\boldsymbol{U}_s(\xi) = \sum_{r=1}^{n_{cp}} R_r^t(\xi)\, \boldsymbol{U}_{sr} \tag{37}$$

where the displacement vector $\boldsymbol{U}_{sr}$ is associated to the control points which are located in the radial scaling direction. The dimension of $\boldsymbol{U}_s$ and $\boldsymbol{U}_{sr}$ is $nds = 3 \cdot n_{bs}$ for 3D problems. n_{bs} is the total number of control points at the boundary of $\partial\Omega_s$. The polynomial degree in the radial scaling direction is denoted as t, and n_{cp} is the total number of control points per line in the radial scaling direction, see Fig. 7. The knot vector $\Xi = [\xi_1, \xi_2, \ldots, \xi_{n_{cp}+t+1}]$ is employed to determine B-spline basis function R_r^t. Here, the radial scaling direction is represented by a straight line. The polynomial degree is $t = 1$ and the corresponding knot vector reads $\Xi = [0, 0, 1, 1]$. These are taken as the start values for further refinement by knot insertion or/and order elevation. In principle h-, p-, and k-refinement can be applied (Cottrell et al. 2009). Consequently, the number of control points n_{cp} is increased.
Rearranging all control point vectors $\boldsymbol{U}_{sr}$ in the vector $\boldsymbol{U}_{s\xi}$, the displacement Eq. (37) and the virtual displacement in the radial scaling direction could be rewritten as

$$\boldsymbol{U}_s(\xi) = \underbrace{\begin{bmatrix} R_1^t & 0 & 0 & \cdots & R_2^t & 0 & 0 & \cdots & R_3^t & 0 & 0 & \cdots \\ 0 & R_1^t & 0 & \cdots & 0 & R_2^t & 0 & \cdots & 0 & R_3^t & 0 & \cdots \\ 0 & 0 & R_1^t & \cdots & 0 & 0 & R_2^t & \cdots & 0 & 0 & R_3^t & \cdots \\ \vdots & \vdots & \vdots & \ddots & \vdots & \vdots & \vdots & \ddots & \vdots & \vdots & \vdots & \ddots \end{bmatrix}}_{\boldsymbol{N}_\xi(\xi)} \boldsymbol{U}_{s\xi},$$
$$\delta\boldsymbol{U}_s(\xi) = \boldsymbol{N}_\xi(\xi)\delta\boldsymbol{U}_{s\xi}. \tag{38}$$

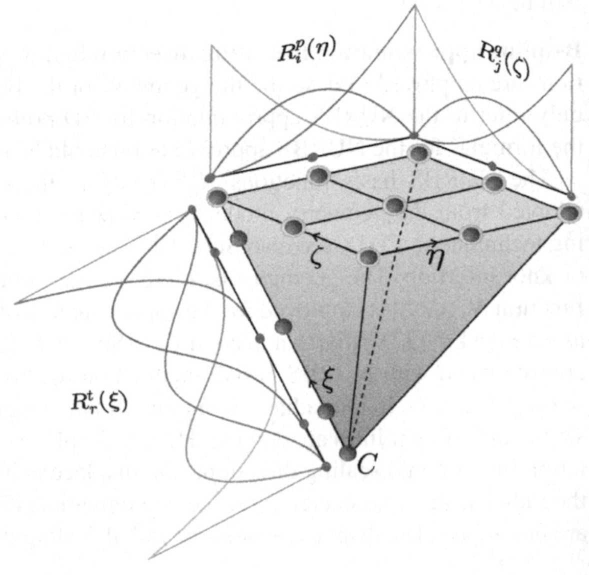

Fig. 7 Illustration of NURBS basis functions in the parameter space for 3D problems. $n_{cp} = 5$ control points are used for the interpolation in the radial scaling direction. The B-splines $R_r^t(\xi)$ with $t = 3$ are shown only along one line in the radial scaling direction. All others radial lines are identical. $n_{bs} = 3 \times 3$ control points are employed in η and ζ directions. The boundary NURBS basis functions $R_i^p(\eta)$ and $R_j^q(\zeta)$ with $p = 2$ and $q = 2$ are shown

An example of the interpolation in the radial scaling direction and also in the circumferential direction is illustrated in Fig. 7 for a 3D problem. Taking into account the high continuity of the NURBS, the governing field equation (36) is approximated as

$$\underset{s=1}{\overset{nsec}{\mathbf{A}}} \left(\sum_{r=1}^{n_{cp}} \left[\xi^2 \boldsymbol{k}_{11} R_r^{t\,''} + \xi \, (2\boldsymbol{k}_{11} + \boldsymbol{k}_{12} - \boldsymbol{k}_{21}) \, R_r^{t\,'} + (\boldsymbol{k}_{12} - \boldsymbol{k}_{22}) \, R_r^t \right] \boldsymbol{U}_r \right) + \underset{s=1}{\overset{nsec}{\mathbf{A}}} \left(\xi \boldsymbol{T}_1 + \xi^2 \boldsymbol{T}_2 \right) = \mathbf{0} \quad \text{in } \Omega \, , \tag{39}$$

where the abbreviation $(\dots)' = \frac{\partial(\dots)}{\partial \xi}$ is used. The approximation of the remaining Neumann boundary conditions in Eq. (35) is given by

$$\underset{s=1}{\overset{nsec}{\mathbf{A}}} \left(\sum_{r=1}^{n_{cp}} (\boldsymbol{k}_{11} R_r^{t\,'} + \boldsymbol{k}_{12} R_r^t) \boldsymbol{U}_r - \boldsymbol{F}_s \right) = \mathbf{0} \quad \text{on } \partial_t \Omega \, . \tag{40}$$

Collocation In the scope of the NURBS-HCGM, the collocation method is employed to solve Equation (39) numerically. One collocation point per control point is sufficient in the framework of the NURBS-based collocation method, which can be interpreted as a rank sufficient one point quadrature as observed by Auricchio et al. (2012) and Schillinger et al. (2013). In the proposed method, only a one-dimensional ODE (Eqs. (35) and (36)) needs to be solved. In this case , the NURBS-based collocation method has been proved to be numerically stable (Auricchio et al. 2012), which motivates the use of NURBS for the approximation in scaling direction. The potential of the collocation method to solve differential equations has been widely investigated in the context of NURBS-based isogeometric analysis (for example, by Auricchio et al. 2010, 2012 and De Lorenzis et al. 2015 as well as Kiendl et al. 2015 and also Reali and Gomez 2015 and most recently Gomez and De Lorenzis 2016). Due to the above features, the NURBS-based collocation method is utilized to solve Eq. (39). The Greville abscissae is employed to define the collocation points. They are related to the knot vector $\Xi = [\xi_1, \xi_2, \dots, \xi_{n_{cp}+t+1}]$ as

$$\hat{\xi}_k = \frac{\xi_{k+1} + \xi_{k+2} + \dots \xi_{k+t}}{t} \quad \text{for} \quad k = 1, \dots, n_{cp} \, . \tag{41}$$

The first collocation point is located at the scaling center C with $\hat{\xi}_1 = 0$, while the last one is at the boundary with $\hat{\xi}_{n_{cp}} = 1$. As stated previously, the proposed method is a boundary-oriented method. Scaling the boundary surfaces yields the 3D solid. The unknown variables are the boundary degrees of freedom. If the first collocation point is chosen exactly at the scaling center C ($\hat{\xi}_1 = 0$), then numerical instabilities will arise. The physical explanation is that several control points will collapse to a single point (the scaling center C). This entails a rank deficiency of the final matrices in the collocation method. Equation (36) is a second-order Euler-type ordinary differential equation (ODE) and its approximations is presented in Eq. (39). If the first collocation point is chosen exactly at the scaling center C ($\hat{\xi}_1 = 0$), then Eq. (39) reduces to

$$\mathop{\mathbf{A}}_{s=1}^{nsec} \left((\boldsymbol{k}_{12} - \boldsymbol{k}_{22})\, R_r^t \boldsymbol{U}_r \right) = \mathbf{0}\,. \tag{42}$$

Considering the property of the NURBS basis functions, $R_r^t(\hat{\xi}_1 = 0) = 1$ holds for the first control point. As a result, Eq. (42) can be rewritten to the homogeneous equation

$$(\boldsymbol{k}_{12} - \boldsymbol{k}_{22})\, \boldsymbol{U}_r(\hat{\xi}_1 = 0) = \mathbf{0}\,. \tag{43}$$

It will lead to either zero solutions or an infinite number of solutions at the scaling center C. However, both the solutions contradict to the prerequisite of finite solutions at the scaling center C. Hence, numerical instability will occur if the first collocation point coincides with the scaling center C ($\hat{\xi}_1 = 0$). To obviate this effect, a tolerance (tol) is introduced for the analysis and the first collocation point is defined as $\hat{\xi}_1 = 0 + tol$. The influence of the choice of the tolerance has been investigated by Chen et al. (2015). It has been observed that the results of the NURBS-HCG depend only very slightly on the choice of the first collocation point $\hat{\xi}_1$. There is only a marginal dependence between the accuracy of the approach and the choice for $\hat{\xi}_1$. Thus, the shifting of the first collocation point $\hat{\xi}_1$ can be allowed from a numerical point of view and the influence of this slight modification on the results can be neglected. In general, the first collocation point can be specified, for example, as $\hat{\xi}_1 = 0 + tol = 10^{-4}$.

The NURBS basis functions employed in the radial scaling direction are abbreviated as collocation NURBS. The displacement vector at the collocation points reads

$$\boldsymbol{U}_s(\hat{\xi}_k) = \sum_{r=1}^{n_{cp}} R_r^t(\hat{\xi}_k) \boldsymbol{U}_r = \sum_{r=1}^{n_{cp}} R_{rk}^t \boldsymbol{U}_r\,. \tag{44}$$

The approximated Eq. (39) is reformulated for each collocation point except the one at the boundary. A system of $k = 1, \ldots, n_{cp} - 1$ equations of the type

$$\begin{aligned} &\mathop{\mathbf{A}}_{s=1}^{nsec} \left(\sum_{r=1}^{n_{cp}} \left[\hat{\xi}_k^2 \boldsymbol{k}_{11} R_{rk}^{t\;\prime\prime} + \hat{\xi}_k \,(2\boldsymbol{k}_{11} + \boldsymbol{k}_{12} - \boldsymbol{k}_{21})\, R_{rk}^{t\;\prime} + (\boldsymbol{k}_{12} - \boldsymbol{k}_{22})\, R_{rk}^t \right] \boldsymbol{U}_r \right) \\ &+ \mathop{\mathbf{A}}_{s=1}^{nsec} \left(\hat{\xi}_k \boldsymbol{T}_1 + \hat{\xi}_k^2 \boldsymbol{T}_2 \right) = \mathbf{0}. \end{aligned} \tag{45}$$

is established. At the boundary collocation point ($\hat{\xi}_{n_{cp}} = 1$), the Neumann boundary conditions defined in Eq. (40) are enforced by

$$\mathop{\mathbf{A}}_{s=1}^{nsec} \left(\sum_{r=1}^{n_{cp}} (\boldsymbol{k}_{11} R_{rn_{cp}}^{t\;\prime} + \boldsymbol{k}_{12} R_{rn_{cp}}^t) \boldsymbol{U}_r - \boldsymbol{F}_s \right) = \mathbf{0} \quad \text{on } \partial_t \Omega\,. \tag{46}$$

For a compact notation, the abbreviations $\bar{\boldsymbol{k}}_{rn_{cp}} = \boldsymbol{k}_{11}(R^t_{rn_{cp}})' + \boldsymbol{k}_{12}R^t_{rn_{cp}}$ and $\hat{\boldsymbol{k}}_{rk} = \hat{\xi}_k^2 \boldsymbol{k}_{11} R^t_{rk}{}'' + \hat{\xi}_k\left(2\boldsymbol{k}_{11} + \boldsymbol{k}_{12} - \boldsymbol{k}_{21}\right) R^t_{rk}{}' + \left(\boldsymbol{k}_{12} - \boldsymbol{k}_{22}\right) R^t_{rk}$ are introduced. After assembling over all sections of the domain Ω, $\boldsymbol{T}_1$ vanishes. The body forces $\boldsymbol{T}_2$ are neglected for the sake of simplicity. The system of equations constituted by Eqs. (45) and (46) is reformulated to

$$\begin{aligned} &\mathop{\mathbf{A}}_{s=1}^{nsec}\left(\sum_{r=1}^{n_{cp}} \hat{\boldsymbol{k}}_{rk}\boldsymbol{U}_r\right) = \boldsymbol{0} \qquad k = 1, \ldots, n_{cp} - 1, \\ &\mathop{\mathbf{A}}_{s=1}^{nsec}\left(\sum_{r=1}^{n_{cp}} \bar{\boldsymbol{k}}_{rn_{cp}}\boldsymbol{U}_r - \boldsymbol{F}_s\right) = \boldsymbol{0} \qquad \text{on } \partial_t\Omega\,. \end{aligned} \tag{47}$$

The degrees of freedom located at the interior nodes are eliminated by static condensation. The degrees of freedom, located at the boundary $\partial\Omega$ are denoted as $\boldsymbol{U}$, while those associated with the interior of Ω are referred to as $\hat{\boldsymbol{U}}$. Let $\boldsymbol{F} = \mathop{\mathbf{A}}_{s=1}^{nsec} \boldsymbol{F}_s(\xi = 1)$, the system of equations reads

$$\begin{bmatrix} \hat{\boldsymbol{K}}_i & \hat{\boldsymbol{K}}_{ib} \\ \bar{\boldsymbol{K}}_{bi} & \bar{\boldsymbol{K}}_b \end{bmatrix} \begin{bmatrix} \hat{\boldsymbol{U}} \\ \boldsymbol{U} \end{bmatrix} - \begin{bmatrix} \boldsymbol{0} \\ \boldsymbol{F} \end{bmatrix} = \boldsymbol{0}\,, \tag{48}$$

where the subscripts b and i denote the matrices related to the boundary and the interior of Ω, respectively. The vector $\hat{\boldsymbol{U}}$ is eliminated from Eq. (48) by a static condensation. It results in the reduced system of equations

$$\boldsymbol{K}_\Omega \boldsymbol{U} + \boldsymbol{P}_\Omega = \boldsymbol{0} \tag{49}$$

with $\boldsymbol{K}_\Omega = \bar{\boldsymbol{K}}_b - \bar{\boldsymbol{K}}_{bi}\hat{\boldsymbol{K}}_i^{-1}\hat{\boldsymbol{K}}_{ib}$ and $\boldsymbol{P}_\Omega = -\boldsymbol{F}$. Equation (35) represents the weak form of the Neumann boundary conditions. De Lorenzis et al. (2015) observed that the imposition of the Neumann boundary conditions in a weak sense produces significantly lower error levels in comparison to a collocation-based evaluation of the Neumann boundary conditions. The Dirichlet boundary conditions are directly applied to the control points at the boundary. Hence, simply columns and rows of Eq. (49) are deleted. Solving Equation (49) yields the displacement $\boldsymbol{U}$ of the control points at the boundary. All nodal displacements $\boldsymbol{U}_s$ of a section can be determined using Eqs. (48) and (37). The displacement vector of an arbitrary point in Ω_s is given by Eq. (27). The strains and stresses are identified by Eqs. (5) and (6). It is noted that the displacement $\boldsymbol{U}$ of the boundary control points is the essential variable in the whole algorithm. All other variables can be derived from it. Hence, the displacement solution at the boundary can be employed to evaluate the performance of this approach. In general, NURBS-HCGM solves the ODE defined by Eqs. (35) and (36) numerically due to the NURBS approximation in the radial scaling direction. The approximation of the displacement in the radial scaling direction allows the analysis of both linear and nonlinear problems. For linear problems, the ODE is directly solved by the collocation method. While for nonlinear analysis, the Newton–Raphson scheme can be employed to solve the equilibrium equations iteratively. The presented formulation is suitable for the analysis of star-shaped solids. To deal with complex geometries, the finite element discretization could be employed to discretize the solid into numerous star-shaped macro elements. The stiffness matrix and the right-hand side for each macro element can be derived as in Eq. (49). This provides a macro element formulation for the general analysis of solids.

4.4 NURBS-Based Galerkin Method

In this Section, the NURBS-based Galerkin method (NURBS-G) will be presented. In the scope of this approach, the weak form of equilibrium equation is applied in all parametric directions of the solid. Hereafter, the NURBS approximation and the derivation of the weak form of the equilibrium equation will be addressed.

NURBS Approximation In this approach, NURBS basis functions are employed to describe the geometry of the boundary, see Sect. 4.1. The approximation is done analogously to the NURBS-HCGM, see Sect. 4.3, which means that the interpolation at the boundary is done with the NURBS basis functions of the geometry whereas in the interior it is done with B-Splines. An example of the interpolation in the radial scaling direction and also in the circumferential direction is illustrated in Fig. 7 for a 3D problem. Here, it is worthwhile to note that the stiffness matrix of the NURBS-G can be alternatively obtained by modifying the geometry of a rectangular patch to a triangle. An example is illustrated in Fig. 8 for the 2D case. In the following, the derivation of the stiffness matrix with the original geometry of the boundary as the starting point for the analysis will be demonstrated.

Weak Form of Equilibrium Equation In this Section, the weak form of equilibrium is derived for the 3D case. For 2D problems, the weak form of equilibrium equation could be obtained analogously.The difference to the procedure in Section 4.3 is that the weak form is employed in all parametric directions. Also here, we employ the principle of virtual work to derive the Eq. (28). Consequently, the first term of Eq. (28) represents the internal virtual work, the second term is the external virtual work

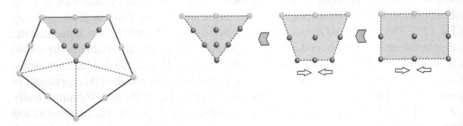

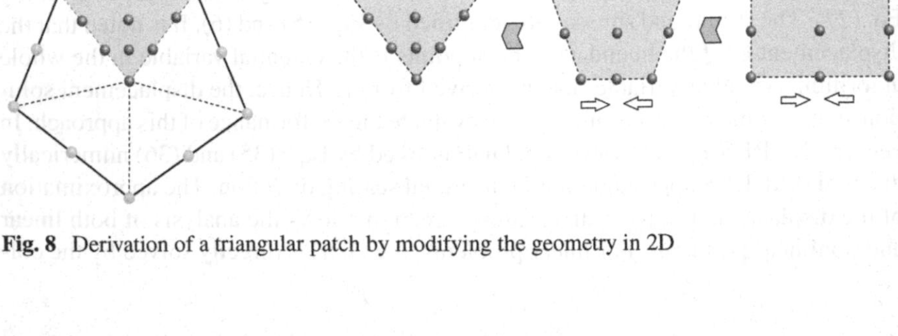

Fig. 8 Derivation of a triangular patch by modifying the geometry in 2D

done by the boundary tractions, and the third term is the external virtual work done by the body forces. The virtual strains are given also as $\delta\boldsymbol{\varepsilon} = \mathbf{D}^T \delta\boldsymbol{u}(\xi, \eta, \zeta)$. The approximation of the displacements and the virtual displacements on the sectional domain Ω_s is defined according to Eq. (27). Note that the sectional domain Ω_s is approximated with NURBS on the boundary as well as in the radial scaling direction. The stress vector is computed by $\boldsymbol{\sigma} = \mathbb{C}\mathbf{D}^T\boldsymbol{u}$ using Eqs. (5) and (6). The first term in Eq. (28) is rewritten by expanding the integral

$$\begin{aligned} \int_{\Omega_s} \delta\boldsymbol{\varepsilon}^T \boldsymbol{\sigma}\, \mathrm{d}V &= \int_0^1 \delta\boldsymbol{U}_s^T{}_{,\xi}\, \xi^2 \boldsymbol{k}_{11} \boldsymbol{U}_{s,\xi}\, \mathrm{d}\xi + \int_0^1 \delta\boldsymbol{U}_s^T{}_{,\xi}\, \xi \boldsymbol{k}_{12} \boldsymbol{U}_s\, \mathrm{d}\xi \\ &+ \int_0^1 \delta\boldsymbol{U}_s^T \xi \boldsymbol{k}_{21} \boldsymbol{U}_{s,\xi}\, \mathrm{d}\xi + \int_0^1 \delta\boldsymbol{U}_s^T \boldsymbol{k}_{22} \boldsymbol{U}_s\, \mathrm{d}\xi \end{aligned} \tag{50}$$

with $(\ldots)_{,\xi} = \frac{\partial(\ldots)}{\partial\xi}$. The coefficient matrices are given in Eq. (30). The second term in Eq. (28) is rewritten by considering Eqs. (A.4)–(A.6) to Eq. (31). The coefficient matrices for the right-hand side are defined in Eq. (32). With the help of Eq. (3), the third term in Eq. (28) is reformulated to Eq. (33). Substituting Eqs. (50), (31), and (33) into Eq. (28) leads to the weak form of equilibrium equation for 3D problems

$$\begin{aligned} &\sum_{s=1}^{nsec} \left(\int_0^1 \delta\boldsymbol{U}_s^T{}_{,\xi} \left(\xi^2 \boldsymbol{k}_{11} \boldsymbol{U}_{s,\xi} + \xi \boldsymbol{k}_{12} \boldsymbol{U}_s\right) \mathrm{d}\xi + \int_0^1 \delta\boldsymbol{U}_s^T \left(\xi \boldsymbol{k}_{21} \boldsymbol{U}_{s,\xi} + \boldsymbol{k}_{22} \boldsymbol{U}_s\right) \mathrm{d}\xi \right) \\ &- \sum_{s=1}^{nsec} \left(\delta\boldsymbol{U}_s^T \boldsymbol{F}_s \Big|_{\xi=0}^{\xi=1} - \int_0^1 \delta\boldsymbol{U}_s^T \xi^2 \boldsymbol{T}_1\, \mathrm{d}\xi \right) = \mathbf{0}. \end{aligned} \tag{51}$$

If the stress resultants in Eq. (28) are replaced with the Cauchy stress, the formulations here are suitable for the geometrical nonlinear analysis. Also, the stress–strain constitutive relation is flexible in these equations, thus, material nonlinearities can be considered (Chasapi and Klinkel 2018).

Substituting Eq. (38) into (51) yields the compact form of the weak form of equilibrium equation for 3D problems. For the sake of simplicity the body forces $\boldsymbol{T}_2$ are neglected. The system of equations is written as

$$\delta\boldsymbol{U}_\xi^T \left(\boldsymbol{F}_\Omega - \boldsymbol{K}_\Omega \boldsymbol{U}_\xi\right) = \mathbf{0} \quad \text{with} \quad \boldsymbol{U}_\xi = \mathop{\mathbf{A}}_{s=1}^{nsec} \left(\boldsymbol{U}_{s\xi}\right) \tag{52}$$

where $\mathop{\mathbf{A}}_{s=1}^{nsec}$ is the assembly operator to assemble the variables over all the sectional domains. $\boldsymbol{K}_\Omega$ is the stiffness matrix of the entire domain. $\boldsymbol{U}_\xi$ is the nodal displacement in the entire domain. The coefficient matrices are given as

$$\boldsymbol{F}_{\Omega} = \overset{nsec}{\underset{s=1}{\mathbf{A}}} \begin{bmatrix} \mathbf{0} \\ \boldsymbol{F}_s\,(\xi = 1) \end{bmatrix}$$

$$\boldsymbol{K}_{\Omega} = \overset{nsec}{\underset{s=1}{\mathbf{A}}} \left(\int_0^1 \left[\boldsymbol{N}_{\xi}^T{}_{,\xi} \left(\xi^2 \boldsymbol{k}_{11} \boldsymbol{N}_{\xi}{}_{,\xi} + \xi \boldsymbol{k}_{12} \boldsymbol{N}_{\xi} \right) + \boldsymbol{N}_{\xi}^T \left(\xi \boldsymbol{k}_{21} \boldsymbol{N}_{\xi}{}_{,\xi} + \boldsymbol{k}_{22} \boldsymbol{N}_{\xi} \right) \right] \mathrm{d}\xi \right) \tag{53}$$

For arbitrary test functions $\delta \boldsymbol{U}_{\xi}$ in Eq. (52), the global system of equilibrium equations can be obtained as

$$\boldsymbol{K}_{\Omega} \boldsymbol{U}_{\xi} - \boldsymbol{F}_{\Omega} = \mathbf{0}. \tag{54}$$

The degrees of freedom located at the interior nodes are eliminated by static condensation from Eq. (54). The degrees of freedom, located at the boundary $\partial\Omega$, are denoted as $\boldsymbol{U}$, while those associated with the interior of Ω are referred as $\hat{\boldsymbol{U}}$. Let $\boldsymbol{F} = \overset{nsec}{\underset{s=1}{\mathbf{A}}} \boldsymbol{F}_s\,(\xi = 1)$, the system of equations is given in Eq. (48). The vector $\hat{\boldsymbol{U}}$ is eliminated from Eq. (48) by a static condensation. This results in a reduced system of equations, which only relates to the boundary degrees of freedom, see Eq. (49).

Until now, all the formulations for the isogeometric analysis of solids in boundary representation are available. To sum up, the solid is divided in the analysis into several sectional domains Ω_s. This division is in principle flexible. For the following numerical examples, C_0 continuity at the boundary is, however, employed to divide the solid. NURBS basis functions are employed for the description of the boundary geometry as well as for the approximation of the displacements at the boundary, see Eqs. (24) and (27). The interior of the domain is described by a scaling center and a radial scaling parameter. The scaling center is chosen in a zone from which the total boundary of the domain is visible (Song and Wolf 1997). The scaling center will, in general, be located inside the domain. It could be defined as the geometric center of the domain if it is convenient to obtain. Or it could be defined as the average coordinate of the control points which are used to define the total boundary of the domain. The displacement in the radial scaling direction is approximated by a one-dimensional B-spline basis function, which is the main difference to the semi-analytical approach where the analytical solution in the scaling direction is employed. The approximation of the displacement in the radial scaling direction allows for the analysis of both linear and nonlinear problems. Applying the weak form to the governing partial differential equation of elasticity, the global system of equilibrium equation is obtained, see Eq. (47). The Galerkin or the collocation method can be employed in the radial scaling direction to solve the ODE. The Dirichlet boundary conditions are directly applied to the control points at the boundary. Hence, simply columns and rows of Eq. (48) are deleted. Solving Equation (48) yields the displacement $\boldsymbol{U}$ of the control points at the boundary. All nodal displacements $\boldsymbol{U}_s$ of a section can be determined by using Eqs. (38) and (47). The displacement vector at an arbitrary point in Ω_s is given by Eq. (27). The strains and stresses are identified by Eqs. (5) and (6). The above- presented analysis procedures are surface-oriented methods. They are suitable for the numerical analysis if only the geometry of the boundary surfaces is given. This is the case with solids, which are designed in CAD with the boundary representation modeling technique. The choice of the method depends on the problem under investigation. In the following, the methods are discussed in terms of accuracy and efficiency. A comparison between the presented formulations and also with standard numerical methods is provided.

5 Numerical Examples

In this Section, five numerical examples related to 2D and 3D elastic problems will be presented. All examples are employed to demonstrate the capabilities of the NURBS-based hybrid Galerkin-collocation method (NURBS-HCGM) and the NURBS-based Galerkin method (NURBS-G). The first two examples are employed to demonstrate the performance in terms of accuracy. Hence, an extensive comparison between both methods as well as a comparison to standard isogeometric analysis (IGA) is considered. The last three examples are presented to illustrate the capability of the methods. Therefore, a comparison with standard FEM and IGA are presented. For all examples, each sectional domain is modeled with the same NURBS basis functions, which employ identical polynomial degree and knot vector. However, it should be noted that the choice of the polynomial degree and the knot vector to approximate each boundary is in principle flexible. Moreover, all 3D computational meshes considered are conforming, which means that adjacent surface patches employ the same polynomial degree and knot vector along the shared edge. However, the NURBS description of boundary surfaces is in principle flexible. Methods for the coupling of nonconforming meshes which could be employed for the analysis are given by Apostolatos et al. (2014), Ruess et al. (2014), and Dornisch et al. (2015). A further extension could be the treatment of trimmed boundary surfaces (Schmidt et al. 2012; Breitenberger et al. 2015).

In the linear analysis, the problems can be solved analytically by employing the eigenvalue approach (Song and Wolf 1997). Here, the eigenvalue solution is used as an optimal solution to evaluate the accuracy of the NURBS-G and NURBS-HCG. In the eigenvalue method, the unknown variables are the displacements $\bar{\boldsymbol{U}}$ of boundary control points. For the numerical examples, we will mainly focus on the error investigation of the boundary displacement $\bar{\boldsymbol{U}}$.

Declarations for the description of solids are summarized in Table 1. The NURBS basis functions employed to describe the boundary are termed as boundary NURBS. The corresponding control points are denoted as boundary control points. Analogously, the NURBS basis functions used in the radial scaling direction are abbreviated as radial or collocation NURBS and the control points are denoted as radial control points or collocations points in case of the NURBS-HCGM. $\boldsymbol{u}_\alpha$ $(\alpha = eg, an, cl)$ represents the displacement obtained by the eigenvalue method, the analytical solution and the NURBS-G or NURBS-HCGM, respectively. The vectors $\boldsymbol{u}_\alpha^g$ $(\alpha = eg, an, cl)$ denote the displacement of boundary Gauss integration points used in the integrals of Eq. (30).

Table 1 Nomenclature to define the numerical models

Parameter	Description
p_B	Polynomial degree of boundary NURBS ($p_B = p = q$)
N_B	Total number of boundary control points
$p_C = t$ $N_C = n_{cp}$	Polynomial degree of radial or collocation NURBS, Eq. (37) number of control points or collocation points per line in the radial scaling direction, Eq. (37), abbreviated as number of radial control points or collocations points

The relative error of displacement response is computed with the aid of L^{∞}-norm. With respect to the displacements at the boundary, the norm is defined as

$$\|\boldsymbol{v}\|_{L^{\infty}(\partial\Omega)} = \max |\boldsymbol{v}| \qquad \forall \boldsymbol{v} \in L^{\infty}(\partial\Omega) \ . \tag{55}$$

This error measure is employed in convergence studies, where mesh refinement and order elevation are considered. It is distinguished between a refinement for the radial scaling direction and for the boundary. The influence of the polynomial degree of radial NURBS and the number of radial control points on the accuracy is investigated for a fixed boundary discretization. An optimal solution is gained by using the eigenvalue method (Song and Wolf 1997). In the next step, the influence of boundary description is discussed. The polynomial degree of boundary NURBS and the number of boundary control points affect the accuracy of the displacement response. Here, optimal rules for the choice of all parameters are provided. Moreover, the capability of the analysis procedures is illustrated by comparison to standard numerical methods. Within the numerical examples, analytical solutions serve as a reference to evaluate the error.

5.1 *Infinite Plate with Circular Hole*

The aim of this example is to study the convergence behavior of the eigenvalue method, the NURBS-G and the NURBS-HCGM for 2D problems. The geometry and boundary conditions of the plate are illustrated in Fig. 9. Due to the symmetry, only one-quarter of the plate is modeled. The exact traction from the analytical solution is imposed at the free boundary (Apostolatos et al. 2014). The material properties are considered with the Young's modulus $E = 100\ \mathrm{N/m^2}$, the Poisson's ratio $\nu = 0.3$, and the thickness $h = 1$ m. The eigenvalue method as well as the NURBS-HCGM and NURBS-G are employed in the analysis. Here, the plate is divided into five sections with respect to the scaling center, which are bounded by dashed lines, see Fig. 9. The scaling center C is defined by the average of all control points at the boundary. The boundary of each section Ω_s is discretized with the initial polynomial degree $p_B = 2$ and with the initial knot vector $\boldsymbol{H} = [0, 0, 0, 1, 1, 1]$. The polynomial

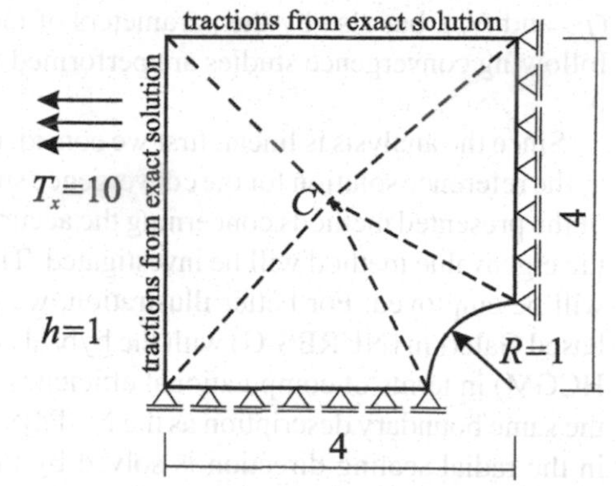

Fig. 9 Infinite plate with circular hole: problem definition

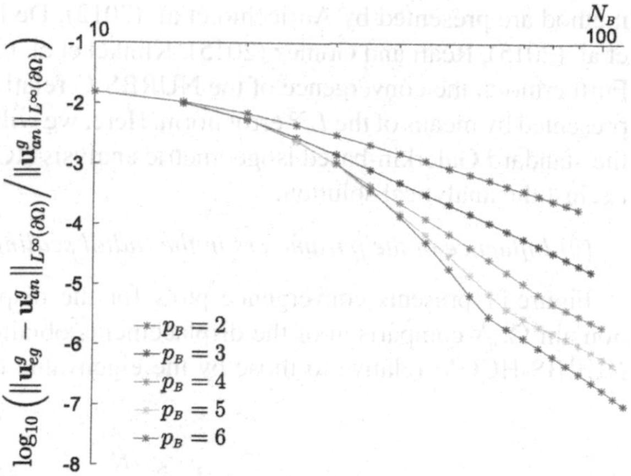

Fig. 10 Relative error of the displacement at the boundary $\partial\Omega$ for eigenvalue method

degree is increased by order elevation. The number of elements is increased for a fixed polynomial degree by h-refinement of the open knot vector. Correspondingly, the total number of boundary control points N_B is increased.

Solution of the eigenvalue method A solution for a given discretization of the boundary is calculated by the eigenvalue method (Song and Wolf 1997). As this method leads to an analytical solution for linear problems, it serves as a reference solution for the evaluation of the NURBS-G and NURBS-HCG. In this method, the decisive parameters which influence the accuracy are the polynomial degree p_B and the total number of boundary control points N_B. Here, the convergence of the displacement at the boundary $\partial\Omega$ in the L^∞-norm is investigated, see Fig. 10. As it can be seen in the figure, the eigenvalue method performs accurately. It leads to the exact solution with increasing polynomial degree of boundary NURBS and total number of boundary control points.

Solution of the NURBS-HCGM and NURBS-G The accuracy of the presented numerical methods is not only determined by the parameters of boundary NURBS

(p_B and N_B), but also by the parameters of radial NURBS (p_C and N_C). Hence, the following convergence studies are performed for different choices of p_B, N_B, p_C and N_C.

Since the analysis is linear, first we consider the solution of the eigenvalue method as the reference solution for the convergence study. Within this study, the performance of the presented methods concerning the accuracy of the displacement with respect to the eigenvalue method will be investigated. The L^∞ error norm for the displacement will be employed. For better illustration, we present a comparison of the NURBS-based Galerkin (NURBS-G) with the hybrid collocation-Galerkin method (NURBS-HCGM) in terms of computational efficiency and accuracy. In the NURBS-HCGM, the same boundary description as the NURBS-G is employed, however, the equation in the radial scaling direction is solved by the NURBS-based collocation method. In the frame of NURBS-based analysis, extensive studies regarding the collocation method are presented by Auricchio et al. (2012), De Lorenzis et al. (2015), Kiendl et al. (2015), Reali and Gomez (2015), Klinkel et al. (2015), and Chen et al. (2015). Furthermore, the convergence of the NURBS-G relative to the analytical solution is presented by means of the L^∞ error norm. Here, we will compare the NURBS-G with the standard Galerkin-based isogeometric analysis (IGA) in terms of their accuracy against the analytical solution.

(a) Influence of the parameters in the radial scaling direction

Figure 11 presents convergence plots for the displacement at the boundary of domain Ω. A comparison of the displacements obtained by the NURBS-G and the NURBS-HCGM relative to those by the eigenvalue method is presented. The L^∞

$$N_B = 30 \quad p_B = 3 \left(\log_{10} \left(\left\| u_{eg}^g - u_{an}^g \right\|_{L^\infty(\partial\Omega)} / \left\| u_{an}^g \right\|_{L^\infty(\partial\Omega)} \right) \right) = -3.0$$

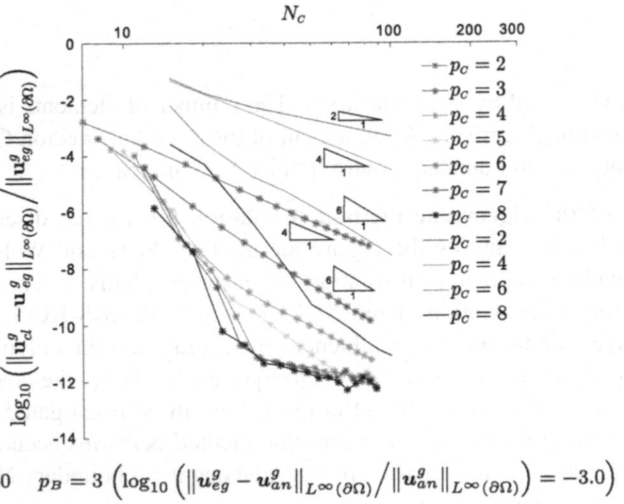

Fig. 11 Relative error of the displacement at the boundary $\partial\Omega$ for NURBS-G denoted with lines & star and NURBS-HCGM denoted with solid lines: different polynomial degrees of radial NURBS and number of radial control points are concerned

error norm for the displacement at the boundary Gauss integration points is considered in the comparison. For a fixed boundary description (i.e., p_B and N_B are fixed), different polynomial degrees of radial NURBS p_C and number of radial control points N_C are utilized in the convergence study. In the figures, the results of the NURBS-G are denoted as line with stars, while the results of the NURBS-HCGM are represented as solid lines. For the NURBS-HCGM, only the results of even polynomial degree of radial NURBS are presented, because the best possible convergence rates are attained for even degrees in the NURBS-HCGM (Klinkel et al. 2015; Schillinger et al. 2013). In addition, to illustrate the best possible accuracy under current boundary description, the relative error between the eigenvalue method and the analytical solution is shown in the caption of Fig. 11. It can be seen, that the accuracy of both methods increases with increasing polynomial degree and number of radial control points. The error level of the proposed NURBS-G is comparably lower than that of the NURBS-HCGM in terms of the control points and the polynomial degree of radial NURBS. For a specified level of accuracy within the NURBS-HCGM, the polynomial degree of radial NURBS should satisfy $p_C \geq even\ (p_B)$ and the number of radial points $N_C \geq N_B$. A further advantage of the NURBS-G is that there is no singularity arising at the scaling center compared to the NURBS-HCGM, hence the solution procedure is stable (see also Klinkel et al. 2015; Chen et al. 2015).

(b) Influence of the parameters in the circumferential direction

The rate of convergence is independent of the polynomial degree of boundary NURBS in the presented NURBS-G as well as NURBS-HCGM method. Greater difference between p_B and p_C will lead, however, to more accurate results for a given number of radial points. The same holds also for the total number of boundary control points, when the same polynomial degree of radial NURBS is concerned. Here also, greater difference between N_B and N_C will yield better results provided that the same number of radial points N_C is employed in the computation. The reader is referred to the studies of Klinkel et al. (2015) and Chen et al. (2015, 2016) for further numerical results.

(c) Comparison of the NURBS-G with the isogeometric analysis (IGA)

Here, the boundary of each section Ω_s is described by NURBS basis functions with identical polynomial degree p_B. It employs $p_B = 2$ with the initial open knot vector $\boldsymbol{H} = [0, 0, 0, 1, 1, 1]$, and extends to $p_B = 3, 4, 5$, and 6, respectively. Under each polynomial degree of boundary NURBS, the number of elements along the boundary of each section is initially $n = 1$, and extends to $n = 2, 3, 4, 5, 6, 7$, and 8, separately. The h-refinement of the open knot vector is employed to generate new open knot vectors. The total number of boundary control points N_B is increased, respectively. To illustrate the capability of the NURBS-G, we provide a comparison to the isogeometric analysis (Dornisch et al. 2013). In Fig. 12, the different meshing strategies of both approaches are depicted. For the NURBS-G model, the scaling center C is defined by the average of the coordinates of the boundary control points, which are denoted as red dots in the figure. The plate is modeled by five sections,

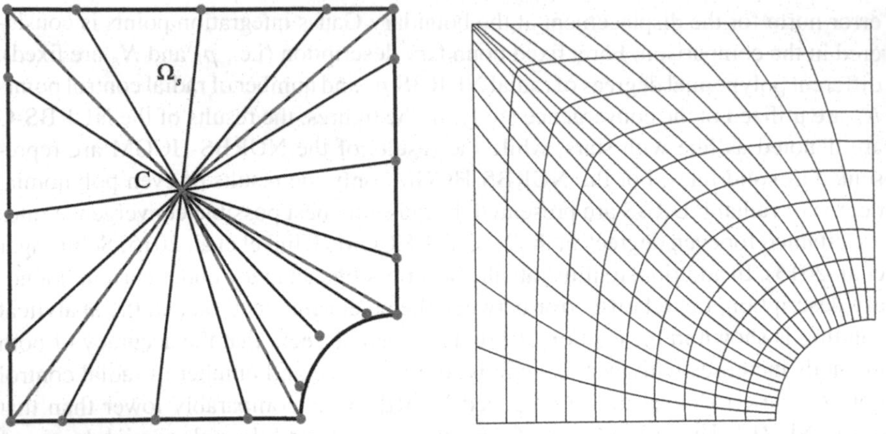

Fig. 12 On the left-hand side, the mesh of the NURBS-based Galerkin method and on the right-hand side the mesh of the isogeometric Galerkin approach are depicted

Fig. 13 L^∞ error norm of the displacement at the boundary $\partial\Omega$: different polynomial degrees of boundary NURBS and total numbers of Gauss integration points are considered

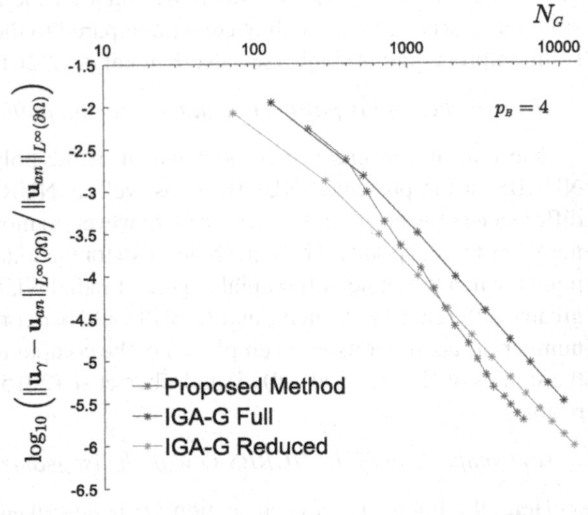

where each section Ω_s is bounded by red colored dashed lines. For the IGA, k-refinement of the open knot vector is employed to generate new open knot vectors.

In Fig. 13, the L^∞ error norm for the displacement at the boundary is employed to demonstrate the accuracy of the NURBS-G approach. In the figure, $\boldsymbol{u}_\gamma$ denotes the displacement solution obtained from the NURBS-G and the IGA approach, respectively. For the NURBS-G, the polynomial degree of radial NURBS is defined as $p_C = p_B$. The number of radial control points is set as $N_C = ceil(N_B/4) + p_C$. Here, $ceil(\cdot)$ denotes the round-toward-infinity function. In the NURBS-G, the reduced quadrature with $ceil(p_C/2)$+1 Gauss points per element is employed to perform the integration. In the IGA approach, two integration strategies are employed for the

Gauss quadrature integration: full quadrature with $p_B + 1$ Gauss points per element and reduced quadrature with $ceil(p_B/2)$+1 Gauss points per element (Hughes et al. 2010). In the figures, the relative L^∞ error norm is plotted versus the total number of Gauss points N_G employed in the NURBS-G and IGA approach, respectively.

Concerning the accuracy of the NURBS-G, it can be seen in Fig. 13 that the method performs accurately. It leads to the exact solution with the rise of polynomial degree of boundary NURBS and total number of Gauss points. Additionally, the proposed NURBS-G and IGA yield error levels in the same range in terms of total integration points. For coarse discretizations, the IGA performs better than the NURBS-G, for finer discretizations the NURBS-G slightly outperforms the IGA.

5.2 *Solid Sphere Under Hydrostatic Pressure*

The aim of this example is to investigate the rate of convergence of the NURBS-G and NURBS-HCG for 3D problems. Due to the symmetry of the system, only one-eighth of the solid sphere is modeled. The geometry is shown in Fig. 14. The radius of the sphere is $R_b = 10$ m. In the analysis, the solid sphere is modeled by four sections, which are bounded by the curved boundary surfaces as shown in Fig. 14c. One section Ω_s is shown in Fig. 14c and indicated by dashed lines. The scaling center C is defined by the centroid of the sphere. The boundary surface of each sectional domain Ω_s is initially described by the polynomial degree $p_B = 2$ and the knot vectors $\boldsymbol{H} = \boldsymbol{Z} = [0, 0, 0, 1, 1, 1]$. The polynomial degree is increased by using k-refinement. The number of elements is increased for a fixed polynomial degree by h-refinement of the open knot vector. An example of the control polygon and mesh of the boundary surfaces is presented in Fig. 15, which employs the knot vector $\boldsymbol{H} = \boldsymbol{Z} = [0, 0, 0, 0.5, 1, 1, 1]$ and the polynomial degree $p_B = 2$. The system is loaded by hydrostatic pressure, which is imposed at the external spherical surfaces. The analytical solution for the displacement and the stress of the solid sphere is given

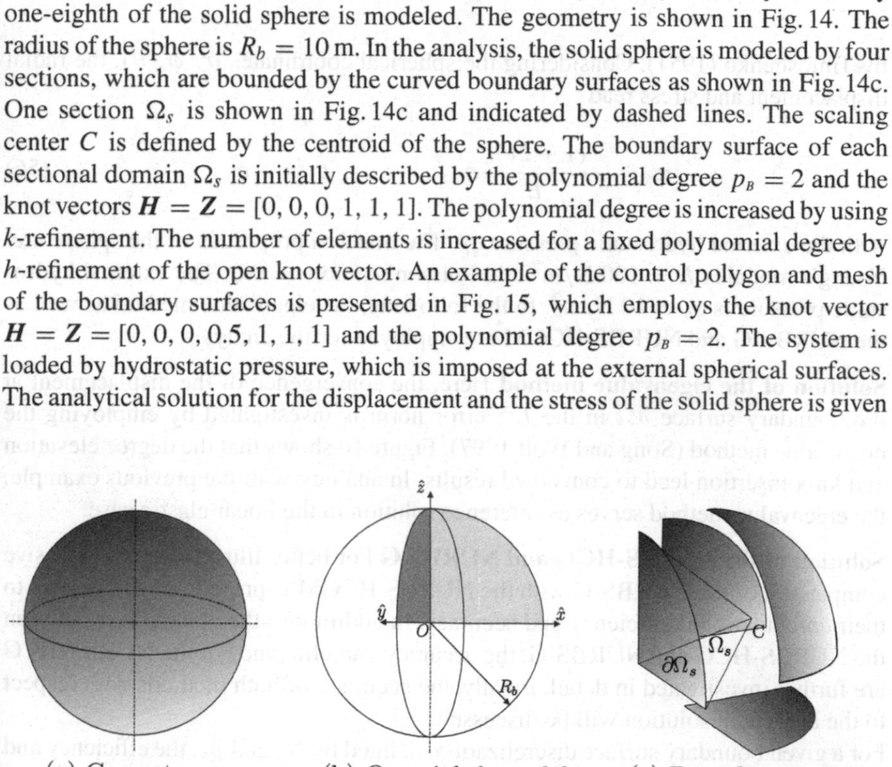

(a) Geometry (b) One eighth model (c) Boundary surfaces

Fig. 14 Solid sphere: problem definition and boundary description

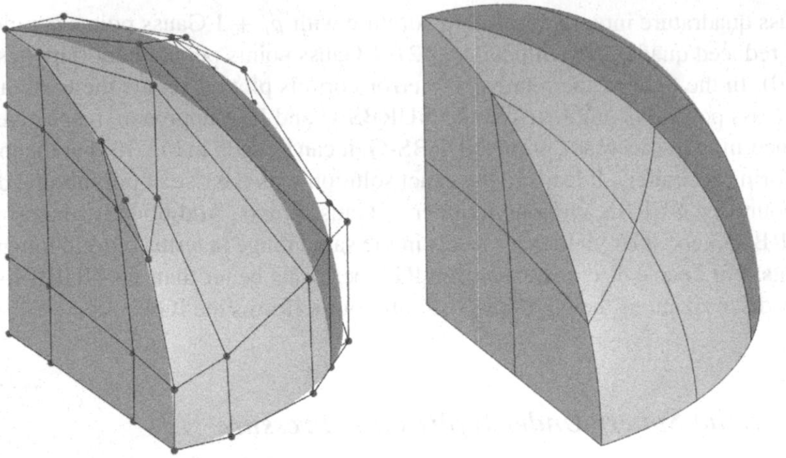

Fig. 15 Control polygon and mesh on the boundary surfaces of the solid sphere ($p_B = 2$ and 2 elements per parametric direction)

by Timoshenko (1951). Considering the spherical coordinates $(r,\ \phi,\ \theta)$, the radial displacement and stress read

$$u_r = -\frac{(1-2\nu)r}{E} q_b \qquad\qquad \sigma_r = -q_b \tag{56}$$

with the external hydrostatic pressure q_b. The material properties of the sphere are: Young's modulus $E = 100\ \mathrm{N/m^2}$ and Poisson's ratio $\nu = 0.0$. The external hydrostatic pressure is $q_b = 10\ \mathrm{N/m^2}$. In the following, the eigenvalue method as well as the NURBS-G and NURBS-HCGM are employed for the analysis.

Solution of the eigenvalue method Here, the convergence of the displacement at the boundary surface $\partial\Omega$ in the L^∞ error norm is investigated by employing the eigenvalue method (Song and Wolf 1997). Figure 16 shows that the degree elevation and knot insertion lead to converged results. In analogy with the previous example, the eigenvalue method serves as reference solution in the linear elastic case.

Solution of the NURBS-HCG and NURBS-G For better illustration, an extensive comparison of the NURBS-G with the NURBS-HCGM is presented with respect to their computational efficiency and accuracy. In addition to the comparison between the NURBS-HCG and NURBS-G, the accuracy and efficiency of the NURBS-HCG are further investigated in detail. Finally, the accuracy of both methods with respect to the analytical solution will be discussed.
For a given boundary surface discretization, defined by N_B and p_B, the efficiency and accuracy of the methods also depend on the polynomial degree of radial NURBS p_C and the number of radial control points N_C. A convergence study will be, therefore, performed for different choices of p_B, N_B, p_C, and N_C in analogy with the previous 2D example. In the following, the L^∞ error norm for displacements will be employed.

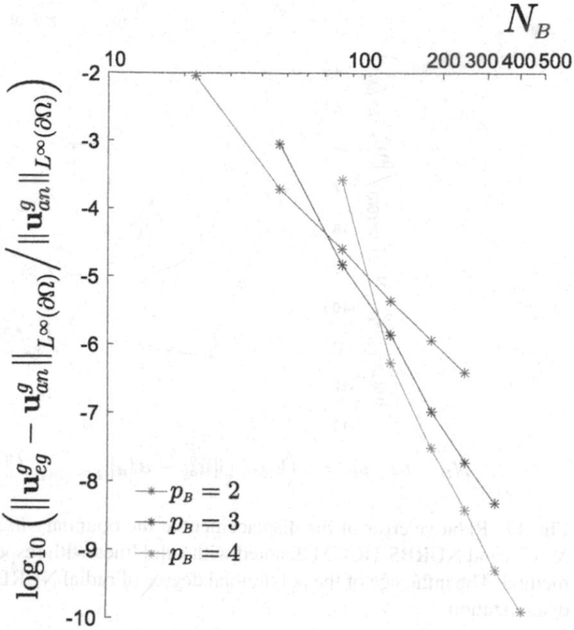

Fig. 16 Relative error of the displacement at the boundary $\partial\Omega$ of the solid sphere (solution of the eigenvalue method)

Figure 17 presents the convergence plots for displacements at the boundary Gauss integration points. For a fixed boundary description, different polynomial degree of radial NURBS p_c and number of radial control points N_c are employed. The solution of the eigenvalue method is set as the reference solution. Thus, the relative error between the eigenvalue method, the NURBS-G and the NURBS-HCGM is displayed in Fig. 17. In the figures, the lines with stars denote the results of the NURBS-G, while solid lines represent the results of the NURBS-HCGM. For the NURBS-HCGM, only the results of even polynomial of radial NURBS are presented in order to achieve best possible convergence rates (Chen et al. 2015). The best possible accuracy under current boundary description is given as the relative error between the eigenvalue method and the analytical solution with the L^∞ error norm, see Fig. 17.
It can be seen in the figure that both the NURBS-HCG and NURBS-G converge with increasing p_c. The error level of the NURBS-G is comparably lower than that of the NURBS-HCGM. In the figure, it can be observed that a rate of convergence $2p_c$ is attained in the L^∞-norm, which can be considered as the best possible rate of convergence. Although the solution for high polynomial degrees indicates instabilities for fine discretizations, note that these results are already converged at the machine precision under consideration of the conditioning of the matrix.

(a) Influence of the parameters in radial scaling direction on the accuracy of the NURBS-HCGM

Figure 18 presents convergence plots for the displacement at the boundary Gauss integration points, which illustrate the influence of collocation. For a fixed polynomial

$$N_B = 82 \quad p_B = 4 \left(\log_{10} \left(\| u^g_{eg} - u^g_{an} \|_{L^\infty (\partial \Omega)} / \| u^g_{an} \|_{L^\infty (\partial \Omega)} \right) = -3.5 \right)$$

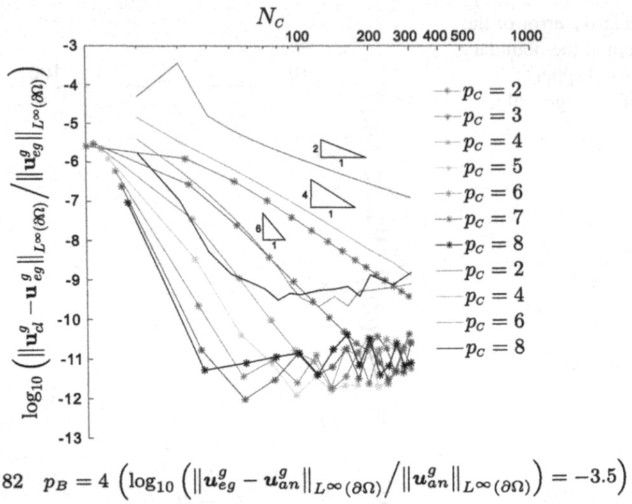

Fig. 17 Relative error of the displacement at the boundary $\partial \Omega$ for NURBS-G denoted with lines & stars and NURBS-HCGM denoted with solid lines with respect to the solution of the eigenvalue method: The influence of the polynomial degree of radial NURBS is examined for a fixed boundary discretization

$$N_B = 127 \quad p_B = 2 \left(\log_{10} \left(\| u^g_{eg} - u^g_{an} \|_{L^\infty (\partial \Omega)} / \| u^g_{an} \|_{L^\infty (\partial \Omega)} \right) = -5.4 \right)$$

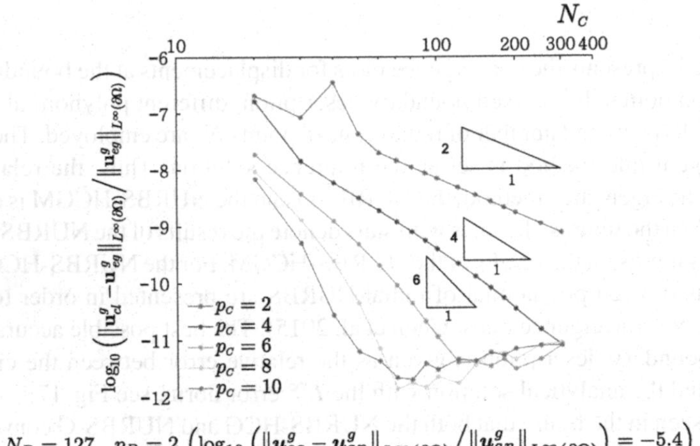

Fig. 18 Relative error of the displacement at the boundary $\partial \Omega$ (solution of the NURBS-HCGM compared to the eigenvalue method): The influence of the polynomial degree of the collocation NURBS is examined for a fixed boundary discretization

degree of boundary NURBS and a fixed total number of boundary control points, different polynomial degrees of collocation NURBS p_C and numbers of collocation points N_C are utilized. Here, the solution of the eigenvalue method is set as the reference solution. Thus, the relative error between the eigenvalue method and the NURBS-HCGM is displayed in Fig. 18. The L^∞-norm of the error is employed for the comparison. In the figure, only the results of even polynomial degrees of collocation NURBS are presented for better illustration. The results given in Fig. 18 show that the displacements converge for increasing polynomial degrees p_C. A rate of convergence p_C is attained for even degrees.

This is consistent with the observations of Auricchio et al. (2010) and can be referred to as the best possible convergence rates. Although unstable results are attained for very high polynomial degrees and number of collocation points, note that these results are already converged at the machine precision under consideration of the conditioning of the matrix.

Figure 18 demonstrates that also an increase of the number of collocation points N_C results in a higher accuracy. However, unstable results are also obtained on the convergence line before it attains the corresponding rate of convergence. The turning point is approximately observed at $N_C \approx N_B/2$. The following two rules are, therefore, suggested to achieve a stable collocation with the best possible convergence behavior:

1. The number of collocation points should satisfy $N_C \geq N_B/2$ to avoid unstable results on the convergence lines.
2. The polynomial degree of collocation NURBS should be set to $p_C \geq even\ (p_B)$, where $even\ (A)$ rounds A to the nearest even number greater than or equal to A. Higher difference between p_C and p_B will lead to more accurate results.

These rules may underestimate the accuracy of the NURBS-HCGM for some lower polynomial degrees of collocation NURBS and number of collocation points. However, they provide the best possible accuracy of the method.

(b) Influence of the parameters in the circumferential direction of the boundary on the accuracy of the NURBS-HCGM

As the same polynomial degree of collocation NURBS p_C is concerned, the rate of convergence is identical for the different choices of p_B, which implies that the rate of convergence is independent of the polynomial degree of boundary NURBS. However, greater difference between p_B and p_C will lead to more accurate results. The results of NURBS-HCG converge in general to the theoretical convergence rates (p_C). This is also the case with increasing number of boundary control points. The results also converge with the theoretical convergence rates (p_C). Moreover, the rate of convergence is independent of the total number of boundary control points. These observations are therefore valid for the 2D as well as the 3D case. For further details on the effect of p_B and N_B, see also the studies of Chen et al. (2015).

(c) Efficiency of the NURBS-HCGM

The efficiency of the NURBS-HCGM is hereafter investigated as it is of fundamental importance and largely determines the potential of the method for the use in engineering applications. Here, for simplicity, we only give a brief discussion about it. For a detailed investigation, the reader can refer to Klinkel et al. (2015). The efficiency of the NURBS-HCGM is mainly determined by the choice of p_B, N_B, p_C, and N_C, see Eq. (48). However, the size of the matrix in Eq. (48) enlarges primarily with the rise of the number of boundary control points N_B and collocation points N_C, which accordingly increases the computation time. Order elevation of the polynomial degrees p_B and p_C plays a minor role for the computation time. Computational costs occur only for the computation of the NURBS basis functions and their derivatives. There is no influence on the dimension of the matrices in Eq. (48). As a result, the total computation time does not change significantly with the rise of p_B and p_C. As already mentioned, however, a specific polynomial degree of boundary NURBS and number of boundary control points should be employed to achieve a high level of accuracy with respect to the exact solution. In addition, it has been already observed that a higher polynomial degree of collocation NURBS and larger number of collocation points should be applied to ensure the accuracy of the NURBS-HCGM. Hence, the optimal choice of p_B, p_C and N_B, N_C is significant for an efficient and accurate computation. To meet this need, we suggest the following two rules for the application of the NURBS-HCGM in the analysis:

1. Initially apply order elevation for the polynomial degree of boundary NURBS p_B. Thereafter, increase the number of boundary control points N_B. These steps ensure the accuracy with respect to the exact solution.
2. Apply order elevation of the polynomial degree of collocation NURBS p_C and increase the number of collocation points N_C in order to achieve high accuracy of the NURBS-HCGM. Note, that order elevation is computationally more efficient than increasing the number of collocation points as discussed above.

(d) Accuracy in respect to the exact solution

It has been observed that the convergence behavior of the NURBS-HCGM is equal to the convergence behavior of the eigenvalue method with respect to the analytical solution by Timoshenko (1951). This holds both for low and high polynomial degrees of boundary NURBS (Chen et al. 2015). Order elevation or *h*-refinement of the boundary NURBS entails more accurate results. The very good agreement between the results of the eigenvalue method and the NURBS-HCGM results certifies also the validity and rationality of the aforementioned rules for the best possible convergence behavior.

Similar results have been observed by comparing the NURBS-G with the eigenvalue method with respect to the analytical solution. According to Chen et al. (2016), the convergence behavior of the NURBS-G is equal to the convergence behavior of the eigenvalue method, both for low and high polynomial degrees of boundary NURBS. Order elevation or *h*-refinement of the boundary NURBS entails more accurate results as is also the case for the NURBS-HCGM.

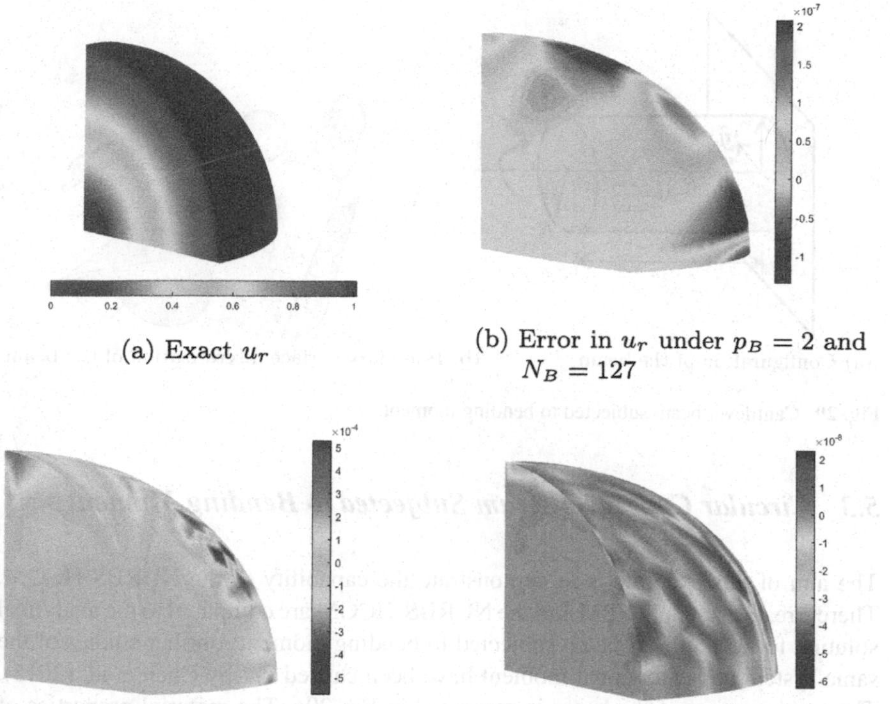

(a) Exact u_r

(b) Error in u_r under $p_B = 2$ and $N_B = 127$

(c) Error σ_r under $p_B = 2$ and $N_B = 127$

(d) Error in σ_r under $p_B = 4$ and $N_B = 407$

Fig. 19 u_r and σ_r plots of the 3D solid sphere under hydrostatic pressure. Here, for the radial NURBS, the polynomial degree is $p_C = p_B$ and the number of radial control points is $N_C = ceil(N_B/2) + p_C$

Finally, we will present the contour plot of the analytical solution and the errors in the radial displacement (u_r) and stress (σ_r) at the boundary surfaces ($\xi = 1$) obtained by the NURBS-G. Here, the error is defined as the difference between the numerical solution and the analytical solution. It should be noted that the radial stress σ_r is homogeneous as shown in Eq. (56). Hence, we will only show the error of the radial stress under two different boundary descriptions. The results are shown in Fig. 19. It can be seen in the figures that the error level of the NURBS-G is quite low, which implies that the method performs accurately.

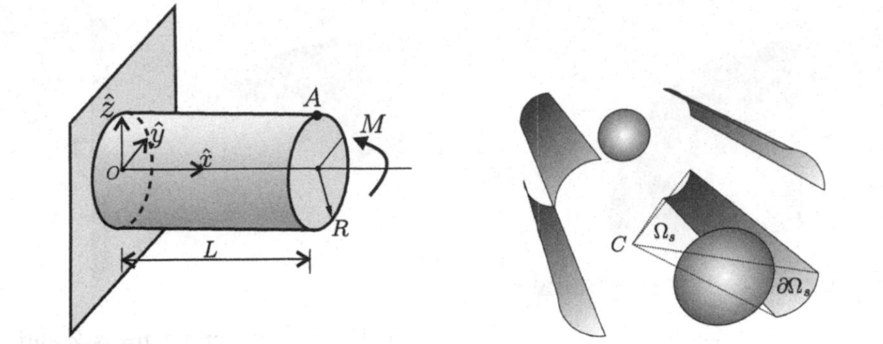

(a) Configuration of the beam (b) Boundary surface discretization of the beam

Fig. 20 Cantilever beam subjected to bending moment

5.3 Circular Cantilever Beam Subjected to Bending Moment

The aim of this example is to demonstrate the capability of the NURBS-HCGM. Therefore, the standard FEM and the NURBS-HCGM are compared to the analytical solution for a cantilever beam subjected to bending moment. Further studies of the same system under torsional moment have been carried out by Chen et al. (2015). The configuration of the beam is presented in Fig. 20a. The material properties of the beam are: Young's modulus $E = 100\ \text{N/m}^2$, Poisson's ratio $\nu = 0.0$. The beam has a length of $L = 50\,\text{m}$ and radius of $R = 5\,\text{m}$. The external bending moment is $M = 1000\ \text{N} \cdot \text{m}$. The scaling center C is defined as the center of the beam, $(\hat{x}_0,\ \hat{y}_0,\ \hat{z}_0) = (25\,\text{m},\ 0,\ 0)$. Thus, with respect to the scaling center the domain Ω is partitioned into 6 sections Ω_s bounded by the boundary surfaces $\partial\Omega_s$, see Fig. 20b. An analytical solution is given by Timoshenko (1951) and is considered here as the reference solution. The vertical displacement w ($\hat{z}$-direction) of the cantilever beam subjected to the tip bending moment M is given by

$$w = \frac{M}{2EI}(\hat{x}^2 + \nu\hat{y}^2 - \nu\hat{z}^2)\,. \tag{57}$$

The standard FEM and the NURBS-HCGM are employed to solve the problem. The boundary surface of each section Ω_s is initially described by NURBS basis functions with polynomial degree $p_B = 2$ and the open knot vector $\boldsymbol{H} = \boldsymbol{Z} = [0, 0, 0, 0.5, 1, 1, 1]$. The polynomial degree is elevated to $p_B = 3$ and 4. For each polynomial degree of boundary NURBS, the number of elements per parametric direction is initially $n = 2$, and is increased to $n = 3, 4, 5$, and 6 using h-refinement. The control polygon and the element mesh of the boundary surfaces of the beam are presented in Fig. 21. The contour plot of the displacement is presented in Fig. 22. Note that the contour is very smooth. The L^∞-norm for the error of the displacement at point A (Fig. 20a, $\boldsymbol{x}_A = (50\,\text{m},\ 0,\ 5\,\text{m})$) is employed to demonstrate the accuracy

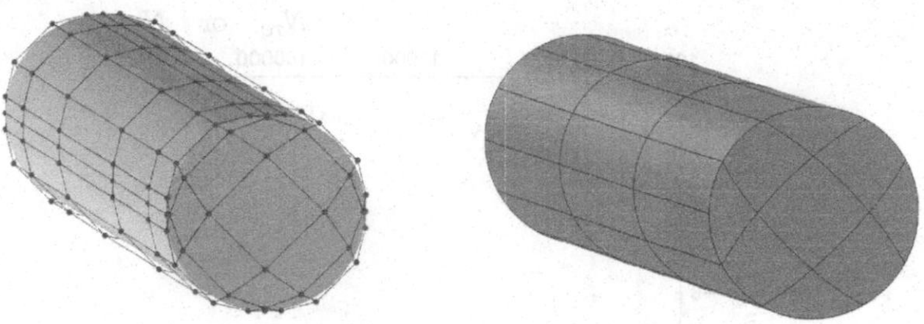

Fig. 21 Control polygon and mesh on the boundary surfaces of the beam ($p_B = 3$ and 3 elements per parametric direction)

Fig. 22 Displacement contour ($p_B = 3$ and 3 elements per parametric direction)

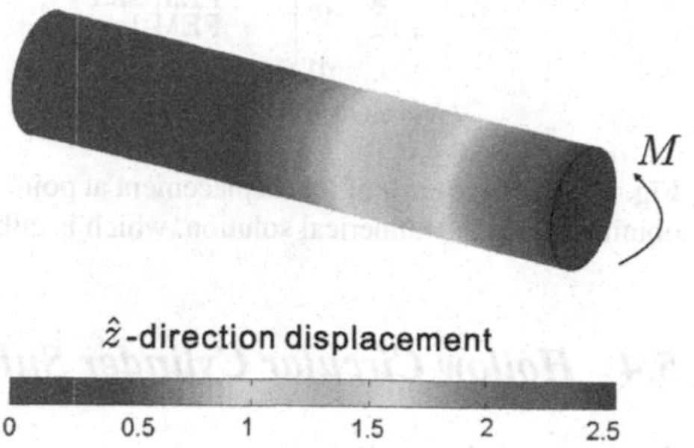

of both methods, see Fig. 23. In the figure, $w_\gamma{}^A$ denotes the deformation obtained from the numerical solution, which is either NURBS-HCGM or standard FEM. For the NURBS-HCGM, the rules proposed in the previous example are employed. The polynomial degree of collocation NURBS is defined as $p_C = even\ (p_B + 6)$. The number of collocation points is determined as $N_C = N_B$. The standard FEM employs full Gauss quadrature (Hughes 2000). Both linear and quadratic shape functions are used and are denoted as FEM-1st and FEM-2nd, see Fig. 23. Here, the error norm is plotted versus the total number of collocation points for NURBS-HCGM given by $N_{TC} = N_C \times N_B$, and versus the total number of nodes employed in the standard FEM, which is denoted by N_F.

Figure 23 shows that the NURBS-HCGM yields accurate results on the basis of the proposed rules. It approaches the analytical solution with increasing polynomial degree of the boundary NURBS and with increasing total number of collocation points. The NURBS-HCGM and the standard FEM yield error levels in the same range with respect to the total number of collocation points or FEM-nodes, respectively. In case of a coarse discretization the standard FEM performs better than the NURBS-HCGM, for finer discretizations the NURBS-HCGM outperforms the standard FEM.

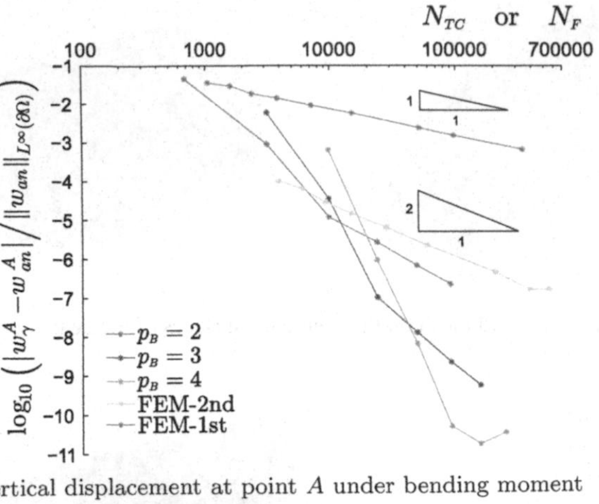

Fig. 23 Relative error of the displacement at point A. Here, $w_\gamma{}^A$ and u_γ^{AR} denote the deformations obtained from the numerical solution, which is either NURBS-HCGM or the standard FEM

5.4 Hollow Circular Cylinder Subjected to Internal Pressure

The aim of this numerical example is to investigate the capability of the NURBS-HCGM. Therefore, the standard IGA and the NURBS-HCGM are compared with the analytical solution for a hollow circular cylinder subjected to internal pressure The geometry of the cylinder is shown in Fig. 24. Plane strain conditions are assumed in the axial direction. A constant pressure is applied at the inner surface of the cylinder. Here, the inner and outer radius are $R_a = 1$ m, and $R_b = 2$ m, respectively. The properties of the cylinder are defined by a Young's modulus $E = 40\ \text{N/m}^2$ and a Poisson's ratio $\nu = 0.0$. The magnitude of the inner pressure is $p = 20\ \text{N/m}^2$. Considering symmetry only a quarter of the cylinder is modeled, see Fig. 24b. The scaling center C is defined as the average coordinate of all control points at the boundary. With respect to the scaling center, the domain Ω is partitioned into 6 sections Ω_s bounded by the boundary surfaces $\partial\Omega_s$, see Fig. 24c.

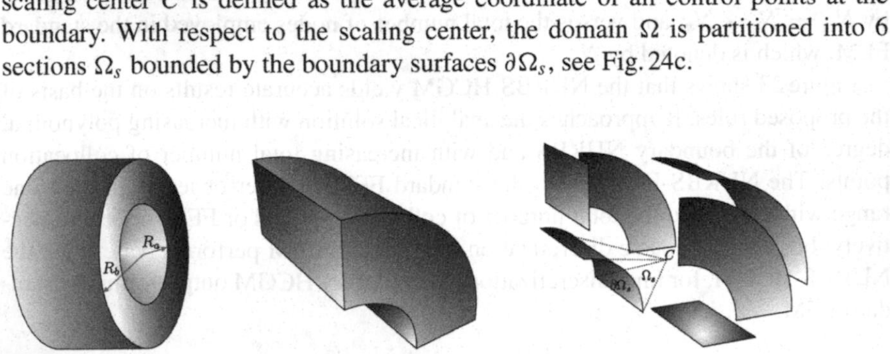

(a) Configuration (b) One quarter model (c) Boundary surface discretization

Fig. 24 Thick cylinder subjected to internal pressure

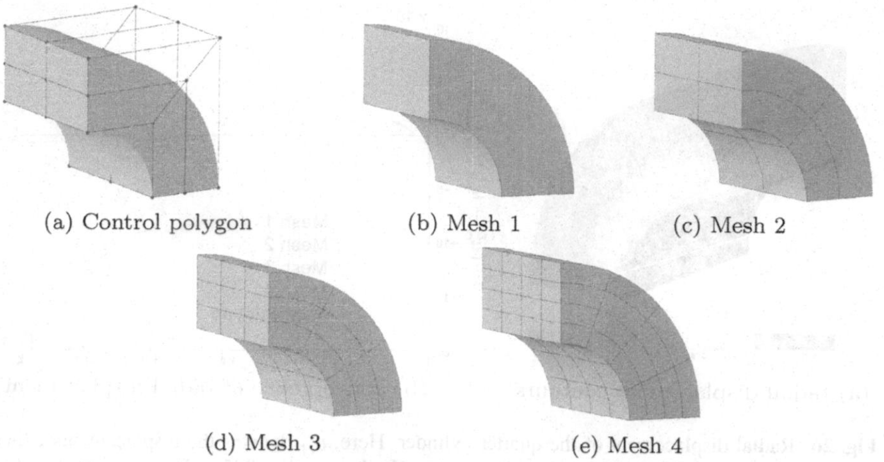

Fig. 25 Control polygon (1 element per parametric direction) and mesh on the boundary surfaces of the quarter cylinder ($p_B = 2$)

An analytical solution for the displacement response is given by Hughes et al. (2005). A plot of the control polygon is presented in Fig. 25a. Element meshes attained by h-refinement are shown in Fig. 25b–e. The polynomial degree is $p_B = 2$ in all cases. The initial open knot vector is $\boldsymbol{H} = \boldsymbol{Z} = [0, 0, 0, 1, 1, 1]$. The rules proposed in the previous examples are employed for the NURBS-HCGM. The polynomial degree of the collocation NURBS is defined as $p_C = even\,(p_B + 6)$, and the number of collocation points is determined by $N_C = N_B$.
Results of the displacement solution for each mesh are presented in Fig. 26. The contour plot of the radial displacement of the cylinder given in Fig. 26a is clearly apparent. Errors in the radial displacement are plotted in Fig. 26b. For better illustration, the result taken from Hughes et al. (2005) for Mesh 1 is also presented in the figure. Hughes et al. (2005) employed the isogeometric approach for the analysis. The maximum error of this approach is slightly below that of the proposed method (Mesh 1). The maximum error through the cylinder thickness attained with the NURBS-HCGM is: for Mesh 1 approximately 1.5%, for Mesh 2 approximately 0.25%, for Mesh 3 0.08%, and for Mesh 4 0.01%. Higher accuracy of the displacement solution can be achieved on all meshes by increasing the polynomial degree p_B for the boundary NURBS.

5.5 Solid with Free Form Geometry and Arbitrary Number of Boundary Surfaces

The last numerical example is employed to illustrate the capability of the NURBS-G to deal with 3D solids bounded by an arbitrary number of boundary surfaces.

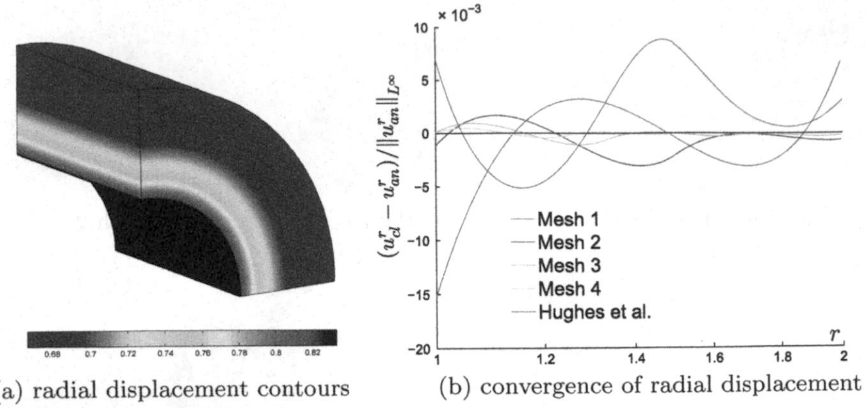

Fig. 26 Radial displacement of the quarter cylinder. Here, u^r_{cl} denotes the displacements either computed with the NURBS-HCGM or taken from Hughes et al. (2005). u^r_{an} represents the displacements obtained from the analytical solution

Therefore, the standard FEM and the NURBS-G are compared for a solid loaded by surface tractions. The system is depicted in Fig. 27a, b. The material properties of the solid are defined by Young's modulus $E = 100\ \text{N/m}^2$ and Poisson's ratio $\nu = 0.0$. The initial geometry of the boundary surface is described by NURBS basis functions of polynomial degree $p_B = 2$ and open knot vectors $\boldsymbol{H} = \boldsymbol{Z} = [0, 0, 0, 1, 1, 1]$. For the response analysis, order elevation and h-refinement are performed to generate boundary NURBS with higher polynomial degree and a larger number of elements. For the NURBS-G, the polynomial degree of radial NURBS is defined as $p_C = p_B$. The number of radial control points is set as $N_C = ceil(N_B/4) + p_C$. In the NURBS-G, the reduced quadrature with $ceil(p_C/2)$+1 Gauss points per element is employed to perform the integration in Eq. (53).

The geometry of elastic cube with circular hole is defined by the length of cube $B = L = H = 40$ m and the radius of circular hole $R = 10$ m. The magnitude of surface traction is $p = 10\ \text{N/m}^2$. Due to the symmetry of the problem, only one-eighth of the cube is considered with the symmetric boundary conditions, see Fig. 27b. In the analysis, the scaling center C is defined as the center of the one-eighth cube. With respect to the center, the domain Ω is partitioned into 7 sections Ω_s bounded by the boundary surfaces $\partial\Omega_s$, see Fig. 27c.

Sample plots of control polygon and mesh on the boundary surfaces of the one-eighth cube are presented in Fig. 28. An analytical solution for this problem in terms of displacements is not available. Thus, the comparison is made between the standard FEM and the NURBS-G. In the standard FEM, quadratic shape function (C3D20) and full Gauss quadrature integration are employed (Hughes et al. 2010). The computation is performed in Abaqus (2007) with 55273 elements and 235171 nodes. The contour plots of the vertical displacement for both methods are presented in Fig. 29. As it can be seen in the figure, good agreement is achieved.

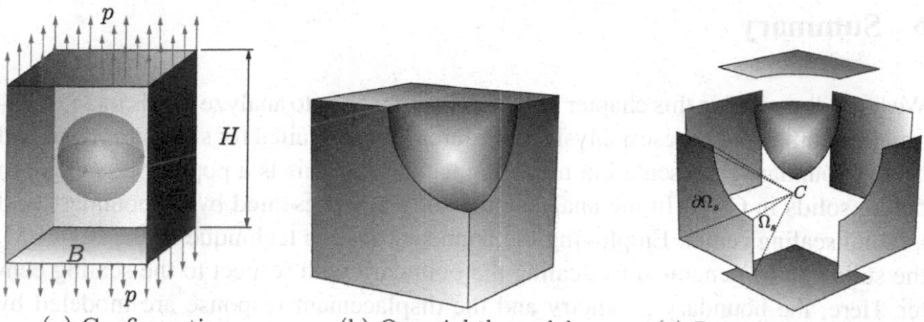

(a) Configuration (b) One eighth model discretization (c) Boundary surface

Fig. 27 Elastic cube with circular hole in tension regime

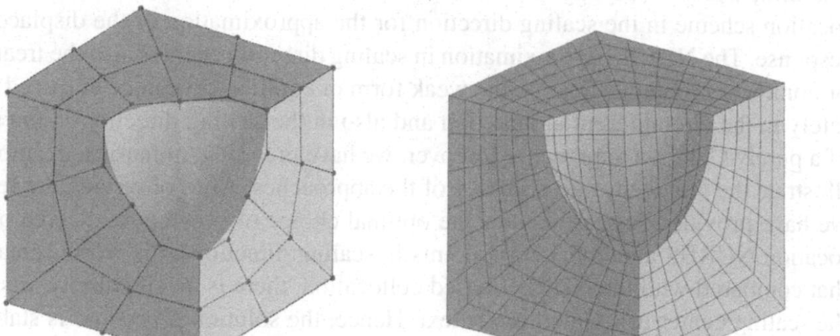

(a) Control polygon (b) Mesh

Fig. 28 Control polygon ($p_B = 2$ and 1 elements per parametric direction) and mesh ($p_B = 6$ and 6 elements per parametric direction) on the boundary surfaces of the one-eighth cube

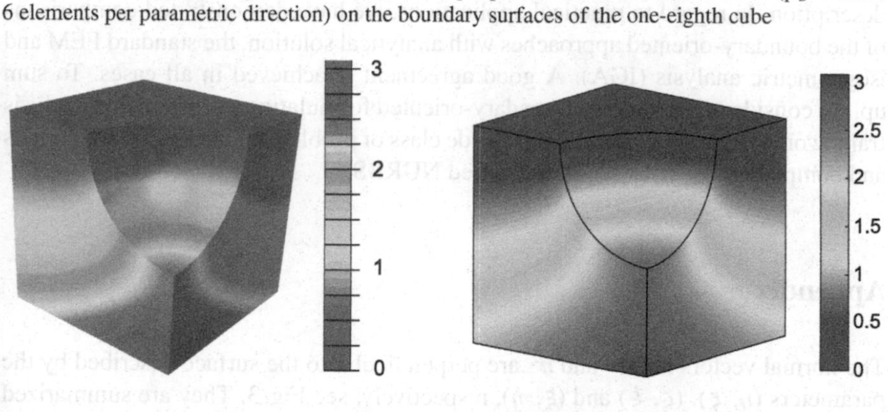

(a) Standard FEM (b) NURBS-G (see mesh Figure 28b)

Fig. 29 Contours of the vertical displacement for the one-eighth cube

6 Summary

We have discussed in this chapter numerical approaches to analyze solids in an isogeometric framework. These analysis procedures are well suited for structures designed by the boundary representation modeling technique. This is a popular technique to define solids in CAD. In the analysis, the solid is represented by its boundary and a radial scaling center. Employing the boundary scaling technique of the SB-FEM, the solid can be generated by scaling the boundary with respect to the scaling center. Here, the boundary geometry and the displacement response are modeled by the NURBS basis functions, which are the same functions used for the definition of the geometry. For the approximation of the displacement response in the interior domain the eigenvalue method can be employed for linear problems. Alternatively, one-dimensional NURBS basis function is introduced in combination with a collocation scheme in the scaling direction for the approximation of the displacement response. The NURBS approximation in scaling direction enables also the treatment of nonlinear problems. Finally, the weak form of equilibrium can be enforced separately in the circumferential direction and also in the scaling direction in the scope of a purely Galerkin approach. Moreover, we have presented numerical examples to illustrate the efficiency and accuracy of the approaches. After observing the results, we have provided suggestions for the optimal choice of polynomial degree of collocation NURBS and collocation points in scaling direction. It is worth remarking that compared with the NURBS-based collocation, there is no singularity arising at the scaling center in a Galerkin context. Hence, the solution procedure is stable. In addition, the error level of a Galerkin approach is lower than that of the NURBS-based collocation and higher rate of convergence is achieved. Furthermore, we have seen that the rate of convergence for both approaches is primarily dependent on the NURBS description in the scaling direction and it is independent of the boundary description. In regard to practical applications, we have demonstrated comparisons of the boundary-oriented approaches with analytical solution, the standard FEM and isogeometric analysis (IGA). A good agreement is achieved in all cases. To sum up, we consider the presented boundary-oriented formulations as promising analysis frameworks that can be extended to a wide class of problems including nonlinearities and complex geometries such as trimmed NURBS.

Appendix

The normal vectors $\boldsymbol{n}^{\xi}$, $\boldsymbol{n}^{\eta}$ and $\boldsymbol{n}^{\zeta}$ are perpendicular to the surface described by the parameters $(\eta,\ \zeta)$, $(\zeta,\ \xi)$ and $(\xi,\ \eta)$, respectively, see Fig. 3. They are summarized as

$$\boldsymbol{n}^{\xi} = [n^{\xi}_{\hat{x}}, n^{\xi}_{\hat{y}}, n^{\xi}_{\hat{z}}]^T = \frac{\hat{\boldsymbol{x}}_{s,\eta} \times \hat{\boldsymbol{x}}_{s,\zeta}}{\|\hat{\boldsymbol{x}}_{s,\eta} \times \hat{\boldsymbol{x}}_{s,\zeta}\|} = \frac{\boldsymbol{x}_{s,\eta} \times \boldsymbol{x}_{s,\zeta}}{\|\boldsymbol{x}_{s,\eta} \times \boldsymbol{x}_{s,\zeta}\|}$$

$$= \frac{1}{g^{\xi}} \begin{bmatrix} y_{s,\eta}\, z_{s,\zeta} - z_{s,\eta}\, y_{s,\zeta} \\ z_{s,\eta}\, x_{s,\zeta} - x_{s,\eta}\, z_{s,\zeta} \\ x_{s,\eta}\, y_{s,\zeta} - y_{s,\eta}\, x_{s,\zeta} \end{bmatrix}, \tag{A.1}$$

$$\boldsymbol{n}^{\eta} = [n^{\eta}_{\hat{x}}, n^{\eta}_{\hat{y}}, n^{\eta}_{\hat{z}}]^T = \frac{\hat{\boldsymbol{x}}_{s,\zeta} \times \hat{\boldsymbol{x}}_{s,\xi}}{\|\hat{\boldsymbol{x}}_{s,\zeta} \times \hat{\boldsymbol{x}}_{s,\xi}\|} = \frac{\boldsymbol{x}_{s,\zeta} \times (\boldsymbol{x}_s - \hat{\boldsymbol{x}}_0)}{\|\boldsymbol{x}_{s,\zeta} \times (\boldsymbol{x}_s - \hat{\boldsymbol{x}}_0)\|}$$

$$= \frac{1}{g^{\eta}} \begin{bmatrix} (z_s - \hat{z}_0) y_{s,\zeta} - (y_s - \hat{y}_0) z_{s,\zeta} \\ (x_s - \hat{x}_0) z_{s,\zeta} - (z_s - \hat{z}_0) x_{s,\zeta} \\ (y_s - \hat{y}_0) x_{s,\zeta} - (x_s - \hat{x}_0) y_{s,\zeta} \end{bmatrix}, \tag{A.2}$$

$$\boldsymbol{n}^{\zeta} = [n^{\zeta}_{\hat{x}}, n^{\zeta}_{\hat{y}}, n^{\zeta}_{\hat{z}}]^T = \frac{\hat{\boldsymbol{x}}_{s,\xi} \times \hat{\boldsymbol{x}}_{s,\eta}}{\|\hat{\boldsymbol{x}}_{s,\xi} \times \hat{\boldsymbol{x}}_{s,\eta}\|} = \frac{(\boldsymbol{x}_s - \hat{\boldsymbol{x}}_0) \times \boldsymbol{x}_{s,\eta}}{\|(\boldsymbol{x}_s - \hat{\boldsymbol{x}}_0) \times \boldsymbol{x}_{s,\eta}\|}$$

$$= \frac{1}{g^{\zeta}} \begin{bmatrix} (y_s - \hat{y}_0) z_{s,\eta} - (z_s - \hat{z}_0) y_{s,\eta} \\ (z_s - \hat{z}_0) x_{s,\eta} - (x_s - \hat{x}_0) z_{s,\eta} \\ (x_s - \hat{x}_0) y_{s,\eta} - (y_s - \hat{y}_0) x_{s,\eta} \end{bmatrix} \tag{A.3}$$

where g^{ξ}, g^{η} and g^{η} are considered according to Chen et al. (2015, 2016). The transformation of an infinitesimal surface element dS is derived by employing Eqs. (1) and (A.1)–(A.3) as

$$\mathrm{d}S^{\xi} = |\hat{\boldsymbol{x}}_{s,\eta} \times \hat{\boldsymbol{x}}_{s,\zeta}|\, \mathrm{d}\eta\, \mathrm{d}\zeta = |\xi \boldsymbol{x}_{s,\eta} \times \xi \boldsymbol{x}_{s,\zeta}|\, \mathrm{d}\eta\, \mathrm{d}\zeta = \xi^2 g^{\xi}\, \mathrm{d}\eta\, \mathrm{d}\zeta, \tag{A.4}$$

$$\mathrm{d}S^{\eta} = |\hat{\boldsymbol{x}}_{s,\zeta} \times \hat{\boldsymbol{x}}_{s,\xi}|\, \mathrm{d}\zeta\, \mathrm{d}\xi = |\xi \boldsymbol{x}_{s,\zeta} \times (\boldsymbol{x}_s - \hat{\boldsymbol{x}}_0)|\, \mathrm{d}\zeta\, \mathrm{d}\xi = \xi g^{\eta}\, \mathrm{d}\zeta\, \mathrm{d}\xi, \tag{A.5}$$

$$\mathrm{d}S^{\zeta} = |\hat{\boldsymbol{x}}_{s,\xi} \times \hat{\boldsymbol{x}}_{s,\eta}|\, \mathrm{d}\xi\, \mathrm{d}\eta = |(\boldsymbol{x}_s - \hat{\boldsymbol{x}}_0) \times \xi \boldsymbol{x}_{s,\eta}|\, \mathrm{d}\xi\, \mathrm{d}\eta = \xi g^{\zeta}\, \mathrm{d}\xi\, \mathrm{d}\eta. \tag{A.6}$$

References

Abaqus. (2001). *6.7. User's manual*. Dassault Systemes.

Apostolatos, A., Schmidt, R., Wüchner, R., & Bletzinger, K. U. (2014). A Nitsche-type formulation and comparison of the most common domain decomposition methods in isogeometric analysis. *International Journal for Numerical Methods in Engineering*, *97*(7), 473–504.

Auricchio, F., da Veiga, L. B., Hughes, T. J. R., Reali, A., & Sangalli, G. (2010). Isogeometric collocation methods. *Mathematical Models and Methods in Applied Sciences*, *20*(11), 2075–2107.

Auricchio, F., da Veiga, L. B., Hughes, T. J. R., Reali, A., & Sangalli, G. (2012). Isogeometric collocation for elastostatics and explicit dynamics. *Computer Methods in Applied Mechanics and Engineering*, *249*, 2–14.

Bazilevs, Y., Long, C. C., Akkerman, I., Benson, D. J., & Shashkov, M. J. (2014). Isogeometric analysis of lagrangian hydrodynamics: Axisymmetric formulation in the rz-cylindrical coordinates. *Journal of Computational Physics*, *262*, 244–261.

Behnke, R., Mundil, M., Birk, C., & Kaliske, M. (2014). A physically and geometrically nonlinear scaled-boundary-based finite element formulation for fracture in elastomers. *International Journal for Numerical Methods in Engineering*, *99*, 966–999.

Breitenberger, M., Apostolatos, A., Philipp, B., Wüchner, R., & Bletzinger, K. U. (2015). Analysis in computer aided design: Nonlinear isogeometric B-Rep analysis of shell structures. *Computer Methods in Applied Mechanics and Engineering*, *284*, 401–457.
Chasapi, M., & Klinkel, S. (2018). A scaled boundary isogeometric formulation for the elasto-plastic analysis of solids in boundary representation. *Computer Methods in Applied Mechanics and Engineering*, *333*, 475–496.
Chen, L., Dornisch, W., & Klinkel, S. (2015). Hybrid collocation-Galerkin approach for the analysis of surface represented 3D-solids employing SB-FEM. *Computer Methods in Applied Mechanics and Engineering*, *295*, 268–289.
Chen, L., Simeon, B., & Klinkel, S. (2016). A NURBS based Galerkin approach for the analysis of solids in boundary representation. *Computer Methods in Applied Mechanics and Engineering*, *305*, 777–805.
Cottrell, J. A., Hughes, T. J. R., & Bazilevs, Y. (2009). *Isogeometric analysis: Toward integration of CAD and FEA*. John Wiley & Sons.
De Lorenzis, L., Evans, J. A., Hughes, T. J. R., & Reali, A. (2015). Isogeometric collocation: Neumann boundary conditions and contact. *Computer Methods in Applied Mechanics and Engineering*, *282*, 21–54.
Dornisch, W., Klinkel, S., & Simeon, B. (2013). Isogeometric Reissner-Mindlin shell analysis with exactly calculated director vectors. *Computer Methods in Applied Mechanics and Engineering*, *253*, 491–504.
Dornisch, W., Vitucci, G., & Klinkel, S. (2015). The weak substitution method - an application of the mortar method for patch coupling in NURBS-based isogeometric analysis. *International Journal for Numerical Methods in Engineering*, *103*, 205–234.
Düster, A., Parvizian, J., Yang, Z., & Rank, E. (2008). The finite cell method for three-dimensional problems of solid mechanics. *Computer Methods in Applied Mechanics and Engineering*, *197*(45), 3768–3782.
Gomez, H., & De Lorenzis, L. (2016). The variational collocation method. *Computer Methods in Applied Mechanics and Engineering*, *309*, 152–181.
Hughes, T. J. R. (2000). *The finite element method: Linear static and dynamic finite element analysis*. Courier Dover Publications.
Hughes, T. J. R., Cottrell, J. A., & Bazilevs, Y. (2005). Isogeometric analysis: CAD, finite elements, NURBS, exact geometry and mesh refinement. *Computer Methods in Applied Mechanics and Engineering*, *194*, 4135–4195.
Hughes, T. J. R., Reali, A., & Sangalli, G. (2010). Efficient quadrature for NURBS-based isogeometric analysis: Computational Geometry and Analysis. *Computer Methods in Applied Mechanics and Engineering*, *199*, 301–313.
Kiendl, J., Auricchio, F., da Veiga, L. B., Lovadina, C., & Reali, A. (2015). Isogeometric collocation methods for the Reissner-Mindlin plate problem. *Computer Methods in Applied Mechanics and Engineering*, *284*, 489–507.
Klinkel, S., Chen, L., & Dornisch, W. (2015). A NURBS based hybrid collocation-Galerkin method for the analysis of boundary represented solids. *Computer Methods in Applied Mechanics and Engineering*, *284*, 689–711.
Lin, G., Zhang, Y., Hu, Z., & Zhong, H. (2014). Scaled boundary isogeometric analysis for 2D elastostatics. *Science China Physics, Mechanics and Astronomy*, *57*(3), 286–300.
Lin, Z., & Liao, S. (2011). The scaled boundary FEM for nonlinear problems. *Communications in Nonlinear Science and Numerical Simulation*, *16*(1), 63–75.
Natarajan, S., Wang, J. C., Song, C., & Birk, C. (2015). Isogeometric analysis enhanced by the scaled boundary finite element method. *Computer Methods in Applied Mechanics and Engineering*, *283*, 733–762.
Ooi, E., Song, C., & Tin-Loi, F. (2014). A scaled boundary polygon formulation for elasto-plastic analyses. *Computer Methods in Applied Mechanics and Engineering*, *268*, 905–937.
Piegl, L. & Tiller, W. (1997). *The NURBS book*. Monographs in visual communications. Springer.

Rank, E., Ruess, M., Kollmannsberger, S., Schillinger, D., & Düster, A. (2012). Geometric modeling, isogeometric analysis and the finite cell method. *Computer Methods in Applied Mechanics and Engineering*, *249*, 104–115.

Reali, A., & Gomez, H. (2015). An isogeometric collocation approach for Bernoulli-Euler beams and Kirchhoff plates. *Computer Methods in Applied Mechanics and Engineering*, *284*, 623–636.

Ruess, M., Schillinger, D., Özcan, A. I., & Rank, E. (2014). Weak coupling for isogeometric analysis of non-matching and trimmed multi-patch geometries. *Computer Methods in Applied Mechanics and Engineering*, *269*, 46–71.

Schillinger, D., Evans, J. A., Reali, A., Scott, M. A., & Hughes, T. J. R. (2013). Isogeometric collocation: Cost comparison with Galerkin methods and extension to adaptive hierarchical NURBS discretizations. *Computer Methods in Applied Mechanics and Engineering*, *267*, 170–232.

Schmidt, R., Wüchner, R., & Bletzinger, K. U. (2012). Isogeometric analysis of trimmed NURBS geometries. *Computer Methods in Applied Mechanics and Engineering*, *241–244*, 93–111.

Song, C. (2004). A matrix function solution for the scaled boundary finite-element equation in statics. *Computer Methods in Applied Mechanics and Engineering*, *193*(23), 2325–2356.

Song, C., & Wolf, J. P. (1997). The scaled boundary finite-element method–alias consistent infinitesimal finite-element cell method–for elastodynamics. *Computer Methods in Applied Mechanics and Engineering*, *147*, 329–355.

Song, C., & Wolf, J. P. (1998). The scaled boundary finite-element method: analytical solution in frequency domain. *Computer Methods in Applied Mechanics and Engineering*, *164*(1–2), 249–264.

Stroud, I. (2006). *Boundary representation modelling techniques*. Springer.

Temizer, I., Wriggers, P., & Hughes, T. J. R. (2012). Three-dimensional mortar-based frictional contact treatment in isogeometric analysis with NURBS. *Computer Methods in Applied Mechanics and Engineering*, *209*, 115–128.

Timoshenko, S. (1951). *Theory of elasticity*. Engineering societies monographs: McGraw-Hill.

Printed in the United States
By Bookmasters